Macroevolution in Human Prehistory

Anna Marie Prentiss · Ian Kuijt · James C. Chatters
Editors

Macroevolution in Human Prehistory

Evolutionary Theory and Processual Archaeology

Springer

Editors
Anna Marie Prentiss
Department of Anthropology
University of Montana
Missoula, MT
USA

Ian Kuijt
Department of Anthropology
University of Notre Dame
Notre Dame, IN
USA

James C. Chatters
Department of Earth & Environmental Sciences
California State University
2576 E. San Ramon Ave.
Fresno, CA
USA

ISBN 978-1-4419-0681-6 e-ISBN 978-1-4419-0682-3
DOI 10.1007/978-1-4419-0682-3
Springer New York Dordrecht Heidelberg London

Library of Congress Control Number: 2009934687

Cover illustration: (Figure 3.5 from book). Atlatl stabilizers ("winged objects") of the Old Bering Sea and Punuk cultures (from Fitzhugh and Crowell 1988). Upper, Punuk from Nunagiak (Ford 1959:62), measures 15 cm width; Middle, Punuk from Grave M4, Meruchta midden, Dovelayaq Bay (Hoffman-Wyss 1987:plate 4) measures 15 cm in width; Bottom, Old Bering Sea from Burial 133, Ekven measures 20 cm in width (Arutiunov and Sergeev 2006b:166).

Printed on acid-free paper

Springer is part of Springer Science+Business Media (www.springer.com)

Acknowledgements

This book project began as a Society for American Archaeology symposium titled "Macroevolution and Archaeology: Concepts and Applications" held at the 70th annual meeting in Salt Lake City. The excitement generated by the session, reflected in audience interest and participant enthusiasm, led us to realize that this would make a significant edited volume. Consequently, we invited all participants to contribute chapters. In addition, we added Marcy Rockman to the lineup given the synergism between her research and many of the topics discussed elsewhere in this volume. Over the past several years, exchange of papers and subsequent discussions have significantly refined all of the works to the point where we can feel confident that despite variation in individual contributions, the chapters now offer a unified message. We hope readers will agree!

We thank all contributors to the volume for their thoughtful consideration of issues associated with cultural macroevolutionary theory and their work developing the chapters. The volume has benefited from thorough peer review comments offered by Hector Neff, Loukas Barton, and one anonymous reviewer. We thank them for their efforts and hope that the final work meets and exceeds their expectations. We thank Teresa Krauss, Springer acquisitions editor, for believing in our project and moving it through peer review. Finally, we thank Katie Chabalko at Springer for her editorial and administrative help.

Prentiss thanks her colleagues and graduate students (particularly those of ANTH 595, Seminar in Darwinian Archaeology) at The University of Montana for feedback on cultural macroevolutionary theory and her family for their patience and support during the writing and editing of chapters in this book. She thanks her coeditors, Ian Kuijt and Jim Chatters, for their friendship, enthusiasm, and hard work on this project. Finally, she thanks her parents, Bill and Sallie Prentiss, who encouraged her interest in ancient life and evolution from a very young age. Sorry about the dinosaur drawings on the marble coffee table!

Kuijt is deeply thankful for the professional support and kindness provided by the members of the Department of Anthropology, University of Notre Dame, and the Institute for Scholarship in the Liberal Arts, University of Notre Dame, who provided funding support to attend the SAA meetings and move this project forward. He would like to specifically thank Anna Prentiss for her drive, organization, and ability to firmly herd a diverse group of academics to a major problem and question.

Thanks also to Meredith Chesson for her constant support, Kaatje for bringing light to our lives, Diane and Ralph Chesson for being great in-laws, and my father Job Kuijt, who as a botanist has always urged me to think about change through time and the natural and social mechanisms behind evolution.

Chatters wishes to thank Anna Prentiss for her diligence in seeing this volume to its completion, many stimulating conversations, and always thoughtful comments on previous drafts of his paper. Thanks also to Robert Dunnell for first directing my thinking toward evolutionary process, and to Jenny Elf and Claire Chatters for tolerating the many hours of isolation this project required.

Contents

Contributors

Robert L. Bettinger Department of Anthropology, University of California, Davis, CA, USA

James C. Chatters Applied Paleoscience; AMEC Earth and Environmental, Inc, Bothell, WA; Department of Earth and Environmental Sciences, California State University, Fresno, CA, USA

Niles Eldredge The American Museum of Natural History, NY, USA

Ian Kuijt Department of Anthropology, University of Notre Dame, Notre Dame, IN, USA

Michael Lenert Department of Anthropology, University of California, Los Angeles, Los Angeles, CA, USA

Owen K. Mason Geoarch Alaska, Anchorage, AK, USA

Anna Marie Prentiss Department of Anthropology, The University of Montana, Missoula, MT, USA

Marcy Rockman Cotsen Institute of Archaeology, UCLA, Los Angeles, CA, USA

Michael Rosenberg Department of Anthropology, University of Delaware, Newark, DE, USA

Charles S. Spencer Division of Anthropology, American Museum of Natural History, New York, NY, USA

Melinda A. Zeder Department of Anthropology, National Museum of Natural History/National Museum of Man, NMNH, NW, Smithsonian Institution, Washington, DC, USA

Introduction

Anna Marie Prentiss, Ian Kuijt, and James C. Chatters

Evolution is the proposition that all organisms on earth, past, present, and future are descended from a common ancestor...
(Eldredge 1989:1)

Cultural evolution, much like general evolution, works from the assumption that cultures are descendent from much earlier ancestors. Indeed, it is now becoming clear that the world's cultures may even derive from a specific common ancestor, perhaps originating in the vicinity of the Upper Nile in East Africa over 50,000 years ago (Bar-Yosef 1998). If this is the case, cultures have since diversified on a scale eclipsed only by the array of biological species. Human culture manifests itself in forms ranging from the small bands of hunter-gatherers that in many ways probably charactorize that original culture, through intermediate scale complex hunter-gatherers and farmers, to the high-density urban settlements and complex polities that characterize much of today's world. A major implication is that cultural evolution was partially (if not substantially), as originally argued by Darwin (1859) for biological species, a historically branching process (e.g., Collard et al. 2006).

Developing an understanding of the history and processes of cultural evolution has been a paramount goal of archaeology. Archaeology has made substantial strides toward this goal, identifying and explaining the major transitions leading to today's cultural patterns (e.g., intensive agricultural systems and associated complex societies) and, recently, considering the effects of cultural practices on landscapes. Advances have come so fast that we seem to be on the brink of a new generation of questions about our past. Yet, as lamented by Schiffer (1999) and Upham (2004), archaeology suffers from insufficient unity in its theoretical approaches and practices. This is particularly true in reference to the development of evolutionary theory in archaeology. As noted by Upham (2004:8), "archaeology ... has yet to find meaningful accommodation with evolutionary theory, a body of knowledge that

---

J.C. Chatters (✉)
Applied Paleoscience; AMEC Earth and Environmental, Inc, Bothell, WA; Department of Earth and Environmental Sciences, California State University, Fresno, CA, USA
e-mail: james.chatters@amec.com

A.M. Prentiss et al. (eds.), *Macroevolution in Human Prehistory*,
DOI 10.1007/978-1-4419-0682-3_1,

includes the most powerful ideas in science." Particularly detrimental has been the schism between proponents of processual archaeology, with its emphasis on ecology and cultural systems, and those of evolutionary archaeology, with its NeoDarwinian interest on culture as populations of information units reflected in artifacts.

During the 1960s and 1970s paleontologists settled a similar debate through the development of paleobiology, which focuses on macroevolutionary processes (Gould 1995). This facilitated new collaborative research, greater discussion, and a shared set of scientific questions and goals (compared to archaeology) focusing research on some of the most important of evolutionary problems, including the evolution of diversity, variation in the frequency of evolutionary innovation, effects of global environmental perturbations on the biosphere, and impacts of biological systems on the earth's surface (Jablonski 1999). The primary distinction between paleobiology and archaeology (other than the obvious difference in subject matter) comes in its degree of theoretical integration, which enables it to address the complex and fascinating interaction between micro- and macroevolutionary processes that give rise to the history of life.

The chapters in the volume examine the dynamic interaction between the micro- and macroscales of cultural evolution, explicating a theoretical approach to the archaeological record that has been termed evolutionary processual archaeology (Spencer 1997). We believe that this approach offers significant opportunity to successfully integrate positive elements of both evolutionary and processualist schools of thought. The approach offers novel insights into traditional processualist areas of study that include the processes of cultural emergence, stasis, collapse, and extinction in a wide array of cultural contexts. Authors draw on Darwinian macroevolutionary theory to address topics that include socioeconomic change, evolution of social inequality, and the development of new sociopolitical entities in contexts ranging from the High Arctic to the arid valleys of Oaxaca, Mexico, and the Near East. Consequently, these contributions form a stepping-off point for a significant new range of cultural evolutionary and socio-ecological studies.

In this chapter, we provide background information to help contextualize the chapters in the book. We introduce macroevolutionary thinking in anthropology and subsequently argue for a workable and inclusive macroevolutionary model for archaeology. Finally, we close with a discussion of directions in cultural evolutionary research.

Macroevolution and Archaeology

The promise of archaeology lies in its ability to examine cultural evolution across great spans of time (Binford 1962; Lyman and O'Brien 1998; O'Brien and Lyman 2003). While it is impossible to track change on a short-term ethnographic timescale, archaeologists are in an optimal position to assess the long-term histories of cultures and cultural elements. Consequently, archaeology is aligned with paleontology as a macroevolutionary science and is excluded from the kind of microevolutionary study that must be undertaken exclusively by ethnographers and molecular biologists. Of course this does not excuse archaeologists from understanding the processes of cultural microevolution, since they form, in many ways, the backbone of macroevolution.

Evolutionary archaeologists (O'Brien and Lyman 2000 2003; see also Mesoudi et al. 2005) promote archaeology as a macroevolutionary science but with an emphasis on the long-term histories of artifacts, in essence, focused on artifact-based cultural phylogeny. Assuming artifacts represent extended human phenotypes (per Dawkins 1990), they argue that by examining change in artifacts they are monitoring the general evolution of *Homo sapiens*. If they are tracking evolution on this extended organismic level, then this specifies a particular form of macroevolution defined by Eldredge (1989:11) as organismic macroevolution or "the accumulation of phenotypic (and underlying genotypic) change in organisms that results from their living in aggregates called *populations* or *demes*" (italics in original). This is the historically standard form of macroevolutionary study in which investigators examine adaptive relationships between organismic phenotypes and their environments (Simpson 1944). The synthetic model of Darwinian evolutionary process extended to the long term is organismic macroevolution.

Taxic Macroevolution

Paleobiologists study another form known as taxic macroevolution that emphasizes "(monophyletic) taxa as ... branches ... of the phylogenetic tree of life" (Eldredge 1989:12–13). With a focus on evolutionary entities at higher scales, such as species and genera, paleobiologists examine such phenomena as relationships between extinction, speciation, and radiation of taxa or the complex relationships between taxa and ecosystems. Most critically, they recognize that law-like evolutionary processes may act exclusively on higher scales that cannot be recognized by microevolutionists. When combined with genic and organismic evolution, taxic macroevolution offers an inclusive approach to the history of life (Eldredge 1989, 1995, 1999; Gould 2002; Jablonski 1999; Stanley 1979, 1998).

Taxic macroevolution is predicated on the ability to define and recognize evolutionary individuals above the scale of organism. Taxic macroevolutionists rely on the Biological Species Concept (BSC) of Dobzhansky (1937) and Mayr (1942), which asserts that species are real in the sense that they are reproductively isolated populations (or interacting populations) of organisms (debates over this concept are reviewed in Eldredge [1989] and Gould [2002]). If species are in some empirical sense "real," they can also be defined as evolutionary individuals (Ghiselin 1974; Hull 1976) or entities that have parts, names, and are spatiotemporally bounded (in contrast to classes without bounds) (Eldredge 1989). Within this framework, the species *Homo sapiens* is recognized as a bounded entity with a birth, life span, and eventual death in the same way that individual organisms are bounded entities. Eldredge (1989) argues that this is confirmed in the fossil record where species are individuated and sorted in a manner consistent with the actions of genetic drift and natural selection. If the birth (and death) of species is a critical element in evolution, it implies that the pattern of evolution is not necessarily always continuous and gradual but perhaps often punctuated in its form (Eldredge 1989; Eldredge and Gould 1972; Gould and Eldredge 1977; Stanley 1979). Further, if species are real and can evolve as units, then it also means that membership within a particular taxon must confer some degree of fitness beyond the simple abilities of the individual organism (Vrba and Eldredge 1984).

For archaeologists seeking to build a similarly comprehensive and inclusive evolutionary model for culture, it is essential to define multiple levels of cultural organization, from basic informational units to complex entities. While the existence of basic units of information and their artifactual counterparts is not controversial (O'Brien and Lyman 2003), the existence and identification of higher level cultural entities have been subject to debate. This recognition of higher level cultural entities is critical if we are to develop an inclusive macroevolutionary approach to archaeology. If we cannot define a defensible hierarchy of evolutionary entities, it will be impossible to develop an understanding of how the higher level—or more complexly integrated—entities interact with more basic units of culture. There are several critical issues that must be faced. First, what is a complex cultural entity, and how do we recognize it? Second, could such a construct be a bounded entity? Finally, how does such a thing evolve? We return to these questions at a later point in this chapter.

Evolutionary Anthropology

As has been amply documented elsewhere (Dunnell 1980; O'Brien and Lyman 2000; Maschner 1996; Rambo 1991; Shennan 2001), contemporary views of cultural evolution in anthropology are greatly varied. Three prominent approaches are the culture history (Lyman et al. 1997), neo-evolutionary (Harris 1979; Steward 1955; White 1949), and Darwinian evolutionary models. We view much of classic processual archaeology (e.g., Binford 1968, 1985) within the broader rubric of neo-evolutionary anthropology. Neo-evolutionists and processualists offer the influential idea that cultures can be defined as functioning systems (much like ecosystems) bounded from other such systems by differences in fundamental socioeconomic strategy (Spencer 1997). The chapters in this volume, however, make it clear that this "space-like" ecology-inspired formulation is incomplete without a means to examine variability in evolutionary "tempo and mode" of these multi-scalar cultural entities. The integration of Darwinian evolutionary theory with systemic elements of processual archaeology offers a potentially powerful approach to the study of cultural evolution.

The evolutionary synthesis in biology (in which the sciences of genetics and naturalism were linked to form the modern or "neo" Darwinian theory of evolution) influenced the development of four Darwinian perspectives on culture: sociobiology (Alexander 1979; Irons 1976; Wilson 1975a,b), human behavioral ecology (Winterhalder and Smith 1992), gene-culture coevolution (Boyd and Richerson 1985; Cavalli-Sforza and Feldman 1981; Durham 1976, 1979), and cultural macroevolution.

The study of gene-culture coevolution has become increasingly influential in archaeology as indicated by a range of recent studies (e.g., Bettinger and Eerkens 1997; Eerkens et al. 2006; Mesoudi and O'Brien 2008). However, other researchers have studied artifact histories under the assumption that the mode of cultural transmission is less important than the overwhelming effects of natural selection and cultural drift (e.g., Darwent and O'Brien 2006; Dunnell 1980, 1989; Neiman 1995; O'Brien et al. 1994, 2008). Archaeologists have now expanded the role of Darwinian thinking in archaeological contexts to focus on cultural entities existing on organizational scales well above that of the artifact.

While Diener (1974, 1980; Diener et al. 1976, 1980) and Marks and Staski (1988) offered early attempts at framing an inclusive macroevolutionary approach, Rosenberg (1994) was the first to offer an archaeological model that integrated microevolutionary and macroevolutionary theory into a comprehensive whole. Rosenberg's approach defines cultures as self-replicating informational systems characterized by a design structure or *Bauplan* (see also Spencer 1997). Whereas Marks and Staski substantially rejected the role of basal units and microevolutionary processes in macroevolution, Rosenberg outlined their central importance, particularly as a means of maintaining a reservoir of knowledge useful on all scales of cultural formations. Rosenberg's pioneering study was followed by several additional papers building on the same themes.

Spencer (1997) recognized the evolution of complex cultural entities and developed a sophisticated approach to the explanation derived from processual archaeology and paleobiology. Prentiss and Chatters (2003a, b; Chatters and Prentiss 2005) constructed a model of cultural evolution based upon tenets of punctuated equilibrium. Exploration and modeling of higher level cultural variation as seen from the standpoint of phylogenetics is now common in evolutionary anthropology (e.g., Boyd et al. 1997; Collard et al. 2005; Holden 2006; Holden and Shennan 2005; Jordan and Mace 2005; Mace and Holden 2005). Evolutionary archaeologists now recognize that adaptive strategies exist as evolutionary entities (Mesoudi and O'Brien 2008; O'Brien et al. 2008). These studies indicate that evolutionary processes acting on such higher scales could potentially override forces at lower levels (e.g., microevolution).

Units, Boundaries, and Heritability

Formulations of organizational strategies, political units, and the like as evolutionary individuals such as those of Diener, Marks and Staski, Rosenberg, and Spencer have not been without critics. Rambo (1991) argues that these ideas are not drawn from Darwinian thinking. He suggests that evolutionary units on the scale of cultural systems or societies do not exist in large enough populations to sustain rates of extinction typical of biological evolution. Finally he rejects the idea that such entities have any empirical status, arguing that no clear boundaries can be identified. Early writings by evolutionary archaeologists (selectionists) were generally negative on this idea as well, typically arguing that such entities are essentialist constructs useful only as heuristic tools for description of variation (Leonard and Jones 1987; Lyman and O'Brien 1998; O'Brien and Lyman 2000).

The critiques of macroevolutionary thinking (beyond the organismic variety) raise serious issues regarding the definition of cultural evolutionary units. First, the boundaries critique (e.g., Rambo 1991) is unavoidable and common in anthropology. If cultural communication is continuous, surely ideas (basal cultural units) are freely exchanged between cultural groups, thereby negating any possibility of a higher level cultural construct. Second, even if boundaries can be defined, it stretches the imagination to suggest that an entire culture could be transmitted between individual people. It is certainly unlikely that a single person could hold all defining knowledge associated with a given culture and even less likely that he/she could transmit it to another.

Definition of appropriate analytical entities has been essential to researchers seeking to build phylogenies of culture and language (Mace et al. 2005), who offer potential solutions to the problem of defining higher cultural entities. Boyd et al. (1997) outline and critique the relevance for phylogenetic analysis of four alternative constructs: culture as species, cultures as hierarchical systems possessing core elements, cultures as sets of cultural packages, and cultures as incoherent collections of memes.

The culture-as-species concept is probably best exemplified by the work of Marks and Staski (1988). Boyd et al. reject this concept, asserting that intergroup (and intercultural) communication is constant—so constant that it would be impossible to sustain an analogy to biological systems, for which substantial genetic isolation can be demonstrated. Contemporary macroevolutionary theorists (e.g., Chatters and Prentiss 2005; Rosenberg 1994, this volume; Spencer 1997, this volume) have eschewed this definition in their formulations of higher level cultural units.

Cultural cores include sets of cultural values, beliefs, and procedures that are particularly meaningful to their holders but may also, when used in concert, be evolutionarily adaptive (Holden and Shennan 2005). If they provide the basis for critical adaptations and, as argued by Holden and Shennan (2005:16), are transmitted in a "conservative way," then it may be difficult for outside memes to significantly invade and compromise that hierarchical system, comprising the core's structure. Rosenberg's (1994) model is probably closest to this form of organization, since it recognizes a stable infrastructure and yet allows complex interactions (exchange of memes) around it. Languages may fit the cultural core model best, because they have core vocabularies and grammars and are transmitted in a very conservative manner (Holden and Shennan 2005).

The idea of culture as a set of traits or memes bundled in a nearly endless array of "packages" that may have independent histories but may also be "glued" together is not antithetical to the culture as a hierarchical system model (Boyd et al. 1997). The "package" model helps to explain the complex histories of "borrowing" between adjacent cultural groups in many parts of the world that have resulted in widely variable recombination of cultural practices. Holden and Shennan (2005) observe that the core and package models may have significant relevance for those seeking to understand deeper cultural phylogenies. Recent research suggests that the respective histories of particular cores and packages may be complex and will not always entirely correlate with one another (Jordan and Mace 2005; O'Brien et al. 2008).

The fourth model, cultures as collections of memes, suggests that ideas are too incoherent and short-lived to permit phylogenetic analysis, since they must recombine very quickly and change rapidly in the face of new contexts (Holden and Shennan 2005). Holden and Shennan reject the collection of memes argument, given the overwhelming evidence for persistence of so many kinds of complex integrated cultural traditions (languages, beliefs, technologies, etc.). If correct, this means that culture has not evolved simply as a by-product of competition between memes operating as independent entities analogous to the selfish genes of Dawkins (1976). This does not mean, however, that those smallest units have not undergone evolution, but merely that "memetic evolution" is but one part of a more complex process (O'Brien et al. 2008).

Viewing culture in the genealogical context as a hierarchically structured system of cores, packages, and memes seems to offer significant potential for macroevolutionary-scale analysis, including the construction of phylogenies (but see Eldredge 2000, this volume; Temkin and Eldredge 2007). Such constructions do not require assumptions of firm boundaries, and it is clear that they are heritable, albeit in sometimes complex ways (per Boyd et al. 1997). We also add that they exist in almost endless populations, certainly enough for cultural extinctions to play a role in macroevolution. Consequently, we argue that there is no need to reject cultural macroevolution on the grounds that culture lacks units, boundaries, or heritability.

Explanation and Evolutionary Process

To a cultural evolutionist, explanation requires an understanding of several things: (1) the production of cultural variation (via the action of cultural agents, factions, and the like); (2) the law-like operation of evolutionary principles (e.g., selection, drift, and cultural proxies of selection and drift as occur during cultural transmission); and (3) contingency, or the impact of unpredictable historical incidents (Gould 2002), which combine to create histories of change within taxonomic groups. Evolutionary explanation, thus, requires construction of historical narratives that outline the interaction of these factors on the development and time/space distribution of cultural entities, much like that of biological taxa (e.g., Mayr 1982). These procedures differ from those of ecology, which seeks law-like generalizations about such phenomena as the energetic relationships between entities or how ecosystems unfold.

To the more hard-line evolutionary archaeologists, culture consists of populations of extended human phenotypes (e.g., artifacts), each in some way representing underlying memes that are, per Dawkins (1990), competing for maximum distribution and longevity (O'Brien and Lyman 2000:7–8). Explanation of change must rely on developing a detailed historical understanding of the variable persistence of those memes. For some archaeologists (e.g., Rosenberg, this volume), this argument makes little sense in isolation as a means for understanding the origin of more complexly integrated cultural entities. Others (e.g., Bettinger 1999, this volume; Prentiss, this volume) suggest that artifact evolution could play a significant role in cultural macroevolution, though the actual process could be complicated and, often, indirect.

Evolutionary processualists recognize that evolution acts on multiple scales (meme to integrated package) resulting in complex dynamics across long time spans. Rosenberg argues that change at the highest macroevolutionary level is contingent upon socioeconomic and demographic stress (e.g., infrastructure per Harris 1979). To Rosenberg (1994), stress, such as population pressure (the result of "contracultural behaviors"), is enough to favor higher rates of innovation that lead to changes at that infrastructural level. He also notes that despite transmission of internal or external innovations, some cultural designs (cores) will simply run their course and eventually become "exhausted" (Rosenberg 1994:327). Dramatic decline in the effectiveness of a cultural core (or *Bauplan* in Rosenberg's convention) may lead to disintegration and reemergence of a new system. Following this high-stress period, cultures go through a short period of "crystallization" (e.g., Foster 1960) in which new ideas are integrated as norms in part due to microevolutionary processes like frequency dependence (e.g., Boyd and Richerson 1985). Consequently, they may or may not be the most adaptive, as long as they meet basic needs by not conflicting with newly emerging core infrastructural elements. The resulting macroevolutionary pattern is much like that described by Diener (1980), while the process is different in that it also invokes the actions of transmission and selection-like processes at lower levels of organization.

Spencer (1987, 1990, 1997) also invokes the term *Baupläne* to describe structural elements of cultural systems and, as with Rosenberg, is describing something like a cultural core of Boyd et al., (1997). Spencer's model (1997, this volume) concerns the rapid emergence of state-level societies, namely, Monte Alban in the Valley of Oaxaca, through a strategy he calls extrapolation. To understand extrapolation, one must imagine several sociopolitical units (say complex chiefdoms) of relatively equal size, counterpoised in relation to one another (a heterarchy). One of these polities, through the use of clever military force, extrapolates its own internally hierarchical system onto the regional heterarchy, thus forming a much larger hierarchy and incipient state. Spencer and Redmond (2001; see also Spencer, this volume) demonstrate that this reflects an example of multi-scalar selection forming a "driven trend" (e.g., Gould 1988; McShea 1994) in which the selective payoffs of an emergent state ultimately override those of the former chiefdoms. Kosse (1994) makes a similar though less explicit argument.

Spencer appears to provide a more explicit role for selection during the critical transition period of change than does Rosenberg. However, this may be a matter of definition. Rosenberg asserts that change happens too quickly for selection to have any effect. He is undoubtedly right that strict natural selection on biological phenotypes would be far too slow. If we take selection to mean culturally induced economic advantage (e.g., selection by consequences [Skinner 1981]) that *might* have reproductive impacts in the long term, then his rapid crystallization period could in effect be selection driven. Indeed, if the core change occurs within the infrastructure, it could not help but be driven by selection or some cultural selection proxy (e.g., Boyd and Richerson 1985), and given its hierarchical nature, it is probably also the kind of multilevel selection described by Spencer and Redmond (2001).

As noted by Ames (2004), Prentiss and Chatters (2003a, b; Chatters and Prentiss 2005) argue for the evolutionary origin of cultural entities on the northwest coast of North America using a model drawing from Eldredge and Gould's (1972; Gould and Eldredge 1977) theory of punctuated equilibrium. Drawing from Eldredge (1985), Rosenberg (1994), and Spencer (1997) they propose a hierarchical cultural framework of interacting entities ranging from basal memes (as reflected in artifacts) to more complex socioeconomic "cores" (e.g., Boyd et al. 1997) such as Resource Management Strategies (RMS; similar to Bettinger's "complex adaptive strategies"). As discussed by Prentiss and Chatters (both this volume; Chatters and Prentiss 2005), RMS are best understood as organizational frameworks coordinating a host of activities spanning tool production and use through group positioning (mobility), subsistence practices, and labor organization. Recognition of RMS structure in the archaeological record is not a simple process, requiring consideration of many lines of evidence (e.g., Chatters 1987) to infer the emergent organizational logic underlying integrated human adaptive behavior.

Recent research in community structure and clustering in networks (Clauset et al. 2008; Newman 2008) may offer future opportunities for mapping and quantifying the internal structure of such entities. When that happens, we can then start to explore in rigorous mathematical terms the roles of phylogenesis and ethnogenesis in the evolution of such complex cultural entities. However, while we have far to go, this should not preclude us from theorizing cultural evolution on the scale of RMS and using the concept to aid our understanding of long-term cultural change (e.g., Bettinger, this volume; Riede 2008; Shennan 2001, 2008; Shott 2008).

Prentiss and Chatters argue that RMS can emerge under conditions of social and/or geographic isolation that prevents swamping (cultural equivalent of gene flow) by the "mother culture." However, isolation alone is insufficient, since some groups become isolated and permanently locked into an original strategy (e.g., the pre-Dorset and Dorset cultures in the Canadian Arctic [Prentiss and Lenert, this volume]). Prentiss and Chatters assume that while techno-economic innovation is constantly occurring in human societies, most conditions (and traditional societies) will not reward system-threatening changes. Consequently, resource conditions must be productive enough (resulting in lowered risk and annual uncertainty) that changes do not automatically result in economic catastrophe (or at least are not anticipated to result in disaster). Viewed in terms of Wright's (1931, 1932) adaptive landscape, each RMS sits on an adaptive peak surrounded by maladaptive valleys (see also Bettinger, Spencer, this volume). To achieve a new state, an RMS must reduce the depth of an adjacent valley or the valley itself must become shallow enough for safe crossing. Therefore, major change at the scale of an entire RMS is very rare and associated with a process Chatters and Prentiss (2005) call niche reorganization.

Niche reorganization is similar to Binford's (2001) concept of niche filling but reflects an evolutionary process by which tactical changes, particularly in critical areas such as scheduling, rapidly give rise to changes in the structure of resource acquisition, processing, distribution, and consumption (Chatters, this volume; Chatters and Prentiss 2005; Prentiss, this volume). Niche reorganization and niche filling are terms that essentially describe a more general process identified in biology as niche construction (Laland et al. 2000, 2001; see also Kuijt and Prentiss, this volume), which refers to the propensity of organisms to modify and in some ways "create" their habitats. Prentiss and Chatters argue that once present, the RMS and its implied niche can be transported by its users (see also Rockman, this volume) and their biological descendants or can be transmitted as a package to other groups. Often, although archaeological data may suggest that a process of RMS transmission did occur (Chatters and Prentiss 2005), details of the transmission process require more research (e.g., Jordan and Shennan 2001).

The models of Rosenberg, Spencer, and Chatters and Prentiss have commonalities that point to important future research directions. All rely on the concept of *Bauplan*, a German term translated variously as "ground plan" or "engineering design" (McGhee 1999:6; Seilacher 1970, 1973), to define a basic organizational structure for entire monophyletic clades. While the appropriateness of the term in archaeology has been debated (Lyman and O'Brien 1998; O'Brien and Lyman 2000), the concept does offer utility in helping us to grasp the nature of evolutionary process. In particular it helps us to envision the complex whole of higher evolutionary entities (whether species or Resource Management Strategies) and to recognize that alterations in their development are too complex to be explained entirely piecemeal (Gould and Lewontin 1979).

An important implication of this is that while structures (*Baupläne*) result from prior change, they also constrain future developments. Adaptive change could come about through a kind of traditional Darwinian process that begins with the expansion of favored innovations. It appears to also occur as a product of the kind of multiscalar adaptive processes modeled by Spencer and Redmond (see also Prentiss, this volume; Rosenberg, this volume) favoring emergent characters that only become visible at those higher levels.

Bettinger (this volume) points out that we still have much research ahead before we truly understand how new cultural characters emerge, persist, and expand in human populations. At the most simplistic level, it is possible that innovations simply are developed and become common via their inherent replicative fitness (Leonard and Jones 1987) and potentially reproductive fitness payoffs (O'Brien and Lyman 2000), leading to broader adaptive change. However, the issue is complex from several standpoints. Bamforth (2002) laments the lack evidence for changes in fitness associated with artifact evolution. Eldredge (this volume) demonstrates that artifact evolution can occur as the by-product of interacting cultural forces across multiple dimensions of integration. Chatters (this volume) and Zeder (this volume) make it clear that innovations may not be readily integrated to their full potential and may lie within a cultural repertory until broader adaptive changes enable them to fluoresce. It is also possible that new cultural practices might evolve as correlated traits or as effects of other cultural interactions (Bettinger, this volume). If this is the case, it also implies that some cultural constructs that are not advantageous to all members of a society (e.g., social inequality) could still be permitted to come into being as exaptations (e.g., Prentiss 2010; Rosenberg, this volume).

Macroevolution and Human Prehistory

This collection of essays addresses the need to refine ideas about the processes of cultural evolution at the macro level. Chapters are organized to take the reader through the major issues in macroevolutionary (or evolutionary processual) archaeology. The first group discusses significant issues in macroevolution theory, the second explores processes by which cultural entities emerge, the third examines cultural expansion/radiation, stasis, and extinction in a cultural macroevolutionary context, and the fourth presents closing reflections. The intent is to provide the reader with an introduction to issues in cultural macroevolutionary thought. Toward that end, chapters provide a host of concepts, models, and hypotheses for further study. Phylogenetic analysis is not generally a focus of these contributions, and while this might seem to be a glaring omission, it is more than adequately addressed elsewhere (Lipo et al. 2006; O'Brien and Lyman 2003).

Issues in Cultural Macroevolution

As this introduction has shown, microevolutionary and macroevolutionary processes combine to generate evolution in the long term. This is an inclusive view of the evolutionary process that is at odds with programs that exclude action at higher levels of cultural organization. Rosenberg begins this section by arguing that understanding the evolution of cultures and cultural phenomena will be significantly inhibited by exclusive reliance on highly reductionist schools of thought. His discussion points to the inescapable conclusion that cultural evolution is complex and cannot be reduced to processes occurring exclusively on the smallest scales.

Rockman considers macroevolutionary implications of landscape learning or what happens when groups with a socioeconomic strategy evolved in one geographic area move into a new context. Her paper introduces the concept of "holon" (drawing from the work of Dean [1988] and others), or complex behavioral systems maintained or critically modified during population movements to new environments. This research is essential for understanding not only how cultures remain stable under altered conditions but also how variation could emerge during these transitions.

In the final contribution to this section, Mason considers the "fundamental dichotomy" of artifact style and function (Dunnell 1978) from a cultural macroevolutionary standpoint. In contrast to early arguments from evolutionary archaeology, he demonstrates that art forms (e.g., harpoon designs) are not necessarily adaptively neutral, since they may be critical identifiers of societies with new, adaptive and highly competitive strategies. Indeed, evolution of harpoon heads in Bering Strait may have been highly dependent upon success at more complex levels of integration. Thus, while harpoon evolution could be modeled from the organismic macroevolutionary standpoint, it makes perhaps more sense to examine it within a multi-scalar cultural framework.

Emergence of Cultural Variants

The majority of contributions to this volume explore elements of macroevolutionary process, beginning with issues of emergence. Macroevolutionary theorists recognize that when new higher level phylogenetic entities, like species, come about, they often have characters recognizable only at those higher levels. Those higher level characters may impart some economic and possibly reproductive advantages to their members (or in the case of culture, users). As pointed out by Bettinger (this volume), it may be productive for archaeologists to view these emergent characters as adaptive peaks on complex fitness landscapes (Wright 1931, 1932). Understanding emergence is, therefore, a critical aspect of macroevolutionary modeling.

Prentiss initiates this discussion by considering the emergence of new hunter-gatherer strategies on North America's northwest coast and interior. She argues that new strategies reflect emergent characters sometimes entailing emergent fitness for their users, implying that we cannot fully explain the evolution of the complex societies in this region without consideration of the consequences of differential possession of these characters.

Spencer relies upon Wright's (1931, 1932) concept of fitness landscape and his now familiar extrapolation model to explore the emergence of the Monte Alban state in Oaxaca, Mexico. Spencer argues that agency and selection played joint roles in the complex history of Monte Alban, culminating in the development of this complex state that persisted for several centuries. Like Prentiss, he also finds linkages between emergent characters and fitness, when considering culture at the macroevolutionary level.

Zeder outlines the complex reasoning required to understand the emergence of food producers in the Near East. She notes that while ecological context plays a critical selective role, this role is mediated within the social world of these last foragers and first farmers. Drawing upon Spencer (1997, this volume), she concludes that we will not understand cultural evolution without recourse to examination of multiple variables across different levels. She proposes that while the Neolithic of the Levant could be recognized as a classic example of a punctuated evolutionary event, it is clear that it was also dependent upon innovations accumulated over the previous 10,000 years.

Cultural Expansion, Stasis, and Extinction as Macroevolutionary Processes

Contributors to this book argue that complex cultural entities may have histories that include periods of stasis, competition with other groups, expansion into new geographic areas, and eventual extinctions. Chatters explores these issues in a macroevolutionary perspective on the archaeology of North America. In particular, he sheds new light on such archaeological constructs as the Archaic, Woodland, and Mississippian cultures, arguing that the appearance of these unique patterns often marks the emergence and rapid expansion of new socioeconomic strategies with unique *Baupläne*. He argues that competition, replacements, and extinctions may be common in the long-term record of human cultural history.

Chatters argues that once present, complex cultural entities like Resource Management Strategies can enter periods of little or no change or "stasis" before evolving further or becoming extinct. This conclusion implicates punctuated equilibrium as an essential model for understanding cultural evolutionary process at the highest level. The archaeological record overwhelmingly confirms that a punctuational view is highly relevant for understanding cultural evolution (Arnold 1993; Bar-Yosef and Meadow 1995; Berry 1982; Chatters and Prentiss 2005; Prentiss and Chatters 2003a; Rosenberg 1994).

In their chapter, Prentiss and Lenert examine the pre-Dorset and Dorset cultures of the Canadian Arctic as a case study in cultural stasis as an evolutionary process. They argue that key emergent characters, including the nature of Dorset social structures, intergroup arrangements, and learning strategies, may have played roles in reducing the opportunities for significant variation to emerge, which may also have played a role in their eventual demise. Conversely they also offer a new hypothesis concerning the pre-Dorset–Dorset transition as a by-product of a short-lived cladogenetic event involving cultural diversification and decimation (Prentiss and Chatters 2003a).

Kuijt and Prentiss examine macroevolutionary issues associated with the emergence and decline in complex hunter-gatherer and early horticulturalist cultures of the Near East. They argue for the importance of integrating evolutionary and ecological concepts in exploring the complex long-term patterns of change so obviously visible in the Upper Pleistocene Levant and surrounding environs. Like Zeder, they argue for the utility of the biological concept of niche construction in evolutionary modeling and suggest that it yields novel insights into socio-natural interactions and cultural evolutionary processes. They assert that the earliest Neolithic emerged from within a period of cultural diversification and decimation associated with shifting climates of 10,000–15,000 B.P.

Closing Reflections: Archaeological Approaches to Macroevolution

In their closing reflections, Bettinger and Eldredge comment on the current state of research and recommend future directions for an archaeology focused on macroevolutionary processes. Like Spencer, Bettinger envisions cultural macroevolution from the standpoint of a fitness landscape (per Wright 1931, 1932). This perspective permits him to envision emergent cultural phenomena as peaks characterized by local fitness optima. He then addresses the thorny problem of peak-shifting—how does macroevolutionary change occur? Bettinger points out that there are contexts where "garden variety" microevolution is adequate for understanding evolution, but other contexts with more complex group-level phenomena must be considered. Major peak shifts may represent significant challenges to archaeologists seeking general explanations, particularly where change is historically contingent, thus seemingly defying the general laws sought by processual archaeologists. One important lesson is that change may come about through complex processes whereby characters evolve as by-products of their association with other characters. Dramatic change may also depend on processes of cultural recombination, as illustrated by many contributors to this volume. Another lesson is that cultural changes are often too rapid to be explained by selection acting on individuals or groups. Bettinger asserts that archaeologists must explore new models of social transmission (e.g., Boyd and Richerson 2001) if they are to begin to understand these phenomena.

Eldredge closes the volume with a consideration of some fundamental issues in cultural macroevolution, highlighting his studies of evolutionary change in lineages of musical instruments. Most fundamentally, he demonstrates the impacts of culture as a complex hierarchical system (e.g., "the sloshing bucket") on artifact histories. It is clear from his research that simple style/function dichotomies will typically be too simplistic to explain emergence, persistence, diversification, and extinction in artifact lineages. In reality, individual agency (innovation), economic forces, and social phenomena all will contribute. One obvious implication is that innovation may not immediately lead to enhanced adaptation. Indeed, innovations may remain essentially dormant pending altered conditions where they become more broadly incorporated by human populations and not always for adaptive reasons!

Discussions by Bettinger and Eldredge make it clear that while we have learned much about the nature of cultural change in the very long term, we have far to go before a mature macroevolutionary science of culture comes into existence. Nevertheless, we hope this volume makes a strong contribution in that direction.

References

Alexander, R.D. (1979) Evolution and Culture. In N.A. Chagnon and W. Irons (Eds.), *Evolutionary Biology and Human Social Behavior: An Anthropological Perspective*, pp. 59–89. North Stituate: Duxbury.

Ames, K.M. (2004). Supposing Hunter-Gatherer Variability. *American Antiquity* 69, 364–374.

Arnold, J.E. (1993). Labor and the Rise of Complex Hunter-Gatherers. *Journal of Anthropological Archaeology* 12, 75–119.

Bamforth, D.B. (2002). Evidence and Metaphor in Evolutionary Archaeology. *American Antiquity* 67, 435–452.

Bar-Yosef, O. (1998). On the Nature of Transitions: The Middle to Upper Paleolithic and the Neolithic Revolution. *Cambridge Archaeological Journal* 8, 141–163.

Bar-Yosef, O. and Meadow, R. (1995). Origins of Agriculture in the Near East. In T.D. Price and A.B. Gebauer (Eds.), *Last Hunters-First Farmers*, pp. 39–94. Santa Fe: School of American Research Press.

Berry, M.S. (1982). *Time, Space, and Transition in Anasazi Prehistory*. Salt Lake City: University of Utah Press.

Bettinger, R.L. and Eerkens, J. (1997). Evolutionary Implications of Metric Variation in Great Basin Projectile Points. In C.M. Barton and G.A. Clark (Eds.), *Rediscovering Darwin: Evolutionary Theory and Archaeological Explanation*, pp. 177–191. Washington D.C.: American Anthropological Association, Archeological Papers No. 7.

Bettinger, R.L. (1999). What Happened in the Medithermal? In C. Beck (Ed.), *Models for the Millenium: Great Basin Anthropology Today*, pp. 62–74. Salt Lake City: University of Utah Press.

Bettinger, R.L. (2003). Comment on "Cultural Diversification and Decimation in the Prehistoric Record" by W.C. Prentiss and J.C. Chatters. *Current Anthropology* 44, 48.

Binford, L.R. (1962). Archaeology as Anthropology. *American Antiquity* 28, 217–225.

Binford, L.R. (1968) Post-Pleistocene Adaptations. In S.R. Binford and L.R. Binford (Eds.), *New Perspectives in Archaeology*, pp. 313–341. Chicago:Aldine.

Binford, L.R. (1985). In Pursuit of the Future. In D.J. Meltzer, D.D. Fowler, and J.A. Sabloff (Eds.), *American Archaeology Past and Future: A Celebration of the Society for American Archaeology 1935–1985*, pp. 459–479. Washington, DC: Smithsonian Institution Press.

Binford, L.R. (2001). *Constructing Frames of Reference*. Berkeley: University of California Press.

Boyd, R., Mulder, M.B., Durham, W.H., and Richerson, P.J. (1997). Are Cultural Phylogenies Possible? In P. Weingart, S.D. Mitchell, P.J. Richerson, and S. Maasen (Eds.), *Human by Nature: Between Biology and the Social Sciences*, pp. 355–384. Mahwah: Lawrence Erlbaum Associates.

Boyd, R. and Richerson, P.J. (1985). *Culture and the Evolutionary Process*. Chicago: University of Chicago Press.

Boyd, R. and Richerson, P.J. (2001). Group Beneficial Norms can Spread Rapidly in a Structured Population. *Journal of Theoretical Biology* 215, 287–296.

Boyer, P. (1998). Cognitive Tracks of Cultural Inheritance: How Evolved Intuitive Ontology Governs Cultural Transmission. *American Anthropologist* 100, 876–889.

Cavalli-Sforza, L.L. and Feldman, M.W. (1981). *Cultural Transmission and Evolution: A Quantitative Approach*. Princeton: Princeton University Press.

Chatters, J.C. (1987). Hunter-Gatherer Adaptations and Assemblage Structure. *Journal of Anthropological Archaeology*6:336-375.

Chatters, J.C. and Prentiss, W.C. (2005). A Darwinian Macro-Evolutionary Perspective on the Development of Hunter-Gatherer Systems in Northwestern North America. *World Archaeology* 37, 45–65.

Clauset, A., Moore, C., and Newman, M.E.J. (2008). Hierarchical Structure and the Prediction of Missing Links in Networks. *Nature* 453, 98–101.

Collard, M., Shennan, S.J. and Tehrani, J.J. (2006). Branching Versus Blending in Macroscale Cultural Evolution: A Comparative Study. In C.L. Lipo, M.J. O'Brien, M. Collard, and S.J. Shennan (Eds.), *Mapping our Ancestors: Phylogenetic Approaches in Anthropology and Prehistory*, pp. 53–64. New Brunswick: Aldine/Transaction.

Darwent, J. and O'Brien, M.J. (2006). Using Cladistics to Construct Lineages of Projectile Points from Northeastern Missouri. In C.L. Lipo, M.J. O'Brien, M. Collard, and S.J. Shennan (Eds.), *Mapping our Ancestors: Phylogenetic Approaches in Anthropology and Prehistory*, pp. 185–208. New Brunswick: Aldine/Transaction.

Darwin, C. (1859). *On the Origin of Species by Means of Natural Selection or Preservation of Favored Races in the Struggle for Life*. London: Murray.

Dawkins, R. (1976). *Selfish Gene*. Oxford: Oxford University Press.

Dawkins, R. (1990). *Extended Phenotype: The Long Reach of the Gene* (new ed.). Oxford: Oxford University Press.

Dean, J.S. (1988). A Model of Anasazi Behavioral Adaptation. In G.J. Gumerman (Ed.), *The Anasazi in a Changing Environment*, pp. 25–44. New York: Cambridge University Press.

Diener, P. (1974) Ecology or Evolution? The Hutterite Case. *American Ethnologist* 1, 600–618.

Diener, P., Nonini, D. and Robkin, E.E. (1976). The Dialectics of the Sacred Cow: Ecological Adaptation Versus Political Appropriation in the Origins of India's Cattle Complex. *Dialectical Anthropology* 3, 221–241.

Diener, P. (1979). Quantum Adjustment, Macroevolution, and the Social Field: Some Comments on Evolution and Culture. *Current Anthropology* 21, 423–443.

Diener, P., Nonini, D., and Robkin, E.E. (1980). Ecology and Evolution in Cultural Anthropology. *Man* 15, 3–31.

Dobzhansky, T. (1937). *Genetics and the Origin of Species*. Reprint edition, 1982, New York: Columbia University Press.

Dunnell, R.C (1978). Style and Function: A Fundamental Dichotomy. *American Antiquity*43: 192–202.

Dunnell, R. C. (1980). Evolutionary Theory and Archaeology. In M.B. Schiffer (Ed.), *Advances in Archaeological Method and Theory*, Vol. 6, pp. 35–99. New York: Academic Press.

Dunnell, R. C. (1989).Aspects of the Application of Evolutionary Theory in Archaeology. In C.C. Lamberg-Karlovsky (Ed.), *Archaeological Thought in America*, pp. 35–49. Cambridge: Cambridge University Press.

Durham, W. H. (1976). The Adaptive Significance of Cultural Behavior. *Human Ecology* 4, 89–121.

Durham, W.H. (1979). Toward a Coevolution Theory of Human Biology and Culture. In N.A. Chagnon and W. Irons (Eds.), *Evolutionary Biology and Human Social Behavior: An Anthropological Perspective*, pp. 39–58. North Stituate: Duxbury.

Eerkens, J.W., Bettinger, R.L., and McElreath, R. (2006). Cultural Transmission, Phylogenetics and the Archaeological Record. In C.J. Lipo, M.J. O'Brien, M. Collard, and S.J. Shennan (Eds.), *Mapping our Ancestors: Phylogenetic Approaches in Anthropology and Prehistory*, pp. 169–184. New Brunswick: Aldine/Transaction Books.

Eldredge, N. (1985). *Unfinished Synthesis: Biological Hierarchies and Modern Evolutionary Thought*. New York: Oxford University Press.

Eldredge, N. (1989). *Macroevolutionary Dynamics: Species, Niches, and Adaptive Peaks*. New York: McGraw Hill Publishing Company.

Eldredge, N. (1995). *Reinventing Darwin: The Great Debate at the High Table of Evolutionary Theory*. New York: John Wiley and Sons.

Eldredge, N. (1999). *The Pattern of Evolution*. New York: W.H. Freeman and Company.

Eldredge, N. (2000). Biological and Material Cultural Evolution: Are There any True Parallels? *Perspectives in Ethology* 13, 113–153.

Eldredge, N. and Gould, S.J. (1972). Punctuated Equilibria: An Alternative to Phyletic Radualism. In T.J.M. Schopf (Ed.), *Models in Paleobiology*, pp. 82–115. San Francisco: Freeman, Cooper, and Company.

Foster, G.M. (1960). *Culture and Conquest: America's Spanish Heritage*. New York: Viking Fund Publications in Anthropology 27.

Ghiselin, M.T. (1974). A Radical Solution to the Species Problem. *Systematic Zoology* 23, 536–544.

Gould, S.J. (1988). Trends as Changes in Variance: A New Slant on Progress and Directionality in Evolution (Presidential Address). *Journal of Paleontology* 62, 319–329.

Gould, S.J. (1995). A Task for Paleobiology at the Threshold of Majority. *Paleobiology* 21, 1–14.

Gould, S.J. (2002). *The Structure of Evolutionary Theory*. Cambridge: Belknap Press.

Gould, S.J. and Eldredge, N. (1977). Punctuated Equilibria: The Tempo and Mode of Evolution Reconsidered. *Paleobiology* 3, 115–151.

Gould, S.J. and Lewontin, R.C. (1979). The Spandrels of San Marco and the Panglossian Paradigm: A Critique of the Panglossian Programme. *Proceedings of the Royal Society of London B* 205, 581–598.

Gould, S.J. and Vrba, E.S. (1982). Exaptation – A Missing Term in the Science of Form. *Paleobiology* 8, 4–15.

Harris, M. (1979). *Cultural Materialism: The Struggle for a Science of Culture*. New York: Random House.

Holden, C.J. (2006). The Spread of Bantu Languages, Farming and Pastoralism in Sub-Equatorial Africa. In C.L. Lipo, M.J. O'Brien, M. Collard, and S.J. Shennan, (Eds.), *Mapping our*

Ancestors: Phylogenetic Approaches in Anthropology and Prehistory, pp. 249–268. New Brunswick:Aldine/Transaction.

Holden, C.J. and Shennan, S. (2005). Introduction to Part 1: How Tree-Like is Cultural Evolution? In R. Mace, C.J. Holden, and S. Shennan (Eds.), *The Evolution of Cultural Diversity: A Phylogenetic Approach*, pp. 11–29. London: UCL Press

Hull, D.H. (1976) Are Species Really Individuals? *Systematic Zoology* 25, 174–191.

Irons, W. (1976). Cultural and Biological Success. In N.A. Chagnon and W. Irons (Eds.), *Evolutionary Biology and Human Social Behavior: An Anthropological Perspective*, pp. 257–272. North Stituate: Duxbury.

Jablonski, D. (1999). The Future of the Fossil Record. *Science* 284, 2114–2116.

Jordan, P. and Mace, T. (2005). Tracking Culture-Historical Lineages: Can "Descent with Modification" be Linked to "Association by Descent"? In C.L. Lipo, M.J. O'Brien, M. Collard, and S.J. Shennan (Eds.), *Mapping our Ancestors: Phylogenetic Approaches in Anthropology and Prehistory*, pp. 149–168. New Brunswick:Aldine/Transaction.

Jordan, P. and Shennan, S. (2001). Cultural Transmission, Language, and Basketry Traditions Amongst the California Indians. *Journal of Anthropological Archaeology* 22, 42–74.

Kosse, K. (1994). The Evolution of Large, Complex Groups: A Hypothesis. *Journal of Anthropological Archaeology* 13, 35–50.

Laland, K.N., Odling-Smee, J., and Feldman, M.W. (2000). Niche Construction, Biological Evolution and Cultural Change. *Behavioral and Brain Sciences* 23, 131–175

Laland, K.N., Odling-Smee, J., and Feldman, M.W. (2001). Cultural Niche Construction and Human Evolution. *Journal of Evolutionary Biology* 14, 22–33.

Leonard, R.D. and Jones, G.T. (1987). Elements of an Inclusive Evolutionary Model for Archaeology. *Journal of Anthropological Archaeology* 6, 199–219.

Lipo, C.L., O'Brien, M.J., Collard, M. and Shennan, S. (Eds.) (2006). *Mapping our Ancestors: Phylogenetic Approaches in Anthropology and History*. New Brunswick: Aldine/Transaction.

Lyman, R. L. and O'Brien, M.J. (1998). The Goals of Evolutionary Archaeology: History and Explanation. *Current Anthropology* 39, 615–652.

Lyman, R.L., O'Brien, M.J., and Dunnell, R.L. (1997). *The Rise and Fall of Culture History*. New York: Plenum.

Mace, R. and Holden, C.J. (2005). A Phylogenetic Approach to Cultural Evolution. *Trends in Ecology and Evolution* 20, 116–121.

Mace, R., Holden, C.J. and Shennan, S. (2005). *The Evolution of Cultural Diversity: A Phylogenetic Approach*. London: UCL Press.

Maschner, H.D.G. (ed.) (1996). *Darwinian Archaeologies*. New York: Plenum.

Marks, J. and Staski, E. (1988). Individuals and the Evolution of Biological and Cultural Systems. *Human Evolution* 3, 147–161.

Mayr, E. (1942). *Systematics and the Origin of Species*. New York: Columbia University Press.

Mayr, E. (1982). *Growth of Biological Thought: Diversity, Evolution, and Inheritance*. Cambridge: Belknap Press.

McGhee, G.R. Jr. (1999). *Theoretical Morphology: The Concept and Its Applications*. New York: Columbia University Press.

McShea, D.W. (1994). The Mechanisms of Large Scale Evolutionary Trends. *Evolution* 48, 1747–1763.

Mesoudi, A. and O'Brien, M.J. (2008). The Cultural transmission of Great Basin Projectile Point Technology I: An Experimental Simulation. *American Antiquity* 73, 3–28.

Mesoudi, A., Whiten, A. and Laland, K.N. (2005). Towards a Unified Science of Cultural Evolution. *Behavioral and Brain Sciences* 29, 329–383.

Neiman, F.D. (1995). Stylistic variation in Evolutionary Perspective: Inferences from Decorative Diversity and Interassemblage Distance in Illinois Woodland Ceramic Assemblages. *American Antiquity* 60, 7–36.

Newman, M. (2008). The Physics of Networks. *Physics Today* November, 33–38.

O'Brien, M.J., Holland, T.D., Hoard, R.J., and Fox, G.L. (1994). Evolutionary Implications of Design and Performance Characteristics of Prehistoric Pottery. *Journal of Archaeological Method and Theory* 1, 259–304
O'Brien, M.J. and Lyman, R.L. (2000). *Applying Evolutionary Archaeology*. New York: Kluwer Academic/Plenum.
O'Brien, M.J. and Lyman, R.L. (2003). *Cladistics and Archaeology*. Salt Lake City: University of Utah Press.
O'Brien, M.J., Lyman, R.L., Collard, M., Holden, C., Gray, R.D., and Shennan, S.J. (2008). Transmission, Phylogenetics, and the Evolution of Cultural Diversity. In M.J. O'Brien (Ed.), *Cultural Transmission and Archaeology: Issues and Case Studies*, pp. 39–58. Washington D.C.: The SAA Press.
Prentiss, A.M. (2010). Social Histories of Complex Hunter-Gatherers: Pacific Northwest Prehistory in a Macroevolutionary Framework. In K.E. Sassaman and D.H. Holly Jr. (Eds.), *Hunter-Gatherer Archaeology as Historical Process*. Tucson: University of Arizona Press (in review).
Prentiss, W. C. and Chatters, J.C. (2003a). Cultural Diversification and Decimation in the Prehistoric Record. *Current Anthropology* 44, 33–58.
Prentiss, W. C. and Chatters, J.C. (2003b). The Evolution of Collector Systems in the Pacific Northwest Region of North America. *Senri Ethnological Studies* 63, 49–82.
Rambo, A.T. (1991). The Study of Cultural Evolution. In A.T. Rambo and K. Gillogly (Eds.), *Profiles in Cultural Evolution: Papers from a Conference in Honor of Elman R. Service*, pp. 23–109. Ann Arbor: Anthropological Papers 85, Museum of Anthropology, University of Michigan.
Riede, F. (2008). Maglemosian Memes: Technological Ontogeny, Craft traditions, and the Evolution of Northern European Barbed Points. In M.J. O'Brien (Ed.) *Cultural Transmission and Archaeology: Issues and Case Studies*, pp. 178–189. Washington D.C.: SAA Press.
Rosenberg, M. (1994). Pattern, Process, and Hierarchy in the Evolution of Culture. *Journal of Anthropological Archaeology* 13, 307–340.
Schiffer, M.B. (1999). Return to Holism. *Anthropology Newsletter* (March), 64, 62.
Seilacher, A. (1970). Arbeitskonzept zur Konstruktions-Morphologie. *Lethaia* 3, 393–396.
Seilacher, A. (1973). Fabricational Noise in Adaptive Morphology. *Systematic Zoology* 22, 451–465.
Shennan, S. (2001) *Genes, Memes, and Human History: Darwinian Archaeology and Cultural Evolution*. London: Thames and Hudson.
Shennan, S. (2008). Evolution in Archaeology. *Annual Review of Anthropology* 37, 75–91.
Shott, M. (2008). Darwinian Evolutionary Theory and Lithic Analysis. In M.J. O'Brien (Ed.), *Cultural Transmission and Archaeology: Issues and Case Studies*, pp. 146–157. Washington D.C.: SAA Press.
Simpson, G.G. (1944). *Tempo and Mode in Evolution*. New York: Columbia University Press.
Skinner, B.F. (1981).Selection by Consequences. *Science* 213, 501–504.
Spencer, C.S. (1987). Rethinking the Chiefdom. In R.D. Drennan and C.A. Uribe (Eds.), *Chiefdoms in the Americas*, pp. 369–390. Lanham: University Press of America.
Spencer, C.S. (1990). On the Tempo and Mode of State Formation: Neoevolutionism Reconsidered. *Journal of Anthropological Archaeology* 9, 1–30.
Spencer, C.S. (1997). Evolutionary Approaches in Archaeology. *Journal of Archaeological Research* 5, 209–264.
Spencer, C.S. and Redmond, E.M. (2001). Multilevel Selection and Political Evolution in the Valley of Oaxaca 500–100 B.C. *Journal of Anthropological Archaeology* 20, 195–229.
Stanley, S.M. (1979). *Macroevolution: Pattern and Process*. San Francisco: W.H. Freeman.
Stanley, S.M. (1998). *Macroevolution: Pattern and Process* (Second Edition). Baltimore: Johns Hopkins University Press.
Steward, J. (1955). *Theory of Culture Change*. Urbana: University of Illinois Press.

Temkin, I. and Eldredge, N. (2007). Phylogenetics and Material Cultural Evolution. *Current Anthropology* 48, 146–154.
Upham, S. (2004). Hanging Out: The Status and Position of Archaeology in American Universities. *The SAA Archaeological Record* 4, 6–8.
Vrba, E.S. and Eldredge, N. (1984). Individuals, Hierarchies, and Processes: Towards a More Complete Evolutionary Theory. *Paleobiology* 10, 146–171.
White, L.A. (1949). *The Science of Culture*. New York: Grove Press.
Wilson, D.S. (1975a). A Theory of Group Selection. *Proceedings of the National Academy of Science USA* 72, 143–146.
Wilson, E.O. (1975b). *Sociobiology: The New Synthesis*. Cambridge: Harvard University Press.
Winterhalder, B. and Smith, E.A. (1992). Evolutionary Ecology and the Social Sciences. In E.A. Smith and B. Winterhalder (Eds.), *Evolutionary Ecology and Human Behavior*, pp. 3–24. New York: Aldine de Gruyter.
Wright, S. (1931) Evolution in Mendelian Populations. *Genetics* 16, 97–159.
Wright, S. (1932). The Roles of Mutation, Inbreeding, Crossbreeding and Selection in Evolution. *Proceedings of the VI International Congress of Genetics* 1, 356–366.

Part I
Issues in Macroevolutionary Theory

Chapter 1
Proximate Causation, Group Selection, and the Evolution of Hierarchical Human Societies: System, Process, and Pattern

Michael Rosenberg

Introduction

The reductionism of the physical sciences is elegant in its workings and impressive in its explanatory power, with the properties of a unit at any given level determined by the additive structural properties of its constituent units at the next lower level. Such theoretical power has been a seductive lure for approaches to the study of biological evolution in the natural sciences, fostering both genetic reductionism (e.g., Dawkins 1989) and adaptationism (e.g., see Gould and Lewontin 1979). And, via importation from biology, it has been an equally seductive lure for some evolutionary approaches to the study of culturally patterned behavior in anthropology.

The fact is that not every biological characteristic is directly governed by a specific gene or set of genes; there simply are not enough genes for that to be the case. While some are no doubt a product of pleiotropy, some phenotypic characteristics are not *directly* governed by specific genes at all. That is, they are the product of interactions between independent genetically structured components and not the direct expression of specific genes or sets of genes. For example, as noted by Gould and Lewontin (1979), there is no gene for 'chin'. "Rather, it is the product of interaction between [the alveolar and mandibular] growth fields" (1979:585). One can obviously get selection for chins, because selection operates on organisms through their phenotypic attributes, including chins. But, selection is for the whole system's contribution to fitness, because only the integrated system is visible to selection. The individual structural components are visible to selective forces only in the form of their traded-off contribution to the larger system, and the components themselves are likely themselves emergent interactive properties of (sub-) systems with still lower level traded-off components.

With respect to both culture and culturally patterned behavior, even more so than for biology, different levels of organization have emergent properties that do

M. Rosenberg (✉)
Department of Anthropology, University of Delaware, Newark, DE 19716, USA
e-mail: michaelr@udel.edu

A.M. Prentiss et al. (eds.), *Macroevolution in Human Prehistory*,
DOI 10.1007/978-1-4419-0682-3_2, © Springer Science+Business Media, LLC 2009

not reduce directly to lower level structural determinants. Thus, they cannot be explained by simple reference to their structural reducta. Nowhere is this more pronounced than in the domain of human social structure and social organization. Social structure is always an emergent property of a group, not a property of any single individual. Simply put, one cannot be one's own superior or inferior; a solitary individual plainly cannot be socially stratified. Complex societies simply do not fully reduce to the level of individuals and their evolution cannot be fully explained solely by reference to individuals (see also Chatters, Prentiss, and Zeder, all this volume).

The evolution of human social complexity, particularly the evolution of centralized societies based on institutionalized dominance hierarchies, legitimized status inequalities, and extra-familial authority, has historically been an important area of anthropological inquiry. The past quarter century has seen a flowering of anthropological approaches to the study of cultural evolution that explicitly seek to be firmly rooted in a proper understanding of evolutionary theory (i.e., behavioral ecology, evolutionary archaeology, and dual inheritance theory), as opposed to the progressivistic transformationalism that characterized earlier anthropological and archaeological approaches (e.g., evolutionism, processual archaeology). Yet, for all of their notable successes, these modern, more theoretically rigorous approaches to the evolution of culture have made little substantial progress in explaining the evolution of social complexity. I suggest that that this lack stems directly from the reductionism that constrains these evolutionary approaches. I further suggest that group-selection plays a more prominent role in the process, both directly and indirectly, than has generally been considered.

However, while the above-mentioned evolutionary approaches were developing, the criticisms leveled at processualism (e.g., see Dunnell 1980; Rindos 1985) were also acknowledged by some who nevertheless maintained an interest in the systemic structure of culture and the consequences that such a systemic structure has for the mechanics of cultural evolution. This resulted in attempts to integrate the salvageable elements of processualism with evolutionary theory, on the part of those who neither "want to give up the 'process' [nor] the evolution in processual archaeology" (Stanish and Haley 2005:57; see also e.g., Chatters and Prentiss 2005; Prentiss 2010; Prentiss and Chatters 2003; Rosenberg 1990, 1994; Spencer 1993, 1997). The rehabilitated evolutionary processualism that has come out of these attempts to properly integrate proximate (systemic) and ultimate (evolutionary) causation has continued processualism's traditional focus on the evolution of complex human societies (e.g., see Spencer, this volume).

Unfortunately, evolutionary processualists, like adherents of the other above-mentioned modern evolutionary approaches, have continued the tendency, largely inherited from evolutionism, to focus on social complexity as an adaptation produced by selection for the most prominent functions performed by complex forms of organization. I will also suggest here that while such functions do indeed contribute to success at between-group competition, these highly prominent functions evolved as exaptations[1] and that the evolution of social complexity lies elsewhere, in earlier adaptations that are less *obviously* advantageous to investigators because they rapidly become functionally overshadowed by these exaptations.

Evolution and Multilevel Selection

The concept of group selection dates back to Darwin (1871). In its original formulation, it was applied to those attributes of individuals that were thought to be selected for because they benefited the group to which the individual belonged. The evolution of altruistic behaviors by individuals was the classic case of a characteristic that suggested the possible operation of such (for-the-good-of-the) group selection. However, over time, four powerful arguments (e.g., cheaters will outcompete altruists within the group) were advanced to demonstrate the impotence of such group selection in the face of organismic selection (see Dawkins 1989; Fisher 1958; Hamilton 1971, 1987; Williams 1966; see also Gould 2002:646–647 for a concise summary of these arguments).

Nevertheless, more recently, the concept of group selection was being resurrected in two slightly different ways (e.g., see Brandon 1988, 1990; Damuth and Heisler 1988). With respect to the 'classic' debate, centered on the attributes of individuals that conceivably evolved for-the-good-of-the-group, counter arguments were advanced (see Sober and Wilson 1998; Wade 1978; Wilson and Sober 1994). For example, modeling demonstrated that, under several sets of plausible conditions, the declining intra-demic frequency of individuals with a given (for-the-good-of-the-group) characteristic produced by organismic selection can be overcome by the relative inter-demic success of demes having such members. Thus, the "overall frequency [of individuals with that attribute] may rise within the species even while their frequency within each surviving deme declines" (Gould 2002:648). Moreover, some well-documented patterns in nature seem very difficult to explain without recourse to some significant level of such inter-demic selection (see Wilson and Sober 1994; see also Gould 2002:648–649 for a concise summary).

More importantly for present purposes, the theory put forward that biological evolution is characterized by prolonged periods of stasis punctuated by bursts of relatively rapid speciation (Eldredge and Gould 1972; Gould and Eldredge 1977) fostered a growing interest in macroevolutionary theory, hierarchical entities, and species (as opposed to inter-*demic*) selection/sorting (e.g., Eldredge 1985, 1989, 1999; Gould 2002; Stanley 1981). The key point regarding the demic-species sorting distinction is that the counter arguments against group selection "had legitimate force" in the matter of inter-demic selection, "but could be overcome under [certain] conditions" (see above). However, in the matter of species selection, they either do not apply at all or become irrelevant (Gould 2002:649).

More importantly for present purposes, group selection is explicitly viewed by the macroevolutionists as operating on the properties of the group as a whole—e.g., differential susceptibility to speciation (see Eldredge 1999; Gould 1982, 2002). Thus, in a very real sense, the macroevolutionists are proposing a very different concept of group selection than was and still is the subject of the ongoing "classic" debate. Prior to Fisher's (1958) critique of group selection, the focus was on the aggregate bottom-up, "for-the-good-of" effect of individuals' attributes on the success of the group as a whole. In its resurrected form (i.e., Sober and Wilson 1998; Wade 1978; Wilson and Sober 1994), the "classic" focus has shifted to the top-down effect on individuals of membership in groups enhanced by the presence of such beneficial individuals; but, the focus remains on the individual. In contrast, the macroevolutionists have focused on groups as integrated individuals visible to the operation of selection as higher order individuals and their relative success in what amounts to typical Darwinian competition between groups, at the group level.

As noted by Gould, these "two approaches often yield concordant results for the obvious reason that differential proliferation of higher-level units . . . often defines the group effect that influences the fitness of lower level individuals." But, this need not always be the case, "leading to situations where we could identify group selection by one criterion, but deny [it] by the other" (2002:653; see also Dugatkin and Reeve 1994).

Lastly, there is the issue of just what group selection is said to operate on in the hierarchical, macroevolutionary approach to group selection. (see Grantham 1995; see also Gould 2002:659ff). One view (Vrba 1983, 1984, 1989; Vrba and Eldredge 1984; Vrba and Gould 1986) focuses on "emergent character," which looks to the emergent systemic properties of groups as the locus for the operation of selective forces on groups as higher order individuals (see also Prentiss, this volume). Just as individual organisms are systemic individuals with distinct emergent properties resulting from genetic interactions groups are also systemic individuals with distinct emergent properties resulting from the interactions of individual organisms. And, selection operates on individuals of whatever magnitude based on their attributes as individuals.

The alternate view (Lloyd 1988; see also Arnold and Fristrup 1982; Damuth 1985; Damuth and Heisler 1988; Lewontin 1970) focuses on the more inclusive concept of "emergent fitness," thereby including sum-of-parts (e.g., degree of aggregate variability within a group for a given trait), as well as emergent group-level characteristics. Thus, "Vrba's exclusive [emergent character] domain [constitutes] a subset of 'best cases' in Lloyd's [emergent fitness] formulations" (Gould 2002:659). For present purposes, there is no need to go beyond "best case" emergent group properties to make a reasonable case for human social complexity being a product of group selection (see also Prentiss, this volume), though other aspects of emergent fitness no doubt also play a role in determining the relative success of similarly scaled forms of social complexity (e.g., see issue of embedded complexity in Rosenberg 1994).

Reductionism and Social Complexity

There have been several attempts by adherents of modern evolutionary approaches to deal with the evolution of social complexity (Leonard and Jones 1987; Braun 1990; Boone 1992; Richerson and Boyd 1999), but all suffer from significant deficiencies stemming from their reductionist tendencies.

Evolutionary archaeology draws heavily on the genetic reductionism of Dawkins (1986, 1989), most clearly expressed in what Dunnell (1995) calls "the materialist paradox" (see also Dunnell 1985; O'Brien and Lyman 2000:25). Just as Dawkins tends to view species as nothing more than temporal ephemera—constantly in a state of becoming but never truly existing in any meaningful way—evolutionary archaeology has tended to view cultural phenomena at any level above the individual trait as nothing more than unbounded, ever-changing sets of selection and drift-produced cultural traits. Thus, they are considered irrelevant to the study of selection-produced, cultural-patterned adaptations because they are just *transient* reifications.

To facilitate that position, evolutionary archaeology (see O'Brien and Lyman 2000:8–9) invokes Dawkins's (1989, 1990) concept of the "extended phenotype"[2] and the proposal that all cultural-patterned behavioral complexes are just the variable potential expressions of the human version. Thus, the traits on which selection is said to have operated to produce the selective outcome being "explained" were simply there, because the underlying human genotype makes any and all such (extended) "cultural" appearances equally possible innovations to be selected for without need for further explanation (e.g., see O'Brien and Lyman 2000:206).

Consequently, innovative cultural expressions can be invoked in a systemic vacuum, without consideration being given to possible biases affecting what traits can be said to exist for selection to operate on under any given set of conditions (but see Jones et al. 1995:18). However, there are often perfectly valid structural reasons for concluding that some specific variant should not exist under the cited circumstances without further reference to proximate causation (see Rosenberg 1990); and, if it does not exist, how can it be selected for.

Leonard and Jones's (1987) and Braun's (1990) analyses remain the only two significant attempts within evolutionary archaeology to deal with the evolution of complex societies. Leonard and Jones (1987) focus on several correct, but general, points concerning the variability inherent in the structural traits of such complex societies and the evolutionary implications of such variability. In contrast, Braun (1990:84)[3] does focus on proximate causation, invoking the organizational consequences of persistent group size in excess of the "normal" egalitarian group threshold size for fissioning. However, he also relies on chance (i.e., historical, as opposed to processual) events to produce the organizational traits being selected for and these are group level traits. Overlooked in this reliance on the extended (social) phenotype's capacity to blindly produce the necessary social innovations is the fact that there is good reason to believe that individuals would never accede to (i.e., select) such innovative organizational traits in the absence of other compelling factors, even if they appeared.

The reductionism of behavioral ecology is seemingly more in line with what Sober and Wilson (1998:329) call "methodological individualism." Behavioral ecology is predicated on the premise that the decision-making capacity of human mind is an evolved adaptation that consciously generates fitness-enhancing behaviors by means of what are often called evolved "decision rules" (Krebs 1978) or "conditional strategies" (Smith 2000). Methodologically, it studies such "decision rules" by means of a "*piecemeal approach,*" because "complex socioecological phenomena are fruitfully studied piece by piece—in reductionist rather than holistic fashion" (Smith 2000:29, emphasis in original). However, while cultural contexts are implicitly acknowledged to exist and occasionally referenced when simple enough and directly relevant enough to require invocation (e.g., Smith 1985; Bliege Bird and Bird 1997), the larger part of that sociocultural context is more often ignored for the sake of analytical simplicity and clarity.

The reductionism of behavioral ecology also requires a behavioral generator to, in the absence of structural sociocultural referents, account for the full range of possible behaviors being invoked. Here, it is the concept of "phenotypic plasticity"—succinctly defined as "a phenotype's ... capacity to respond differentially to varying environmental conditions" (Boone and Smith 1998:144). Thus, it views cultural evolution as the product of the rational, decision-driven capacity of individuals to intentionally adapt differentially to short-term variability in environmental conditions by what very much amount to Lamarckian mechanisms (cf. Boone and Smith 1998; Smith 2000)—the underlying Darwinian adaptation being the evolved decision rules that confer the capacity to so adapt differentially to changing conditions.

However, for most of our existence, humans were simple foragers. Thus, those decision rules had to have become geared to optimizing caloric and reproductive returns in the relatively simple sociocultural contexts that characterize such societies, contexts wherein the energetic and socio-reproductive cost-benefit parameters were relatively clearly discernable and decision making was restricted to tradeoffs along relatively few axes, allowing for reasonably reliable rational calculations. Hence the many successes of behavioral ecology in explaining economic decisions in foraging societies (e.g., Blurton Jones 1987; Hawkes 1993) or reproductive decisions in both simple societies and family groups within more complex societies (e.g., Borgerhoff Mulder 1990, 1992), contexts where the cultural filter is minimally opaque as regards true somatic and reproductive costs and benefits.

However, the environmental contexts in which those decision rules come to be applied in larger, more complex societies become orders of magnitude more complex. The environmental picture becomes progressively murkier due to an increasing multiplicity of socioeconomic and sociopolitical costs and benefits, with "hidden" costs and benefits becoming harder to discern and factor into reductionist gaming equations for both "players" and scientific "analysts" alike (e.g., see Low 2000; Kaplan and Lancaster 2000; and Mace 2000 regarding the modern demographic transition to low fertility). Adding to that murkiness is the fact that the increased socioeconomic and sociopolitical complexity is typically accompanied by a commensurate increase in ideological complexity, which may further cloud the cognitive ability of individual "players" to accurately discern objectively accurate biological costs and benefits when making decisions. The decision rules can, no doubt, still be said to operate in such contexts. However, as often as not, they progressively fail in their function of optimizing energetic and reproductive outcomes and instead work to optimize other outcomes culturally defined as desirable, which may in fact not be somatically or reproductively adaptive at all (see Boyd and Richerson 1992; Richerson and Boyd 1992; Laland and Brown 2002:145 ff.). Thus, without substantial reference to the sociocultural environment, in more complex contexts the reductionism of behavioral ecology potentially runs into difficulty distinguishing between adaptive behaviors in culturally complex (i.e., opaque) environmental contexts, seemingly maladaptive behaviors that make adaptive sense only in such contexts, genuinely maladaptive latent behaviors that can develop in such contexts, and anything in- between such absolute states (see Durham 1991:362ff; Richerson and Boyd 1992; see also Smith 2000:32). And, all this does not even take into account the fact that in social situations people may act "irrationally" with respect to costs and benefits based on perceptions of fairness (e.g., see Nowak et al. 2000)

Boone recognized the fundamental problem inherent in any reductionist approach to the evolution of social complexity in his attempt to deal with the subject. He noted that if "groups form out of mutual self-interest," on the part of rational individuals, "how can we explain the development of inequality and exploitation within groups" (1992:301)? The answer he proposed is that "the lack of alternative [optimization] strategies... may promote group affiliation even in the face of *extreme* disadvantage to some, perhaps most, of its members" (1992:302, emphasis added). Boone then proceeded to model the institutionalization of *structured* inequality in terms of typically individualistic resource optimization models and game theory in the context of unequal access to resources. However, there are two fundamental problems with his model.

First and foremost, he equates the development of differential wealth with the development of social hierarchies, but formal hierarchal relationships are based on differential power, not differential wealth. And while differential power tends to generate differential wealth as a result of self-interested actions, differential wealth does not necessarily generate differential power *of the type under consideration*. This is particularly the case in the context of the social and economic leveling mechanisms that would exist in any (even only nominally) egalitarian group, given the documented seriousness with which the individuals in such groups tend to protect their egalitarianism (e.g., Cashdan 1980; Lee 1979; von Furer-Haimendorf 1969; see also Boehm 1999, 2000).

Second, while Boone casts his analysis in purely individualistic terms, he too invokes what amounts to group-level selective considerations in his references to "public goods" (1992:327), which play a prominent role in the model he proposes. Thus, elements of his model should properly be viewed in terms of group selection and the top-down effects of group membership.

The ultimate problem with Boone's model is that it treats the individual as the sole meaningful frame of reference. For reductionism to successfully explain, as it does in the physical sciences and sometimes does in biological evolution, the whole must be accounted for by the sum of the parts. But complex social systems are not simple sums of constituent individuals. Some aspects, notably social structure and organization, are emergent properties of the group as a whole, resulting from interaction between the group's constituent individuals. These properties can affect both the constituent individuals as individuals and the group as a whole.

In the case of Richerson and Boyd's (1999) attempt, the authors do indeed invoke between-group selection in their analysis. However, they concentrate on how specific social instincts—a property of individuals—are harnessed to create such societies and maintain them, the cultural mechanisms that facilitate such harnessing, the direct between-group competitive advantages such groups enjoy over more simply organized groups, and the competitive advantage of centralized societies that can better harness those social instincts over other centralized societies that do so less well. But, selection only operates on what is. How some new organizational variant came to be, and so became visible to the operation of selection, is an independent process—mutation in the case of biological traits and innovation in the case of cultural traits. In other words, they analyze why centralized societies are successful as groups in between-group competition and how cohesion within such groups is fostered in the face of the above-mentioned inequalities. However, they do not address the issue of how and why leadership based on institutionalized coercive power comes into existence in the first place. They more or less simply take it as a given that somehow, somewhere coercive authority came to be invested in some individuals within a given group and proceed from there.

Process, Culture, and Constraint

Evolutionary processual, like earlier processual models, are systemic in approach. But, ironically they suffer from a similar failure to account for systemic properties as do reductionist attempts to explain the evolution of social complexity.

Those that invoke an element of voluntarism for the evolution of social complexity, whether transformational (e.g., Johnson 1978; Service 1975; Wittfogel 1957) or evolutionary (e.g., Stanish 2004; Stanish and Haley 2005), either explicitly or implicitly invoke functional benefits (i.e., organizational benefit for the accomplishment of some collective purpose or public good). However, they too fail to demonstrate how the new mode of organization based on coercive authority developed in pristine contexts beyond explicit or implicit reference to unspecified systemic dynamics (e.g., Flannery 1972) or agency (e.g., Stanish 2004). Thus, of necessity, they are required to assume that persuasive power and institutionalized coercive power represent an easily traversed, unbroken sociopolitical continuum (e.g., Stanish 2004; see also Stanish and Haley 2005) and that the former can gradually develop out of the latter without significant social hindrance. As will be explored, that is a dubious assumption. In addition, they invoke conscious acts on the part of people predicated on the additional dubious assumption that the uncertain future benefits of some hitherto untried mode of organization outweigh its immediate costs in the form of loss of autonomy.

Explanations that invoke social manipulation (e.g., Bender 1990; Ellen 1982) typically revolve around individual agency. The fundamental problem with such models is that, while they tend to pay lip service to social constraints, they implicitly view such constraints as generally capable of being circumvented. That is, manipulation models may invoke the same control-of-wealth/trade-generates-power equation invoked by Boone (1992) for the origins of socially sanctioned power. Alternately, they may invoke control over restricted knowledge to produce the change from "power to" to "power over" (Bender 1990:259). Or, they may invoke the gradual accumulation of power by leaders with initially limited investitures of authority, such as war leaders (e.g., Carneiro 1998). The problem here is that all such processes ignore the documented diligence with which would-be leaders who overstep their mandate are reined in by egalitarian societies—by recourse to murder if necessary —and are predicated on the tenuous assumption that such diligence can be circumvented (see Boehm 1999, 2000).

Explanations that invoke coercion revolve around the violent subordination of groups by other groups (e.g., Carneiro 1970). These suffer from a different systemic problem. To begin with, they are implicitly predicated on some starting level of differential within-group power as a given. Social complexity is not simply a matter of between-group status differences but also within-group status differences, and the latter are arguably a prerequisite for the former. That is, the concepts of extra-familial hierarchy and exploitation must already exist within a cultural system in order for them to be conceivable options for expanding to higher order, intergroup relations. Thus, coerced subordination of groups is a potentially viable explanation for the development of class systems from some starting level of formalized interpersonal inequality but not for the origins of such interpersonal inequality. This is a point recently recognized and addressed by Carneiro (1998), but unfortunately by reference to social manipulation.

In a related vein, they presuppose a level of organization already incorporating some internal degree of parasitic appropriation, such that the desire to absorb others in order to systematically appropriate their labor and/or resources was already a conceptual option. They further presuppose that social and ideological organizational features were already in place to make such appropriation both possible *and worthwhile*. Were they not, the winners would simply kill the losers or drive them from their land and appropriate their land for themselves, because that would arguably be the optimal benefit of victory[4] under a *pre-existing* egalitarian form of organization. If military success in circumscribed environments was all that was required for the emergence of social complexity, then the highly circumscribed New Guinea highlands would have been rife with ranked if not truly hierarchical societies at the time of western contact.

As argued elsewhere (Rosenberg 1994), cultural *Baupläne* constrain the directions of potential change,[5] and a *Bauplan* that incorporates an egalitarian ethos—however nominal—will not allow for the legitimation of formal hierarchy. A culture's *Bauplan* is the system's "organic design" in the form of a conceptual core. It is neither the "essence" of a culture (see also Boggs 2004) nor is it the system's *external* boundary, any more than a building's steel girders constitute its essence or boundaries. It is the structural core of that system and its internal "boundary," in the sense that one cannot simply add to a conceptual "building" in places where the existing conceptual girders do not offer support; doing so despite the lack of support will simply cause it to collapse.

True, individuals compete, maneuver, contest, manipulate, and negotiate for social advantage in a dynamic social environment. In other words, they "game" the social system in any and every way possible. But, that social environment is structured by the *Bauplan*, and they must game it within the constraints of the

Bauplan—one that is structurally egalitarian, even if only nominally so. Metaphorically speaking, one gains no "points" with anyone at all by unilaterally starting to play football during a soccer game and tackling another player. Put another way, if a socially disapproved, potentially self-advancing behavior is engaged in, it only accomplishes the opposite of self-advancement. Attempts to socially go beyond the constraints of the *Bauplan* will meet with reactions that have the opposite effect than was sought by the maneuver. In the case of one that is based on egalitarianism, it is a loss of relative respect, and hence power (see Boehm 1999).

The point is that for authority to become institutionalized the existing Bauplan must be abandoned and a new one forged. This is not something within the capabilities of any single individual—any individual so attempting will simply be branded a deviant and dealt with accordingly by the group. It is a group-level phenomenon capable of being produced in either of two ways. The first as a consequence of numerous individuals independently acting in a generally similar deviant manner *despite* the social sanctions and establishing a new behavioral pattern through deviant behavioral *Force Majeure*, leading to the collapse of the old *Bauplan* and the forging of a new one that is consistent with the newly established behavioral pattern (see Rosenberg 1994). The second is as a voluntary group-level decision, akin to a modern corporate reorganization. This will be suggested here to be the case in the evolution of inequality, in contexts where such a change will yield immediate and easily perceptible net top-down benefits.

Social Complexity and Group Selection

While it is no longer accurate to speak of any specific human attribute as being distinctly human in some absolute sense, human sociality is distinctly human in the extremely high degree of complexity to which it has evolved in our lineage. Moreover, one can make a convincing case that for no other species has group selection—including selection for relative social complexity—been as important as it has for our own.

Human social groups are always characterized by emergent properties, in that all are highly structured on the basis of culturally defined status relationships. Status relationships are always paired and thus cannot be a feature of a solitary individual (e.g., see Goodenough 1981); they only come into being in the context of group formation. Statuses always carry reciprocal rights and obligations and so are also characterized by some form of exchange of rights and obligations, if nothing more, making exchange an equally consistent emergent characteristic of groups.

The sexual division of labor-based exchanges that characterize human family organization are said to be the basic structural building block of human societies. But, a division of labor is an emergent property of a group. Individuals may derive somatic or reproductive benefits as individuals for being members of such groups, but they only do so through the top-down effects of membership in such groups.

While group selection no doubt operates at the family level, at this level of direct one-on-one exchange, in practice one need go no further than invoking some combination of self-interest, kin selection, and organismic selection for reciprocal altruism to adequately explain the evolution of the family as a social adaptation (see also Boehm 1999:210). However, such reductionist explanations begin to lose some of their potency when applied to the more widespread generalized reciprocal exchange of resources within multifamily groups (e.g., the sharing of meat within relatively simply structured egalitarian foraging groups), wherein the problem of cheating is an ever-latent threat (e.g., Blurton Jones 1984, 1987; Boehm 1999).

While the explanation for the evolution of such larger social systems can again be phrased, albeit a bit more weakly, in terms of reciprocal altruism and organismic selection (e.g., see Bleige Bird and Bird 1997; Blurton Jones 1984, 1987), it can just as readily be rephrased without conflict in terms of modern concepts of group selection (see Wilson 1998; see also Dugatkin and Reeve 1994) and the concordance of an emergent group-level attribute's top-down contribution to enhanced individual fitness, together with its contribution to the relative success of the multifamily group as a whole. However, while invoking group selection adds a layer of understanding concerning the evolution of multifamily egalitarian groups, reductionist explanations, including those focusing exclusively on top-down group effects (e.g., Boone 1992), can again be said to suffice.

Generalized reciprocity only generates effective benefits in relatively small groups, simply structured, and characterized by face-to-face relationships. In larger group contexts, it is simply too easy for cheaters to avoid the inevitable social penalties drawn by consistent cheating (e.g., see Boehm 1999). Also, the social aspects of the penalties themselves have little power, if the jeopardized social relationship of the distant potential punisher to the would-be cheater has little value to the latter. Thus, for still more complex social systems, the most successful have been ones in which different kinds of exchanges beyond generalized reciprocity play a prominent role, ones formally structured to make cheating inherently difficult.

Occupational specialization is nothing more than a higher order division of labor at the societal level. Administrators are nothing more than a specific type of still higher order specialist who coordinates the activities of other specialists. Their latent value as specialists develops when social structure reaches levels of "organic" interdependence such that coordination begins to positively affect the efficiency of other individual occupational specializations (e.g., see Tainter 1988). Authority makes possible more efficient administration than does a reliance on just persuasive influence. But, authority is the sanctioned right to make *and enforce* decisions. It is explicitly coercive and its appearance in human social systems constitutes a drastic social innovation from the (at least nominal) egalitarianism that preceded its appearance.

The between-group advantage of more complex social systems over relatively simpler systems can easily and amply be documented in the historical and ethnographic record (e.g., see Diamond 1997), a point recognized by both Boone (1992) and Richerson and Boyd (1999). However, the between-group advantage of greater complexity does not necessarily produce positive top-down effects on individual fitness This is particularly the case for complex, centralized social systems with hierarchical power relationships that invest authority in specific individuals. Stratified redistributive exchange systems play a prominent economic role in such centralized social systems, with socially sanctioned extra-kin authority wielded by hereditary dominants, who appropriate some if not all of the surplus resources and labor of subordinate non-kin individuals—much of it for their own private use in one form or another. In purely objective terms, this conforms to what is usually called parasitism when seen in other species. Moreover, such "parasitism" can potentially be extreme and often enough was.

One must not confuse modern, semi-egalitarian[6] democratic systems, which represent something of a social reversion to the bottom-up system of checks on dominants that characterized earlier pre-authoritarian systems (see Boehm 2000), with earlier despotic systems. These modern systems minimize the parasitism of dominants to some tolerable degree, and the large majority of members likely do end up deriving net tangible top-down somatic benefits from group membership.[7] However, selection operates on the basis of current fitness, not future fitness. The earliest authority-based systems did not evolve with the knowledge, expectation, or even hope on the part of the ancients that eventually more egalitarian sociopolitical systems might again someday be a derivative evolutionary result. More often than not, in such despotic systems, the large majority of members were likely left with little more than the bare minimum of resources required for survival and reproduction,. In contrast, dominant individuals and groups enjoyed significant somatic benefits and reproductive advantages.[8]

The oft-cited, ostensibly beneficial redistributions that characterize such systems were more often than not in the form of services (e.g., protective, resource enhancing) than in the form of direct resources; and, a good case can be made that elites derived significantly greater net benefits from any and all such redistributed services than did the large majority of subordinate group members.

For example, irrigation/terracing/flood control systems provided larger surpluses for elites to appropriate. They likely did not provide significantly increased surpluses for peasants to retain or use, because those would by definition be *surpluses* and thus available to be appropriated for use by elites. Likewise, religious activities and religious edifices validated the elites' parasitic statuses, while providing the large majority with no more supernatural aid than they also provided the elites. In the same vein, militaries provided for the protection and projection of the elites' power (i.e., the ability to appropriate from their own and other groups), while protecting the large majority of members from external threats no more than such militaries also protected the elites themselves. The case for skewed benefits is arguably also possible for any functional benefits from any other such redistributed public services.

Moreover, even direct redistributions of resources in times of crisis would arguably be beneficial to the large majority of members only because their household surplus (which would otherwise have been retained at the household, egalitarian peer group, or kin group level) had already been largely, if not totally, appropriated by elites. Thus, such potential surpluses were unavailable for independent storage and subsequent independent use by households or local community groups in such times of need. In a very real sense, such direct redistributions of resources only serve(d) to minimally mitigate the parasitism of elites by, when necessary, maintaining their subordinates at some minimal level for survival. Bluntly put, a parasite's fitness suffers if it lets its host die, for it too then dies. It serves the interests of elites to maintain their subjects through such direct redistributions when the need arises.

The point is, one cannot plausibly explain the evolution of human societies more complex than at least nominally egalitarian ones in purely reductionist terms, even with added reference to the group effects of such social systems on individuals. That is because most individuals in such social systems arguably enjoyed no greater somatic or reproductive benefits over their ancestors living in early (i.e., simpler) food-producing societies—and likely even fewer such benefits given the labor-intensive nature of intensification (see Boserup 1965). The explanation for the evolution of complex societies lies not in organismic selection, with or without top-down group-effect. Rather, it lies in the direct operation of Darwinian between-group selection on such groups as higher order individuals, based on the emergent socioeconomic and sociopolitical properties of such groups *as social groups*. More complexly organized groups enjoy a competitive advantage *as groups* over less complexly organized groups. They tend to grow at their expense, by either expanding to displace them because they can compete more effectively for territory and the resources therein or, in some cases, absorbing them outright at the lowest parasitized levels of membership (see Carneiro 1970).

However, again, most of the individuals in such sucessful (more complexly organized) groups do not actually derive any top-down benefits from the absorption of competing groups, since the marginal added production of the newly absorbed members would be appropriated by elites. Membership merely prevents some latent even greater loss of fitness posed by the mere existence of such groups and the potential, if not real, competition for resources that they pose. Thus, the main benefit of membership to most individuals is in the form of defense against other groups and the threat of either being displaced, killed outright, or being individually absorbed by other groups and even more severely parasitized (e.g., enslaved) than they may already be in their current hierarchical social level.

Proximate Causation and Exaptation

Nominally egalitarian groups arguably constitute the organizational apex of top-down beneficial effect on individual fitness resulting from organizational properties in primary contexts. The large majority of members of the first such group to cross the threshold to any form of hierarchically coercive social structure would potentially suffer some relative loss of individual fitness by becoming parasitized to at least some degree by emergent elites, without gaining even the nominal benefit of protection from other, more centrally organized groups, because such other groups *do not yet exist* to pose competition. More importantly, even if we accepted the hypothetical proposition that such parasitism was not an initial structural feature of the transition, the majority of members would nevertheless suffer an immediate and perceptible loss of personal autonomy *that is widely understood by members of egalitarian societies to be easily exploitable* (cf. Boehm 1993, 1999, 2000).

True, one could hypothetically argue that unilaterally changing to a centralized, hierarchical form of societal organization would provide an organizational benefit resulting in the enhanced ability to drive less organized competitors off their territory. This would make such appropriated territory available for the entire membership of such a centrally organized group to demographically expand into and so opportunistically enhance the reproductive fitness of *all* the victorious group's members. But, this presupposes the rather dubious proposition (see Rindos 1985:72) that the *future* benefits of a hitherto untried mode of organization, of which a group has no prior knowledge or with which it has no prior experience whatsoever, are so obviously apparent as to negate the *immediate* loss of personal autonomy for the large majority of members and the widely recognized risk of abuse by dominants that such a change makes potentially possible.

The obvious question all this raises is the same one that has been debated all along. Why did social complexity involving any degree of structured sociopolitical inequality evolve in pristine contexts and hence the absence of a social arms race? Why would coercive power be permanently invested in some individuals, given the conscientiousness with which even nominal egalitarianism is maintained in non-hierarchical societies (see Boehm1999, 2000), and given the well-documented awareness of individuals in egalitarian societies of the danger to their autonomy posed by the accumulation of power by would-be dominants (e.g., see Boehm 1999, 2000; Lee 1979). Why would individuals willingly give up their personal autonomy and knowingly make themselves susceptible to potential exploitation by a social system wherein the potential exploiters are obviously allowed to control the level of potential exploitation.

Boehm (1993, 1999, 2000) cogently argues that the egalitarianism of simple societies represents an evolved coalition-based bottom-up check on dominants in our lineage. It is rooted in our tendency as primates to form coalitions, coupled with our high sense of self-awareness and related ability to calculate socially affected somatic self-interest; and, it is expressed by the propensity of subordinates to "act collectively in an 'insubordinate' capacity" (1999:170). In other words, it is rooted in the same ability to calculate cost-benefits that underpins behavioral ecology. The problem is delineating some set of immediate conditions that would plausibly negate or overrule individuals' strong self-interest in maintaining the egalitarian status quo, such that latent dominants could, in practice, achieve blatantly dominant status with the concurrence of the group.

The answer must obviously reside in the social purpose serving as the justification for the *initial* legitimation of authority. However, such a primitive purpose need not necessarily be identical to potentially latent functional benefits that coercive leadership may quickly evolve to take on as exaptations that provide advantage in intergroup competition once formal hierarchical structures come to be established.

Richerson and Boyd (1999) correctly note that the evolution of complex societies involves elements of both the voluntarism and coercion alternately invoked by many of the classic transformational and processual explanations. Stanish (2004; see also Stanish and Haley 2005) implicitly makes the same point in the context of his evolutionary processual approach to the problem. Unfortunately, Stanish conflates evolved function with cause in the matter of proximate causation, as did earlier voluntaristic models; and, the traditionally invoked organizational or integrative functions of voluntaristic explanations only confer benefits to the group in the context of between-group competition. Thus, it seems worthwhile to look elsewhere, on the grounds that these benefits may very well be social exaptations and that the origins of coercive leadership within groups may lie in another, *earlier* selected-for function entirely. This earlier selected-for function no doubt continues to provide benefit in some capacity but rapidly becomes selectively eclipsed by the organizational competitive advantages of coercive leadership, *once it comes to be established*. This primitive benefit also comes to be conceptually overshadowed by the later exaptations, which are more prominently visible to anthropologists and have, therefore, tended to figure more prominently in their explanations for the evolution of social complexity. I suggest that primitive benefit is intra-group conflict resolution.

Conflict, Conflict Resolution, and Proximate Causation

Both de Waal (1982, 1996) and Boehm (1993, 1999) argue that conflict resolution, by active intervention if necessary, is an important social role played by dominants in nonhuman higher primate societies. Conflict resolution is an equally important aspect of leadership in egalitarian human societies—even nominally egalitarian societies that go so far as to display elements of ephemeral authority (e.g., see Maybury-Lewis 1974:203–204). However, in the absence of authority, the ability of primus inter pares type leaders to effectively intervene to resolve conflicts is severely curtailed, because they cannot impose a resolution, resulting in the well-documented tendency for egalitarian groups to fission as a consequence of unresolved conflicts.

Richerson and Boyd (1999) also note the importance of intra-group peacekeeping as a function of leadership. However, peacekeeping is apparently seen by them as a secondary, though crucial, function of leadership, essential for the maintenance of the organizational features that determine relative between-group advantage. I suggest that they have erroneously focused on the evolved functions of the social adaptation that is coercive leadership (i.e., organizational benefit), not its primitive function (conflict resolution). I suggest that it is this need to maintain intra-group peace—and with it group cohesion that serves as the catalyst for the investiture of authority in contexts where the maintenance of group cohesion, and hence relative size, is crucial for the perceptible well-being of *all* its members.

I agree with Richerson and Boyd (see also Keeley 1996) about the prevalence of conflict as an expression of intergroup competition (e.g., see Rosenberg 1998) and also about the importance of group cohesion in the context of conflict. However, I suggest that it is the internal peacekeeping function of leaders in times of conflict, not their value as military leaders (e.g., Carneiro 1998) or their value as organizers (e.g., Stanish 2004), that serves as the initial catalyst for the investiture of authority in them.

The suggested role of conflict in the formation of complex societies was most systematically presented by Carneiro (1970). Simply summarized, he proposed that in circumscribed environments, winning groups would absorb the losers, thus establishing the social hierarchies that characterize complex societies. However, I would turn that around and suggest that the ultimate roots of social complexity lie not in the emergent relationship of the losers to the winners in some conflict but in the relationship of the losers to the friendly groups with whom they are forced to seek refuge when driven off their lands.

No matter how tightly circumscribed any given area may be physically, the individual local groups within it are still further circumscribed socially, in that they are invariably surrounded by multiple neighboring local groups in a dynamic social environment. That is, it is difficult to envision a circumscribed physical environment so small as to be populated by just two small local groups (i.e., hamlets or villages)—any environment so small would simply be too unstable to support any human populations at all on a long-term basis. Such other local groups may be either related at some level or not, and friendly to them, neutral, or hostile. While any conflict between such groups could potentially involve large shifting alliances made up of such multiple local groups, ultimate defeat would still tend to befall one local group at a time (e.g., see Rappaport 1968). Egalitarian societies have neither the numbers nor the organization to displace more than one or at most a very few local groups at a time. The point is that even within a highly circumscribed physical environment, losing local groups would still have other friendly or allied local groups with whom to take refuge, even in the logically unlikely event (see above) that the winners would somehow wish to absorb them as a group in order to exploit them.

While it is possible to envision many contexts wherein such refugee groups would not be at a social disadvantage, it is also possible to envision contexts where they would in fact be disadvantaged. One such context would likely be a social landscape consisting of local groups dominated demographically by single lineages, wherein communal ritualistic power is also strongly tied to place. In such contexts, the refugees would at least initially be at some level of, however friendly or even related, "others" relative to the resident group by virtue of descent. Moreover, they would also be relatively powerless having been dispossessed of the source of their own ritualistic power, while living among another cohesive self-identifying group that still maintained theirs (cf. Bender 1990).

In any such disadvantaged and dependent situation, a refugee group would be highly susceptible to subtle manipulation, if not subtle coercion, by their hosts to the relative economic and reproductive advantage of the hosting group. And, given normal human tendencies to pursue self-interested goals to the degree socially possible, any refugee group of "others" in such a position is highly likely to be so manipulated to at least some extent. The point is that any relative advantage held by a hosting group depends on the maintenance of cohesion within that group; any conflicts of interest producing a split would potentially negate their social advantage. At worst, it could result in a realignment, potentially incorporating the refugees, which produces two new groups with equal access to ritualistic power, by virtue of each having members drawn from the original, now split, hosting group. At best, it would water down their social advantage by reducing the advantaged group's size and concomitant ability to manipulate the refugee group.

Spencer (1993:48) suggested that authority initially comes to be invested in leaders when the "success of individuals [becomes] dependent on the success of the group." I suggest that it is precisely in such advantaged contexts that this occurs. It is a social given that, even while part of a social group, individuals will pursue their own interests and those interests will at times be in conflict. Such conflicts, if not resolved, have the potential to fracture the group; and, for members of an advantaged group, such a fracturing would weaken their position of relative advantage, thus jeopardizing the welfare of the group as a whole and all its members. I suggest that in such advantaged contexts, the members of the advantaged group would formally delegate coercive power to one of their own to facilitate conflict resolution. This would maintain cohesion, and thereby the maintenance of their perceived individual *net* advantage even after their subordination to the new, formal dominant is factored into the equation.

I suggest that it is only in such contexts that individuals would see it as their net advantage to do so. Moreover, to invoke the distinction apparently made by some Native American tribes (e.g., see Lowie 1973), it is the *primus inter pares* leader of the "peace chief" type[9] who is likely to be voluntarily granted formal power on the basis of collective self-interest, not the "war chief" type who somehow transforms his limited (and consciously curtailed) military power into more generally applicable power through coercion or manipulation. Would-be dominants need not manipulate the system in the modes typically invoked by agency models to gain coercive power against the group's natural reluctance to grant it; they are voluntarily vested with it by the group as needed to maintain cohesion and thereby relative group advantage. They only need to maneuver through expressions of generosity and fairness (cf. Stanish and Haley 2005) *within* the pre-existing egalitarian constraints of the social system to be the individual(s) so selected, because they would seem least likely to abuse such power.

In other words, I suggest that it is the same inherited human tendency to form alliances (see de Waal 1982, 1996), which Boehm (1999, 2000) suggests underlies an egalitarian society's ability to maintain its egalitarianism, that also underlies emergent inequalities. In the case of egalitarianism, it is an alliance of the latent "underdogs" against the latent "top dogs" that serves to maintain that egalitarianism and forestall potential disadvantage to the alliance's members. In the case of emergent complexity, it is an alliance of the same latent "underdogs" *with* a latent "top dog" that serves to ultimately grant him formal dominance and so maintain their collective at-this-stage-still informal dominant position for the group as a whole.

However, *any form* of structured inequality is a *radical* departure from a pre-existing egalitarian system. The resultant conflicts and contradictions with the still-existing egalitarian *Bauplan* will result in a rapid (i.e., punctuational) disintegration and reorganization of the entire cultural system (see Rosenberg 1994) around a new *Bauplan* incorporating a broadly and generally applied system of formal social ranking. Such rankings will more likely than not be structured around kinship and supernatural descent, simply because kinship is a structural element that can readily be carried over without conflict from the older, abandoned system. In any case, once any form of inequality is structurally sanctioned, even if only internally, it has the potential to be extrapolated to any and all other applicable circumstances (e.g., intergroup relations), and the dominance of one group over the other(s) inevitably becomes formalized as well.

I further suggest that at this early stage of the process, hierarchically organized complex societies typically do not yet grow by the forced incorporation of other groups in the fashion Carneiro proposed. Rather, I suggest that they more often than not grow by continuing to dispossess other groups of their territory and occupying any such newly acquired land themselves through population growth (i.e., the latent tendency for population to grow, made manifest by the availability of additional land to expand into) and/or the attraction of friendly and likely related groups that voluntarily join them in the context of the social arms race that has now begun. At this stage, the level of resource appropriation engaged in by the dominant subgroup is likely to be still relatively mild, and advertisement in the form of aggrandizing behavior for purposes of attracting members may be a feature (e.g., see Prentiss 2010). Lower ranked groups are not likely to vigorously contest the emerging new social order to the degree that, given their only limited "otherness," they likely identify at some significant level with the dominant individual and subgroup, and their subordinate status is only marginally so.

However, resource appropriation, once institutionalized at even some low internal level is now, like inequality, also culturally available conceptually for extrapolation to social relations marked by even greater degrees of conceived "otherness." Moreover, the temptation to thus appropriate (i.e., get something for "nothing")[10] from groups where the scale would be unchecked by the constraints of group identification can be assumed to be a given.

Authority-based centralized leadership, once established, enhances organization and that affects group-level success in competition with other groups. But, potential organizational improvements are not the precipitant of the social innovation; the need to maintain group cohesion is. Once authority is established in some residential community, group size will remain stable due to effective conflict resolution and can even continue to grow. Thus, there would rarely be a context where the basis for that investiture will expire before it becomes well established as a cultural norm. Once established as an accepted norm, it is capable of even being carried into new environments by relatively small breakaway groups, who do not need the basic "public good" of large group size maintenance (e.g., see Kirch 1984).

Structural Variability, Structural Similarity, and Selection

As noted by Leonard and Jones (1987) and others (e.g., Yoffee 2005), each from their own perspective, complex societies do not readily conform to types. The institutionalization of coercive power is a process and thus an entirely different matter than the organizational structure of such power and its trappings. Both of the latter are highly variable, being strongly influenced by the opportunistic exaptations that then evolve based on the availability of such formalized power after it becomes established. The initial appearance of hierarchy may be processual, but the subsequent evolution of complex societies is contingently historical and shaped by the interaction of opportunity and need. They vary in the historical order of their constituent institutions' development, the rate of their individual development, and the specific forms such developing institutions take. Thus, the form any initial investiture of authority takes is unlikely to be uniform. It could be primarily ritual based, primarily kinship based, or some combination of the two. It could be organized along corporate or network lines, or some combination of the two (see Blanton et al. 1996). It could also potentially be either heterarchical or hierarchical (see Crumley 1987, 1995). However, the prevalence of more hierarchical types of organization in the largest (i.e., most successful) known complex societies suggests that hierarchical organization tends to be selected for at the between-group level over the long term, as may network over corporate organization.

Once coercive authority is institutionalized, the organizational structure of leadership develops to accommodate organizational need, as enhanced organization makes possible new collective endeavors made desirable by environmental factors and/or historical opportunity. It does not follow some predetermined developmental trajectory. The organization of power is a potential product of administrative needs, itself a product of what is being administered (e.g., taxes, tribute, trade). For example, as argued by Spencer (1998), effective projection of political power beyond the boundaries of the core polity, for purposes of exploitation, requires the development of some form of an administrative system characterized by internal specialization and partial delegation of decision making. Those polities that do not develop such administrative systems, if not first outcompeted by those that do, soon collapse under the strain of managing a growing polity that becomes too unwieldy for a still-fully centralized and still-unspecialized administrative system to effectively administer. Those polities that do innovatively develop some such systems have the potential to continue to grow until such time as the opportunistic or structural ability to exploit is no longer sufficient to maintain the cultural institutions that opportunistically came to be built thus far, or until such time as growth is halted by competition with another more powerful or better administered polity.

However, the organization of power is different than its trappings, such as ostentatious displays of culturally sanctioned coercive power and its ideological foundations. As long as a developing complex society's membership remains restricted to a single, however loosely self-defined and self-identifying group, the legitimacy of dominants is implicit in the information system (i.e., culture) shared by the group.

Leaders come to symbolize the group, and followers come to identify with leaders to some degree as embodiments of the group itself (e.g., see Levine 1967). It is in this kind of social context that the mechanisms explored by Richerson and Boyd (1999) operate most effectively. Thus, as long as warfare by such a group is geared to competing with other groups for space into which to expand, highly visible propagandistic displays, ostentatiously legitimating the power of dominants, as opposed to those reinforcing group identity, are only minimally necessary. However, once—for whatever reason—the goal of warfare ceases to be the displacement of other groups for purposes of growth-fueled expansion, and instead becomes the subjugation of groups more definitively "other," for the explicit purpose of exploiting them in one capacity or another, then more prominent and dramatic representations of the validating ideology and the symbols of its power become necessary. They do so in order to more effectively control groups who may bow to, but nevertheless might otherwise vigorously contest the legitimacy of the dominants, because they do not naturally identify with the dominants at any social level.

The shift in goals from displacement to exploitation is not inevitable. The reasons for such a shift in goals no doubt vary but more often than not likely revolve around simple historical opportunity. That is, exploitation is already a feature of social relations by virtue of the new *Bauplan*. Thus, opportunities to more ruthlessly do so to groups deemed sufficiently "other" will be seized upon when they present themselves. This would particularly be the case if the territory occupied by such "others" cannot be as effectively utilized by the winners through demographic expansion (e.g., due to distance or spatial discontinuity) as it can be by leaving the losers in place and exploiting them. It is when such opportunities to do so arise, and are taken, that Carneiro's (1970) mechanisms for more marked class formation and Spencer's (1998, see also Spencer in this volume) for greater decentralization and specialization come into play, and even greater complexity follows.

Yet, for all the structural variability between the individual complex societies that evolved at various times and places in human history, there are broad similarities in many of the evolutionary trajectories of such societies. Such common trends include increasingly greater internal specialization and sociopolitical differentiation, increasingly distant exploitation of peripheral "others," increasingly formal differentiation of social statuses, increasingly elaborate ideologies of power, etc. These broad structural similarities are what prompted earlier generations of scholars to speak of sociopolitical "types" and "stages," and they can be most parsimoniously explained by reference to widespread parallel evolution. That is, the likelihood of pure chance producing such levels of fundamental similarity so many independent times in pristine contexts is minuscule. Thus, while evolution is ultimately historical, in that it based on contingency, such parallel evolution strongly implies the operation of recurrent psychosocial and sociocultural processes at the proximate level, producing a fundamental level of similarity in the social novelties that people produce and on which selection can then act.

Discussion

Social structure and organization are emergent characteristics of a social group and selected for in the context of group selection. Such selection is always between groups, but can also sometimes be at the organismic level, based on the top-down benefits of membership. But selection operates on what the social and physical environment produce in the way of innovative behaviors. Thus, the emergence of social complexity cannot be properly understood without reference to the social environment and the proximate social processes it tends to generate.

Evolutionary archaeology's devastating critique of processualism's transformationalism (e.g., see Dunnell 1980; Rindos 1985) was essentially correct. But, whatever its shortcomings with regard to evolutionary theory, processualism's problem orientation viewed emergent complexity in systemic terms and focused on proximate systemic processes that modern, better theoretically grounded evolutionary approaches have thus far had difficulty doing effectively.

Agency may not be knowable by *sole* reference to the archaeological record, as per Evolutionary Archaeology's critique (e.g., Rindos 1984, 1985). But, it is potentially knowable by recourse to behavioral tendencies documented in the data produced by Psychology and Primatology and testable ethnographically, if not archaeologically. In fact, Prentiss (2010) has recently documented a pattern of change in the Pacific Northwest that conforms remarkably well to the early stage of the model proposed here, as regards the emergence of complexity in the Pacific Northwest. Archaeology has historically borrowed liberally from other disciplines in its quest to explain the cultural past. Why should psychology and primatology be off-limits to evolutionary explanations when biology and geography are not and why should Behavioral Ecology have sole access to ethnographic data with which to evaluate its explanations?

Almost all processual models for the rise of complex societies, whether they meant to or not, implicitly invoked human agonistic behavioral tendencies, in the same manner as do Richerson and Boyd (1999). Such agonistic tendencies are amply documented by primate ethology and humans are, after all, primates. Those agonistic tendencies on the part of individuals were held in reasonable social check for most of modern human existence (e.g., see Boehm 1999, 2000), but at some point in the past, in some societies, they ceased to be and formalized dominance hierarchies developed. Such distinctions were an emergent property of the evolving social system. And, while perhaps the economic, environmental, and other components that dynamically contributed to their development as systems can be explained reductively, the evolution of the systems themselves can only be explained by reference to systemic dynamics.

Where most current explanations, whether evolutionary or processual, generally go wrong is in equating evolved function with cause. These commonly invoked organizational benefits of authoritarian leadership are exaptations. They are built on such leadership once it comes into being for other reasons, and then become factors in between-group selection if and when they do evolve. However, the future benefits of some unknown and untried organizational mode are a poor motivation to individuals to relinquish their personal autonomy to would-be dominants and thus open themselves up to exploitation.

The ultimate outcome of unresolved conflict in egalitarian societies is group fissioning. Thus, in cases where the benefit to individuals of group advantage outweighs the loss of individual advantage, and group advantage is tied to group cohesion, it is the maintenance of group cohesion that is the crucial "public good" whose needed maintenance overcomes the justified reluctance on the part of the majority to allow latent dominants to formally dominate. Task-based investitures of authority are inherently temporary, expiring with the completion of the task. Any attempt to manipulate an essentially egalitarian system in order to expand such temporary and limited mandates would likely meet with negative consequences and, if necessary, potentially drastic ones (see Boehm 1999).

Recent years have seen attempts to rehabilitate processualism by retaining its systemic focus but grounding it in a proper understanding of evolutionary theory. This resurrected evolutionary processualism has the potential to move beyond the explanatory limitations of the now-dominant reductionist approaches to cultural evolution. This is by virtue of its ability to effectively deal with not just microevolutionary mechanics but also macroevolutionary ones such as group selection and punctuational change, which are by definition systemic in nature. Abandoning transformational processualism was both correct and necessary. Abandoning systemic approaches to the explanation of proximate causation amounted to throwing the baby out with the bath water.

Notes

1. Exaptation is a term proposed by Gould and Vrba (1982) for some subsequent (selected-for) functional effect of an adaptation, which however was not the original function that the adaptation performed when it initially evolved in some more primitive form, under an earlier, different set of selective pressures. In other words, an exaptation is an "adaptation" that co-opts or builds upon what used to misleadingly be called a "pre-adaptation" and in which the new effect is selected for by some new set of behavioral and/or environmental conditions. For example, feathered wings are now thought to have originally evolved as a thermo-regulatory and/or display adaptation in a lineage of dinosaurs. The subsequent evolution of feathered wings as flight mechanisms is an example of exaptation, wherein the earlier structure was exapted (i.e., co-opted) for a new effect (gliding/flight) *after* it had already evolved in the context of its original (thermo-regulatory and/or display) function (see also Gould 2002:1231 ff.).
2. The extended phenotype can be simply summarized in general terms as the full range of (however indirect) phenomenological expressions of some starting genetic instruction set (a hermit crab's appropriated mollusc-shell home, a beaver's dam, etc.).

3. Drawing on Johnson (e.g., 1978, 1983)
4. In terms of both how members of the winning group would perceive maximum benefit and objective contribution to evolutionary fitness.
5. See also Ghirlanda et al. 2006 for a mathematical model that reaches essentially the same conclusion.
6. In the sense of equal political say by way of the vote.
7. Hence the net emigration from less democratic to more democratic systems that began with the first appearance of such systems a little over two centuries ago and continues unabated into the present.
8. See Borgerhoff Mulder (1990, 1992) for ethnographic examples of the relationship between wealth and reproductive success for males. See also Ridley (1993:193ff) for a more general summary of some of the historical evidence for the relationship between wealth/power and relative reproductive success of dominant males.
9. Whether or not the distinction is formally recognized in any given society.
10. Warfare, being a pre-existing condition, is no added cost.

Acknowledgments I would like to thank Guillermo Algaze, James Chatters, Niles Eldredge, Stuart Fidel, Ian Kuijt, Anna Prentiss, Robert Graber, and Charles Spencer for taking the time to comment on earlier drafts of this article, sharing their expertise and their constructive criticisms with me, and generally acting as sounding boards for me as the ideas contained herein took shape. I would also like to thank the three reviewers, whose constructive criticisms greatly improved the presentation of those ideas.

References

Arnold, A.J. and Fristrup, K. (1982). The Theory of Evolution by Natural Selection: A Hierarchical Expansion. *Paleobiology* 8: 113–129.

Bender, B. (1990). The Dynamics of Nonhierarchical Societies. In S. Upham (Ed.), *The Evolution of Political Systems: Sociopolitics is Small-Scale Sedentary Societies* (pp. 247–263). Cambridge: Cambridge University Press.

Blanton, R.G., Feinman, G., Kowalewski, S. and Peregrine, P. (1996). A Dual-Processual Theory for the Evolution of Mesoamerican Civilization. *Current Anthropology* 37:1–14.

Bliege Bird, R.L. and Bird, D.W. (1997). Delayed Reciprocity and Tolerated Theft: The Behavioral Ecology of Food Sharing Strategies. *Current Anthropology* 38:49–78.

Blurton Jones, N.G. (1984). A Selfish Origin for Human Food Sharing: Tolerated Theft. *Ethology and Sociobiology* 5:1–3.

Blurton Jones, N.G. (1987). Tolerated Theft: Suggestions About the Ecology and Evolution of Sharing, Hoarding, and Scrounging. *Social Science Information* 26:31–54.

Boehm, C. (1993). Egalitarian Behavior and Reverse Dominance Hierarchy. *Current Anthropology* 34: 227–254.

Boehm, C. (1999). *Hierarchy in the Forest: The Evolution of Egalitarian Behavior*. Cambridge, MA: Harvard University Press.

Boehm, C. (2000). Conflict and the Evolution of Social Control. *Journal of Consciousness Studies* 7:79–101.

Boone, J.L. (1992). Competition, Conflict, and the Development of Social Hierarchies. In E.A. Smith and B. Winterhalder (Eds.), *Evolutionary Ecology and Human Behavior* (pp. 301–337). New York: Aldine de Gruyter.

Boone, J.L. and Smith, E.A. (1998). Is It Evolution Yet? A Critique of Evolutionary Archaeology. *Current Anthropology* 39:S141–S173.

Borgerhoff Mulder, M. (1990). Kipsigis Women's Preferences for Wealthy Men: Evidence for Female Choice in Mammals. *Behavioral Ecology and Sociobiology* 27:255–264.

Borgerhoff Mulder, M. (1992). Reproductive Decisions. In E.A. Smith and B. Winterhalder (Eds.), *Evolutionary Ecology and Human Behavior* (pp. 339–374). New York: Aldine de Gruyter.

Boserup, E. (1965). *The Conditions of Agricultural Growth*. Chicago: Aldine.

Boyd, R. and Richerson, P.J. (1992). Punishment Allows the Evolution of Cooperation (or Anything Else) in Sizable Groups. *Ethology and Sociobiology* 13:171–195.

Brandon, R.N. (1988). Levels of Selection: A Hierarchy of Interactors. In H.C. Plotkin (Ed.), *The Role of Behavior in Evolution* (pp. 51–71). Cambridge, MA: MIT Press.

Brandon, R.N. (1990). *Adaptation and Environment*. Princeton, NJ: Princeton University Press.

Braun, D.P. (1990). Selection and Evolution in Nonhierarchical Organization. In S. Upham (Ed.), *The Evolution of Political Systems: Sociopolitics is Small-Scale Sedentary Societies* (pp. 62–86). Cambridge: Cambridge University Press.

Carneiro, R. (1970). A Theory of the Origin of the State. *Science* 169:733–738.

Carneiro, R. (1998). What Happened at the Flashpoint: Conjectures on Chiefdom Formation at the Very Moment of Conception. In E.M. Redmond (Ed.), *Chiefdoms and Chieftaincy in the Americas* (pp. 18–42). Gainsville, FL: University Press of Florida.

Cashdan, E.A. (1980). Egalitarianism Among Hunters and Gatherers. *American Anthropologist* 82: 116–120.

Chatters, J.C. and Prentiss, W.C. (2005). A Darwinian Macroevolutionary Perspective on the Development of Hunter-Gatherer Systems in Northwestern North America. *World Archaeology* 37:46–65.

Crumley, C.L. (1987). A Dialectical Critique of Hierarchy. In T. Patterson and C. Gailey (Eds.), *Power Relations and State Formation* (pp. 115–159). Washington, DC: American Anthropological Association.

Crumley, C.L. (1995). Heterarchy and the Analysis of Complex Societies. In R.M. Ehrenrich et al. (Eds.), *Heterarchy and the Analysis of Complex Societies* (pp. 1–6). Archeological Papers of the American Anthropological Association 6. Washington, DC: American Anthropological Association.

Damuth, J. (1985). Selection Among Species: A Formulation in Terms of Natural Functional Units. *Evolution* 39:1132–1146.

Damuth, J. and Heisler, J.L. (1988). Alternative Formulations of Multilevel Selection. *Biology and Philosophy* 3:407–430.

Darwin, C. (1871). *The Descent of Man and Evolution in Relation to Sex*. London: Murray.

Dawkins, R. (1986). *The Blind Watchmaker*. Harlow: Longham.

Dawkins, R. (1989). *The Selfish Gene* (new edition). Oxford: Oxford University Press.

Dawkins, R. (1990). *The Extended Phenotype: The Long Reach of the Gene* (new edition). Oxford: Oxford University Press.

Diamond, J. (1997). *Guns, Germs, and Steel: The Fates of Human Societies*. New York: W.W. Norton.

Dugatkin L.A. and Reeve, H.K. (1994). Behavioral Ecology and Levels of Selection. *Advances in the Study of Behavior* 23:101–133.

Dunnell, R.C. (1980). Evolutionary Theory and Archaeology. In M.B. Schiffer (ed.), *Advances in Archaeological Method and Theory, vol.3* (pp. 35–99). New York: Academic Press.

Dunnell, R.C. (1985). Methodological Issues in Americanist Artifact Classification. In M.B. Schiffer (Ed.), *Advances in Archaeological Method and Theory, vol.9* (pp. 149–207). New York: Academic Press.

Dunnell, R.C. (1995). What is it That Actually Evolves? In P.A. Teltser (Ed.), *Evolutionary Archaeology: Methodological Issues* (pp. 33–50). Tucson: University of Arizona Press.

Durham, W.H. (1991) *Coevolution: Genes, Culture, and Human Diversity*. Stanford: Stanford University Press.

Hamilton, W.D. (1971). Selection of Selfish and Altruistic Behavior in Some Extreme Models. In E. Eisenberg and W.S. Dillon (eds.), *Man and Beast: Comparative Social Behavior* (pp. 57–91). Washington, DC: Smithsonian Inst. Press.

Eldredge, N. (1985). *Unfinished Synthesis: Biological Hierarchies and Modern Evolutionary Thought*. New York: Oxford University Press.

Eldredge, N. (1989). *Macroevolutionary Patterns and Evolutionary Dynamics: Species, Niches and Adaptive Peaks*. New York: McGraw-Hill.

Eldredge, N. (1999). *The Pattern of Evolution*. New York: W.H. Freeman.

Eldredge, N. and Gould, S.J. (1972) Punctuated Equilibria: An Alternative to Phyletic Gradualism. In T.J.M. Schopf (Ed.), *Models in Paleobiology* (pp. 82–115). San Francisco: Freeman, Cooper.

Ellen, R. (1982). *Environment, Subsistence, and System: The Ecology of Small-Scale Social Formations*. Cambridge: Cambridge University Press.

Fisher, R.A. (1958). *The Genetical Theory of Evolution (Second Edition)*. New York: Dover.

Flannery, K.V. (1972). The Cultural Evolution of Civilization. *Annual Review of Ecology and Systematics* 3:399–426.

Ghirlanda, S., Enquist, M. and Nakamura, M. (2006). Cultural Evolution Develops Its Own Rules. *Current Anthropology* 47:1027–1034.

Goodenough, W.H. (1981). *Culture, Language, and Society* (second edition). Menlo Park, CA: Benjamin/Cummings.

Gould, S.J. and Vrba, S. (1982). Exaptation – a Missing Term in the Science of Form. *Paleobiology* 8:4–15.

Gould, S.J. (1982). The Meaning of Punctuated Equilibrium and Its Role in Validating a Hierarchical Approach to Macroevolution. In R. Milkman (Ed.), *Perspectives on Evolution* (pp. 83–104). Sunderland, MA: Sinauer.

Gould, S.J. (2002). *The Structure of Evolutionary Theory*. Cambridge, MA: Belknap.

Gould, S.J. and Eldredge, N. (1977). Punctuated Equilibria: The Tempo and Mode of Evolution Reconsidered. *Paleobiology* 3:115–151.

Gould, S.J. and Lewontin, R.C. (1979). The Spandrels of San Marco and the Panglossian Paradigm: A Critique of the Adaptationist Programme. *Proceedings of Royal Society London Series B* 205:581–598.

Grantham, T.A. (1995) Hierarchical Approaches to Macroevolution: Recent Work on Species Selection and the "Effect Hypothesis." *Annual Review of Ecology and Systematics* 26: 301–321.

Hamilton, W.D. (1987). Discriminating Nepotism: Expectable, Common, Overlooked. In D.S.C. Fletcher and C.D. Michener (Eds.), *Kin Recognition in Animals* (pp. 417–437). New York: John Wiley.

Hawkes, K. (1993). Why Hunter-Gathers Work: An Ancient Version of the Problem of Public Goods. *Current Anthropology* 34:341–361.

Johnson, G.A. (1978). Information Sources and the Development of Decision-Making Organizations. In C. Redman et al. (Eds.), *Social Archaeology: Beyond Subsistence and Dating* (pp. 87–112). New York: Academic Press.

Johnson, G.A. (1983). Decision-Making Organization and Pastoral Nomad Camp Size. *Human Ecology* 11:175–199.

Jones, G.T., Leonard, R.D. and Abbott, A.L. (1995). The Structure of Selectionist Explanations in Archaeology. In P.A. Teltser (Ed.), *Evolutionary Archaeology: Methodological Issues* (pp. 13–32). Tucson: University of Arizona Press.

Kaplan H.S. and Lancaster, J.B. (2000). The Evolutionary Economics and Psychology of the Demographic Transition to Low Fertility. In L. Cronk, et al. (Ed.) *Adaptation and Human Behavior* (pp. 283–322). New York: Aldine de Gruyter.

Keeley, L.H. (1996). *War Before Civilization: The Myth of the Peaceful Savage*. Oxford: Oxford University Press.

Kirch, P.V. (1984). *The Evolution of Polynesian Chiefdoms*. Cambridge: Cambridge University Press.

Krebs, J.R. (1978). Optimal Foraging: Decision Rules for Predators. In J.R. Krebs and N.B. Davies (Ed.), *Behavioral ecology: An evolutionary Approach* (pp. 23–63). Oxford: Blackwell.
Laland, K.N. and Brown, G.R. (2002). *Sense and Nonsense: Evolutionary Perspectives on Human Behavior*. Oxford: Oxford University Press.
Lee, R.B. (1979). *The!Kung San: Men, Women, and Work in a Foraging Society*. Cambridge: Cambridge University Press.
Leonard, R.D. and Jones G.T. (1987). Elements of an Inclusive Evolutionary Model for Archaeology. *Journal of Anthropological Archaeology* 6:199–219.
Levine, R. (1967). The Internalization of Political Values in Stateless Societies. In R. Hunt (Ed.), *Personalities and Cultures: Readings in Psychological Anthropology* (pp. 185–203). Garden City, NY: Natural History Press.
Lewontin, R.C. (1970). The Units of Selection. *Annual Review of Ecology and Systematics* 1:1–18.
Lloyd, E.A. (1988). *The Structure and Confirmation of Evolutionary Theory*. New York: Greenwood Press.
Low, B.S. (2000). Sex, Wealth, and Fertility: Old Rules, New Environments. In L. Cronk, et al. (Eds.), *Adaptation and Human Behavior* (pp. 323–344). New York: Aldine de Gruyter.
Lowie, R. (1973). Political Organization among American Indians. In D. Black and M. Mileski (Eds.), *Social Organization of Law* (pp. 278–303). New York: Seminar Press.
Mace, R. (2000). An Adaptive Model of Human Reproductive Rate Where Wealth Is Inherited: Why People Have Small Families. In L. Cronk, et al. (Eds.), *Adaptation and Human Behavior* (pp. 261–281). New York: Aldine de Gruyter.
Maybury-Lewis, D. (1974). Akwe-Shavante Society. London: Oxford University Press.
Nowak, M.A., Page, K.M. and Sigmund, K. (2000). Fairness Versus Reason in the Ultimatum Game. *Science* 289:1773–1775.
O'Brien, M.J. and Lyman, R.L. (2000). *Applying Evolutionary Archaeology: A Systemic Approach*. New York: Kluwer/Plenum.
Prentiss, A.M. (2010). Social Histories of Complex Hunter-Gatherers: Pacific Northwest Prehistory in a Macroevolutionary Framework. In K.E. Sassaman and D.H. Holly, Jr. (Eds.), *Hunter-Gatherer Archaeology as Historical Process*. Tucson: University of Arizona Press (in review).
Prentiss, W.C. and Chatters, J.C. (2003). Cultural Diversification and Decimation in the Prehistoric Record. *Current Anthropology* 44:33–58.
Rappaport, R. (1968). *Pigs for the Ancestors*. New Haven: Yale University Press.
Richerson, P.J. and Boyd, R. (1992). Cultural Inheritance and Evolutionary Ecology. In E.A. Smith and B. Winterhalder (Eds.), *Evolutionary Ecology and Human Behavior* (pp. 61–92). New York: Aldine de Gruyter.
Richerson, P.J. and Boyd, R. (1999). Complex Societies: The Evolutionary Origins of a Crude Superorganism. *Human Nature* 10:253–289.
Ridley, M. (1993). *The Red Queen: Sex and the Evolution of Human Nature*. New York: Penguin Books.
Rindos, D. (1984). *The Origin of Agricultural Systems: An Evolutionary Perspective*. New York: Academic Press.
Rindos, D. (1985). Darwinian Selection, Symbolic Variation, and the Evolution of Culture. *Current Anthropology* 26:65–77.
Rosenberg, M. (1990). The Mother of Invention: Evolutionary Theory, Territoriality, and the Origins of Agriculture. *American Anthropologist* 92:399–415.
Rosenberg, M. (1994). Pattern, Process, and Hierarchy in the Evolution of Culture. *Journal of Anthropological Archaeology* 13:307–340.
Rosenberg, M. (1998). Cheating at Musical Chairs: Sedentism in an Evolutionary Context. *Current Anthropology* 39:653–681.
Service, E.R. (1975). *Origins of the State and Civilization: The Process of Cultural Evolution*. New York: Norton.
Smith, E.A. (1985). Inuit Foraging Groups: Some Simple Models Incorporating Conflicts of Interest, Relatedness, and Central Place Sharing. *Ethology and Sociobiology* 6:27–47.

Smith, E.A. (2000). Three Styles in the Evolutionary Analysis of Human Behavior. In L. Cronk et al. (Eds.), *Adaptation and Human Behavior* (pp. 27–46). New York: Aldine de Gruyter.

Sober, E. and Wilson, D.S. (1998). *Unto Others: The Evolution and Psychology of Unselfish Behavior*. Cambridge, MA: Harvard University Press.

Spencer, C.S. (1993). Human Agency, Biased Transmission, and the Cultural Evolution of Chiefly Authority. *Journal of Anthropological Archaeology* 12:41–74.

Spencer, C.S (1997). Evolutionary Approaches in Archaeology. *Journal of Archaeological Research* 5:209–264.

Spencer, C.S (1998). A Mathematical Model of Primary State Formation. *Cultural Dynamics* 10: 5–20.

Stanish, C. (2004). The Evolution of Chiefdoms: An Economic Anthropological Model. In G.M. Feinman and L.M. Nicholas (Eds.), *Archaeological Perspectives on Political Economies* (pp. 7–24). Salt Lake City: University of Utah Press.

Stanish, C. and Haley, K.J. (2005). Power, Fairness, and Architecture: Modeling Early Chiefdom Development in the Central Andes. In K.J. Vaughn et al. (Eds.), *Foundations of Power in the Prehispanic Andes* (pp. 53–70). Archeological Papers of the American Anthropological Association 14, Arlington, VA.

Stanley, S.M. (1981). *The New Evolutionary Timetable: Fossils, Genes, and the Origin of Species*. New York: Basic Books.

Tainter, J.A. (1988). *The Collapse of Complex Societies*. Cambridge: Cambridge University Press.

von Furer-Haimendorf, C. (1969). *Morals and Merit: A Study of Values and Social Controls in South Asian Societies*. Chicago: University of Chicago Press.

Vrba, E.S. (1983) Macroevolutionary Trends: New Perspectives on the Roles of Adaptation and Incidental Effect. *Science* 221:387–389.

Vrba, E.S. (1984). What is Species Selection? *Systematic Zoology* 33:318–328.

Vrba, E.S. (1989). Levels of Selection and Sorting with Special Reference to the Species. *Oxford Surveys in Evolutionary Biology* 6:111–168.

Vrba, E.S. and Eldredge, N. (1984). Individuals, Hierarchies and Processes: Towards a More Complete Evolutionary Theory. *Paleobiology* 10:146–171.

Vrba, E.S. and Gould, S.J. (1986). The Hierarchical Expansion of Sorting and Selection: Sorting and Selection Cannot be Equated. *Paleobiology* 12:217–228.

de Waal, F.B.M. (1982) *Chimpanzee Politics: Power and Sex among Apes*. New York: Harper & Row.

de Waal, F.B.M. (1996). *Good Natured: The Origins of Right and Wrong in Humans and Other Animals*. Cambridge, MA: Harvard University Press.

Wade, M. (1978). A Critical Review of the Models of Group Selection. *Quarterly Review of Biology* 53:101–114.

Williams, G.C. (1966) *Adaptation and Natural Selection: A Critique of Some Current Evolutionary Thought*. Princeton, NJ: Princeton University Press.

Wilson, E.O. (1975). *Sociobiology: The New Synthesis*. Cambridge: Harvard University Press.

Wilson, D.S. (1997). Hunting, Sharing, and Multilevel Selection. *Current Anthropology* 39:73–97.

Wilson, D.S. and Sober, E. (1994). Re-Introducing Group Selection to the Human Behavioral Sciences. *Behavioral and Brain Sciences* 17:585–654.

Wittfogel, K.A. (1957). *Oriental Despotism: A Comparative Study of Total Power*. New Haven: Yale University Press.

Yoffee, N. (2005). *Myths of the Archaic State: Evolution and the Earliest Cities, States, and Civilizations*. Cambridge: Cambridge University Press.

Chapter 2
Landscape Learning in Relation to Evolutionary Theory

Marcy Rockman

Introduction

Human beings adapt not to their real environment but to their ideas about it, even if effective adaptation requires a reasonably close correspondence between reality and how it is perceived Bruce Trigger (1989)

Landscape learning is a model that addresses how human groups gather and share environmental information from the perspective of colonization (Rockman 2003a, b). It has been noted by many scholars that the situation of a group being unfamiliar with an environment may have had substantial social and economic consequences and so may have had archaeologically visible traces (Anthony 1990, 1997; Beaton 1991; Bogucki 1979; Fedele 1984; Kelly and Todd 1988; Whittaker 1989). However, the process of learning itself and the underlying assumptions of how different types of environmental information may have been developed and transmitted over archaeologically visible time frames by different groups and different economies have not yet been fully examined.

The landscape learning model lays a framework for doing this by presenting a consistent process of information gathering in such a manner that it can be tested for in different historically contingent contexts. The objective of the landscape learning model is to determine whether, when, and how aspects of a previously unfamiliar environment become part of an ongoing social system, in other words, to identify environmentally related cultural change. Tests of the landscape learning model are still few (Rockman 2003b) and it is expected that the model will be expanded and refined in the future. Examination of the model and its underlying concepts to date, however, have shown it to be widely applicable to many contexts (for example, see Anderson 2003; Fiedel and Anthony 2003; Blanton 2003; Hardesty 2003; Roebroeks 2003; Steele and Rockman 2003; Tolan-Smith 2003;) and further that it fills a critical gap in linking the intangible realm of thought, perspective, and memory with the solid, traceable, measurable, and "mappable" archaeological record.

M. Rockman (✉)
Statistical Research, Inc., Cotsen Institute of Archaeology, University of California, Los Angeles, CA, USA
e-mail: marcyrockman@hotmail.com

A.M. Prentiss et al. (eds.), *Macroevolution in Human Prehistory*,
DOI 10.1007/978-1-4419-0682-3_3,

The objectives of this volume are to examine the current state of thinking about cultural evolution. As the editors note in their introduction, archaeology is our sole means of examining the deep human past, and collected research over the history of the discipline has clearly demonstrated the long sequences of changes that have taken place from prehistory leading to our current major forms of social organization. As populations can be seen to continue through different sequential forms of social organization in many cases, the cultures they entail can be seen to have evolved following the concept of "descent with modification" (see also Eldredge, this volume; Spencer, this volume).

The theory of evolution began in the biological sciences with the examination of life forms and their interactions with and roles in their surrounding environments. The concepts of natural selection, survival of the fittest, and adaptation describe the long-term interactions of biological populations with those environments.

The relationship of the environment to cultural evolution is less clear. Anthropological theory has been moving away from the idea that the natural environment can determine cultural forms (see review in Dincauze 2000). Current understanding is that for a very long time humans have had, within parameters, the capacity to design their interactions with the environment.

Understanding how those design decisions are made and what their constraining parameters are is one of the primary objectives of cultural macroevolutionary theory (see Prentiss et al., this volume). As landscape learning addresses the role of thought, perception, and memory—all components of decision making—in human–environment interactions over time, it stands to be a useful tool in the archaeological macroevolutionary kit.

Gathering Environmental Information

Landscape learning occurs through the development of environmental knowledge. While there are many ways of dividing an environment into informational packages, the landscape learning model uses the following scheme of environmental knowledge (after Rockman 2003a):

- *Locational* geographic coordinates and physical characteristics of natural resources
- *Limitational* cycles and constraints of the local environment
- *Social* combination of locational and limitational knowledge into a form that is applied to ongoing practices and communicated among group members and between generations; i.e, share sense of "this is how we live here"

Locational knowledge describes fixed resources and other related geographic information, such as placement of a lithic outcrop, spring, or meadow with rich soils, pathway of a game migration, or distribution of sheltered caves or campsites with good visibility. Because locational knowledge is related to features and environmental characteristics that are generally fixed in space and time, gathering locational knowledge is likely the easiest portion of environmental knowledge to gather and may be learned quickly.

Collection of locational knowledge is contingent, however, on the economic and social needs of the given population. Rosenberg (1994) and other scholars in this volume (see Bettinger; Chatters; Mason; Prentiss; Prentiss et al.; and Rosenberg, this volume) use the term *Bauplan* to describe underlying structure and related constraints. Originally conceived as referencing the form of an organism, in cultural macroevolutionary theory it has been applied to the economic and social structure of a given population and the related opportunities and limitations of that structure. It is also a useful concept here. At the simplest level of consideration, resources or landscape features may only be recognized as such if they fall within the needs and material uses of the group, which are related to its *Bauplan*. For instance, van Andel and Runnels (1995) noted that the Mesolithic societies of Greece do not appear to have used upland plains but that these areas were later brought into use by early Neolithic farmers. At other levels of analysis, the process of identification of previously unused resources may slow occupation and use of a given ecozone by a given socioeconomic strategy (with a specific *Bauplan*), such as the time lag proposed by Tolan-Smith (2003) for the late Upper Palaeolithic colonization of maritime western Scotland, or may limit the range of identified available options in times of stress in common resources.

Limitational knowledge describes the periodicities of cyclical environmental characteristics and the constraints of the given environment for certain uses. For example, how much precipitation falls each year, in what forms, when, and how much does it vary from year to year? How many animals or plants can be hunted harvested each year without exhausting the respective populations. How stable are the game migration routes? How workable and reliable are the lithic materials for the range of tasks for which the given group uses them? Because limitational knowledge references phenomena that occur over time, it cannot be learned or confirmed instantaneously. Rather, the time necessary to develop or accommodate limitational knowledge is some function of the rate of change within its given identified cycle combined with the length of the generations within the subject population (Dean 1988; Minc 1986; Minc and Smith 1989; Smith 1988).

Recognition and use of limitational knowledge depends on the "dynamic scale ranges" of both environmental and social entities. As usefully developed from ecology for archaeological application by Hopkinson (2001, 2007a), dynamic scale range theory addresses the range of scales to which a given phenomenon of interest responds. Environments can be measured on multiple scales. For instance, as developed by Hopkinson (2001), a forest may be considered as either a dynamic entity in its own right, a constant setting within which an organism or population operates, or as inconsequential "noise" in major geological and geomorphological processes. Similarly, rainfall and temperature can be reported as daily, annual, or decadal averages. In turn, environmental entities respond primarily to rates of change that approximate their own reproductive cycle; for example, flower populations tend to reflect high frequency daily to seasonal variation in rainfall and sunshine, while trees are more likely to prosper or fail with longer period drought or fire cycles. However, they can reflect changes at much lower frequencies over the long term. The upper limit of the range is the level that can be considered a non-constraining constant in relation to the level of interest. The lower limit consists of "stochastic noise" that does not deliver a measurable signal to the level of interest. The dynamic scale range, therefore, includes both higher and lower frequency levels than the particular level or phenomenon of interest (Hopkinson 2001, 2007a) With respect to imitational knowledge, each instance of this aspect of landscape learning is most likely to reflect cycles that most closely match the life span of the element interacting with the cycle: annual planting procedures should respond to annual or similar precipitation and temperature variations, while settlement patterns respond to longer cycles, such as floods. Locations of annual planting, however, will likely correspond with at least some aspects of the longer term settlement patterns.

Social knowledge describes appropriate responses to gathered locational and limitational knowledge that have been shared among a group (horizontal transmission, after Boyd and Richerson 1985; Hewlett and Cavalli-Sforza 1986; Richerson and Boyd 2001, 2005) and passed to subsequent generations (vertical transmission) such that they form a collective sense of "this is how we live here." It is in fact such interaction of environmentally derived and socially shared activity that creates the phenomenon known as landscape (after Ashmore and Knapp 1999; Basso 1996; Tolan-Smith 1997). As social knowledge is the result of implementation of component locational and limitational knowledge, the time frame to develop it is the product of the time frames for the component knowledge and may be further modified according to the rates and forms in which it is shared, remembered, and stored in group-appropriate social forms (see Vansina 1985; Smith 1988).

Both locational and limitational knowledge can only be acquired from the environment through individual learning. Their transition into social knowledge occurs through social learning which in turn can be subject to the various biases (content-based or direct, frequency-based, and model-based) described by Richerson and Boyd (see definitions and background in Boyd and Richerson 1985; Henrich and Boyd 1998; Richerson and Boyd 2001, 2005).

Each type of knowledge can influence the development of the other two. Further, each type of knowledge can be updated at any time. For the sake of modeling and analysis, the landscape learning model is designed to analyze development of environmental knowledge with respect to a given resource from the perspective of colonization or first contact with an environment.

Landscape Learning and Colonization

Colonization has an uneasy place in macroevolutionary theory in light of the invoked role of colonization and diffusion in early culture-historical models of cultural change (Anthony 1990; Shennan 2000; Trigger 1989). The theoretical complexity of colonization and the important roles of different types of information flow as well as population movements have been more recently addressed and productively applied to multiple colonization case studies (Anthony 1990, 1997; Chapman and Hamerow 1997; Graves and Addison 1995). Colonization studies tend to present movement in one of two primary forms: point-and-arrow or streaming, which involves directed movement of a population along a relatively narrowly defined pathway over a relatively long distance between an origin a destination, and wave-of-advance (e.g., Ammerman and Cavalli-Sforza 1984), which posits relatively short distance movements across a relatively wide swath of environment. A third type of movement, leap-frogging, can be seen as a sort of combination of the two, involving multiple streaming movements between separated areas of similar environmental characteristics. There are many implications of these forms with respect to both motivating factors and potential for archaeological visibility (Anthony 1990, 1997; Beaton 1991; Jochim et al. 1999; Kelly 2003; Meltzer 2004; Webb 1998).

The point of interest for discussion here is the context they present for development of environmental knowledge when considered from the perspective of initial colonization. Point- and-arrow and leap-frogging forms require some form of predeveloped knowledge, at least to the extent that the target of colonization was identified, but describe a situation in which a wide range of knowledge structures (locational, limitational, social) may need to be updated rapidly or simultaneously. In wave of advance movement, in that it predicts shorter distance movements and greater likelihood of maintaining interactions with some portions of previously occupied environments, development of environmental knowledge can be seen as primarily cumulative, and may entail rapid locational knowledge updates with slower updates of limitational and social knowledge.

The development of environmental knowledge in the context of noninitial colonizations is more complex as it requires consideration of the nature of the relationships between previously established populations and the arriving population. Following a biogeographical approach developed in Rockman (2003a), the environmental knowledge relationships may be conceived as three barriers:

- *knowledge barriers* existence of usable previously collected information
- *population barriers* numerical population density and capacity with respect to the structure of existing and incoming economic systems and residence patterns (*Bauplan*); availability of niche space
- *social barriers* resident population's defense and information storage systems, including language

An initial colonization may have a high knowledge barrier—relatively little previously gathered information—but faces low/nonexistent population and social barriers. Contact between previous and immigrant populations with similar *Baupläne* may present lower knowledge barriers but potentially higher population or social barriers. Contact between populations with different *Baupläne*, such as agriculturalists and hunter-gatherers, will have high knowledge barriers and can be anticipated to have variable population and social barriers as well.

Colonization can also be a useful tool for macroevolutionary analysis because it depicts movement from somewhere to somewhere and therefore facilitates consideration of design decision constraints. Such consideration can be described at the micro-level from the perspective of environmental cognition and the use of schemata. An individual's initial contact with and movement through an environment proceeds by means of wayfinding (Golledge 1987, 2003). There are several means of wayfinding, including homing (continual updating) which may be likened to a wave of advance motion in which movement is perpetually referenced to previous movements and the intended destination; piloting (use of landmarks) which may be likened to point-and-arrow colonization in that it moves toward a known (if not immediately seen) destination with orientation provided by means of landmarks; and chunking, which may be likened to leapfrogging in that it breaks a relatively long route into identifiable and more easily negotiated segments. An individual's internal representation of routes traveled and to be traveled, places of origin, and intended destinations comprise the individual's cognitive map. As cognitive maps represent spatial experience, they are almost never complete and are updated as experience and information are added and the effective environment changes. Similarly, cognitive maps do not develop instantaneously. In instances where cognitive maps are incomplete, generalized information about environmental structure based on previous experiences elsewhere may be drawn upon, termed schemata, templates, frames of reference, scripts, and scenes (Golledge 2003). An example developed in Golledge (2003) from the built environment would be a modern urban dweller's knowledge of where to find an exit or restroom in a public building. A natural environment schemata may be that chert or flint may be found at topographic high points, particularly high erosional features (Rockman 2001b, 2003b).

Schemata, then, are patterns into which environmental information is anticipated to fit. If reality does not match imported schemata, then schemata may be updated or the environmental characteristic may be missed. In some instances, however, schemata or ways of doing things can suit a new environment better than the previous environment in which they were developed or find a new use or expression. Biological terminology calls such instances preadaptations or exaptations (see Gould 1991; Gould and Vrba 1982). Gamble (1993, 1999) has linked exaptation to the capacity shown by early hominids for global dispersal. This approach provides a link to landscape learning at the macro-evolutionary scale.

Landscape Learning at the Macroscale

Landscape learning is possible because of human behavioral flexibility. Behavioral flexibility makes it possible to encounter and adjust to new and different environments at a rate faster than can be allowed by physical evolution alone. In turn, the ability to respond successfully to new environments may have set in motion a feedback loop in which encountering novel environments and developing information about those environments supported more behavioral flexibility and a propensity to disperse and continue to adjust to new environments.

Potts's (1998) variability selection model posits that the uniquely wide range of human behavioral flexibility is an adaptation to environmental fluctuations of variable and increasing magnitude over the past 6 million years. Paleoenvironmental studies of a range of deposits dating from the Oligocene onwards have identified at least twelve cycles of environmental variation with cycles ranging from one to 100,000 years. The meshing and unmeshing of these multiple cycles with differing periodicities appears to have resulted in a particularly variable environment in which the influence of any given environmental cycle may have changed substantially from one cycle to the next, making its particular effects more difficult to track.

The variability selection model outlines the development of variability selection adaptations, which Potts defines as "complex structures or behaviors" that allow novel responses in unpredictable settings. These adaptations allow phenotypic variability that can vary with the environment in which the species finds itself, including versatile locomotor capabilities and dental structures, as well as the neurological capacity to process and respond to a range of external signals and social situations (Potts 1998). Following on this, the dynamic scale range of human physical capacities for cultural change can be defined per the wide range of environmental cycles included by Potts in the variability selection model. In turn, learning can be seen as an integral component of macroscale human physical evolution. The range of scales to which human cultures themselves respond is in need of additional investigation. The capacity to learn, however, can also be seen to have played a role in cultural development from hominid times onward.

The ability to behave flexibly and, in so doing, adjust social practice to local conditions, may have been a key component of the early hominid movements across the Old World. Currently, the number, paths, and rates of hominid dispersals out of Africa remain a focus of debate and study. Several recent studies (Ashton and Lewis 2002; Dennell 2003; Pavlov et al. 2004; Roebroeks 2003) have noted the patchiness of the premodern human record in Europe and argued that scattered sites over a long time frame cannot be taken at face value to indicate successful continuous occupation following a single initial colonization. Rather, it is more likely that there were repeated dispersals into and retractions out of different parts of Europe, perhaps on as coarse a scale as glacial–interglacial cycles, perhaps more frequently than that. Lahr and Foley (1994) have in turn suggested that diversity in the hominid and human populations indicates multiple dispersals of different hominid forms along different pathways within and out of Africa to the rest of the Old World. Dennell (2003), in his discussion of long and short chronologies of hominid movements out of Africa, pointed out that while it is possible with current dates to suggest a rate of 10 km/year for the *Homo erectus* dispersal from East Africa to Java, it is unlikely that colonization happened evenly at such a rate. The actual pattern of dispersal was likely much more punctuated.

The combined picture of these studies is that, given the fragmentary record that we have, it is simply not possible to postulate population pressure (as we are able to calculate it based on contemporaneous technology and paleoenvironmental evidence) as the sole motivation for these many dispersals. In other words, it is not possible to apply a simple wave-of-advance colonization model (after Ammerman and Cavalli-Sforza 1984) such that all intervening space was filled up as hominid colonization moved outward. Likewise, it is also not possible to equate hominid ranges with modern human ranges and related territorial behavior (see discussion of Paleolithic settlement at the end of Roebroeks et al. 1992). Therefore, it is also difficult to apply directly colonization "push-pull" analyses that are possible for the social underpinnings of more recent colonizations, particularly point-and-arrow or streaming-type colonizations (Ammerman and Cavalli-Sforza 1984; Anthony 1997). The impetus for hominid dispersals therefore falls outside traditional colonization models.

A proposal for the use of environmental information by early hominids by Gamble (1993), when combined with the prestige model-based pattern of information transmission from the cultural evolution model by Richerson and Boyd (2005) is a possible explanation for the major and relatively rapid early hominid dispersals. Briefly, the prestige model says that, insofar as imitation of others is a primary means of determining appropriate behavior, those individuals perceived to be successful are much more likely to be chosen as models to be imitated than individuals who are not perceived to be successful. Perception of success is critical, and imitation may focus on the outward, visible trappings or behaviors of the successful individual, not necessarily on the specific behaviors that made the individual successful (after Richerson and Boyd 2005).

Gamble's model is based on the situation of group fissioning to exploit patchy resources and unequal opportunities within the group to explore and acquire information. Gamble proposed that subadults, particularly subadult males, old enough to forage for themselves, would have been under pressure to leave the group temporarily by higher ranking older males to reduce rivalry for mates. These subadults collected important environmental information while "out of the house," which was then a source of information for the group as a whole when they returned. The key point for the expansion of the hominid range, or colonization, is that, as Gamble (1993) noted, "The impetus for exploration [came] not from some sort of adaptive curiosity. Instead it [stemmed] from the nature of the cooperative alliances, negotiated and contested between the core and peripheral members of the social group." In this model, environmental information was the social currency with which higher ranking males managed their role in the group and through which younger males gained their place in the group. A younger male with more and detailed environmental information to contribute might have moved further up the ranks than one with less or poorer information. If the prestige model-based transmission force is added to this, such that higher ranking explorers were imitated, more emphasis would have been placed on activities and skills related to exploration and, in turn, more geographic area may have been explored. Thus a social engine was in place that required exploration of more and more territory in order for the group order and functioning to be maintained. In this way, it was not population pressure but social pressure that may have driven early hominid dispersals (Anthony 1997; Fiedel and Anthony 2003). And the capacity to gather, retain, and transmit environmental information was what allowed this to happen.

Holons and Adaptive Landscape

The human capacity to respond culturally to the environment can be seen as a biological adaptation. In turn the cultural capacity to gather and share environmental information is an exaptation to global dispersal. Landscape learning, therefore, is an important link in the creation of a cultural landscape.

But what role does landscape learning play in the cultural changes of more complex societies with which much of macroevolutionary theory is concerned? The concept of adaptive landscape is applied by several other authors in this volume following on work by Wright (1932) (see Chatters; Prentiss; Prentiss and Lenert; Prentiss et al.; Spencer, this volume). In brief, it describes an array of peaks of variable heights and distribution within a plane of possible social and ecological relationships. Each peak represents a combination of relationships that is better suited—adapted—to its social and cultural environment than surrounding combinations of relationships. Macroevolutionary change occurs when a given population, group, culture, or society changes peaks and establishes a new combination of relationships. The potential benefit of changing peaks is that, ideally, the new peak is higher—more adapted—than the previously occupied peak. The challenge to changing peaks is that doing so may require moving across the valley between the peaks which is comprised of less adapted combinations of relationships.

The adaptive landscape is a useful concept for describing the situations that may precipitate cultural change. However, it does not describe how a given population, group, culture, or society comes to recognize that a change in peak is possible, necessary, or appropriate. Landscape learning is about how individual lines of information about different environmental traits, including fixed resources and cyclical patterns, are acquired. One additional theoretical tool developed from dynamic scalar range theory known as the *holon* is useful in considering and identifying the information and decision feedback loop that may play a role in peak switching and corresponding macroevolutionary change.

Hopkinson (2001, 2007; see also Koestler and Smythies 1969) defined a holon as

> Systematic clusters of strongly interacting components [that] can then be conceived as if they were a single entity from the point of view of the signals that they deliver to other components or clusters of components.

Holon boundaries are not necessarily equivalent to tangible boundaries such as the bark of a tree or edge of a flower patch and can include relative intangibles such as metabolic cycles and gene pools. Although holons are collections of system processes, they are bound by the same scalar rules such that ecosystems or learning systems cannot be comprehended by observations at a single scale or via a single holon (Hopkinson 2001, 2007a). A particularly cogent example of a holon for an archaeological context is the hydrologic changes in the American Southwest that contributed to the abandonment of the southern Colorado Plateau after A.D. 1250, as described by Dean (1988). The Great Drought that appears to have brought about the end of the Anasazi occupation of the area was not a single event but the intersection of low-frequency water table cycles, high-frequency drought cycles, and high-frequency agricultural practices which merged into a much lower frequency effect of devastating aridity.

With this example in mind, I suggest that the holon concept can be adapted for the purpose of investigating landscape learning, such that a collection of behaviors also may be considered a type of holon. A given set of behaviors or activities can generate an overall "signal" or level of adaptedness which then interacts with other groupings of behaviors of the given group, population, or species and with external holons found in the environment.

Drawing on Dean's (1988) example of the American Southwest and the dynamic scalar relationship between behavioral responses and environmental variability, the key point of behavioral holons is that they are a collection of individual practices. Within this collection, some previously established adaptations may work for some time in relatively similar environments. Problems develop for human groups when the outcome of all combined practices is no longer adapted to the new local conditions, and the relative adaptedness of the collective signal of the previous behavioral practices and environmental understandings exceeds local environmental tolerances.

For instance, a given behavior developed in a previous environment may be unsuitable as part of a set of behaviors or practices in a new environment. If, however, the results of the given behavior lie below the carrying capacity for that environment or do not immediately coincide with an unfavorable periodicity or downturn of that environment, then that relatively unsuitable behavior may persist within the overall behavioral system for an extended period of time. In turn, if the given behavior acts to bring a population into immediate and direct conflict with a new environmental periodicity or a periodicity downturn not previously encountered, then that particular behavior or the entire behavioral holon will come under direct and immediate selective learning pressure. As a result, individual practices of a behavioral system may be out of phase with each other with respect to their suitability in a given ecosystem. In other words, while a behavioral holon may appear adapted, individual component practices may not yet have come under selective pressure and so may themselves be poorly adapted. In turn, individual practices may be adapted, but the behavioral holon signal may be unadapted.

This is a critical addition to the landscape learning model as it allows consideration of time gaps between environmental experience and response. Lag time is built into the initial landscape learning model described above, with the essentially sequential movement of information from locational knowledge to limitational knowledge to social knowledge. Social knowledge in place at the time of colonization of an unfamiliar environment or major environmental change is unlikely to fully reflect the new environmental conditions and cannot be updated instantaneously; locational and limitational knowledge must be gathered and shared first. The holon concept brings this full circle and explains why it may appear that past groups did not learn their environments, or respond quickly enough to what may seem now to have been drastic environmental change. Humans have developed the ability to behave flexibly, and the landscape learning model suggests that we have the capacity to learn to live just about anywhere. What the holon concept provides is essentially the "brakes" on that flexible process by establishing thresholds for change. Given the capacities for imitation and transmission, learning and subsequent cultural change can happen at "blinding speed" compared to physical evolution (Richerson and Boyd 2005), but it has not necessarily been continuous. It is the development of multiple sets of social knowledge, interacting with and potentially changing one or more behavioral holons, that may account for the some of the punctuations in long-term cultural change.

Case Studies

As noted above, the landscape learning model is a tool in the macroevolutionary kit. In and of itself, at least insofar as its processes are currently understood, landscape learning does not account for the major transitions in cultural complexity that macroevolutionary theory addresses. To date, the two most detailed case studies of landscape learning describe development of locational knowledge by groups with stable *Baupläne*. These examples are summarized below, along with an additional hominid example as illustrations at different spatial and temporal scales of how the landscape learning process contributed to different groups' understanding of their place within their physical and adaptive landscapes.

Early Hominids

Following the discussion of Potts's (1998) variability selection model and the work of Gamble (1993) regarding early hominid dispersals out of Africa above, landscape learning can be seen to have been going on for 2 million years or more, and so the archaeological traces of landscape learning may be detectable from that time onward. Roebroeks (2001, 2003) has examined behavior and landscape learning in the context of the earliest peopling of Europe in the Middle Pleistocene. This setting presents many challenges: the relative sparseness of the archaeological record, the coarseness of our dating techniques for that time period, and the limitations they present with respect to identifying contemporaneous occupations, and the relative evolutionary "newness" of human behavioral flexibility. Roebroeks concluded that, insofar as landscape learning can be distinguished from physical evolution in this setting, it may be detectable only in the broad patterns of hominid presence and absence across glacial and interglacial cycles. For instance, locational knowledge may be deduced from the raw materials identified in individual or groups of sites (see raw material data compiled in Feblot-Augustins 1997). Development of limitational knowledge may be deduced from reuse of specific lithic raw material sources (see again Feblot-Augustins 1997) and evidence of successful hunting such as the carefully made and successfully used spears found at Shöningen (Dennell 1997; Thieme 1997).

The scope, nature, and persistence of social knowledge are difficult to define at this scale. Roebroeks (2003) noted that the form and content of early European sites suggest that the environmental capacities of the earliest Europeans were quite different from later, modern, landscape learners. Dennell (2003) furthered this view, noting that archaeological presence alone is not evidence of successful colonization. "Long" chronologies of hominid dispersals into Europe and the wider Old World (beginning before 1 million years ago) appear to be built on the record of intermittent events accomplished in favorable climatic intervals. "Short" chronologies (beginning after 1 million years ago) are based on records of more permanent colonization. Pavlov et al. (2004) noted, particularly with respect to early occupation of the northern reaches of Europe, however, that archaeologically visible occupation should be taken to represent at least a modicum of survival capability.

Hopkinson (2007b) proposes that, per these difficulties, the expansion of the Neandertal ranges in Europe transition to include high relief uplands in central and eastern Europe after 200,000 years ago was "not an evolutionary non-event". Rather it was, in other words, a macroevolutionary shift that had significant social landscape components. Strict review of sites dating prior to 200,000 years ago indicates that hominin groups avoided upland regions with broken terrain in Europe with occupation primarily, although not absolutely, distributed along the middle and lower reaches of river drainage systems in western Europe after a 200,000 years ago hominin occupation extended both eastward and upward into high-relief uplands. The distribution cannot be explained entirely by preservation issues or history of research; the shift in geography appears to be real. The eastern European landscape was characterized not only by a highly seasonal climate but also by a coarse-grained spatiotemporal distribution of resources. Millennial scale climatic fluctuations did not allow development of an optimally adapted consistent plant community but rather—similar to the effects of the variability selection forces on hominids—resulted in a patchwork plaid of vegetation patches that, given the regional characteristics of seasonality and topography, placed hominid-preferred resources relatively far apart. Lower Palaeolithic populations had use of fire and demonstrated ability to take medium-sized game, so barriers to the use of these areas were not strictly technological. Rather, they were social, cognitive, or both.

As developed by Hopkinson (2007b), the Neandertal Levallois technique combined two previously discrete lithic techniques: *façconage*, shaping of the core by removal of flakes, and *debitage*, production of sharpened flakes from a core, into a single technology; a change requiring what he terms an "incorporation of difference." The same incorporative process can be seen in the Middle Palaeolithic use of landscape. After 200,000 years ago, Neandertals incorporated "larger scales of heterogeneity into their effective systems of landscape exploitation" and lived life on larger spatiotemporal, conceptual, and practical scales. It is not yet clear whether these changes were cognitively based (physical evolution) or emergent factors due to population growth and transmission of practices. Hopkinson argues that these need not be exclusive; given the physical evolutionary basis of the learning capacity described above, this author concurs. The key point here is that the change in spatiotemporal use of coarse-grained landscape can be seen as a macroevolutionary shift in the capacity of hominid lifeways, represented by change in the distribution of sites and the use of integrated lithic technologies.

Recolonization of Britain at the End of the Last Ice Age

The archaeological record indicates that the British Isles were abandoned for up to 8,000 years during the last glaciation, which peaked at approximately 18,000 years ago. Hunter-gatherers returned to Britain by approximately 13,000 years ago, most likely from the direction of northern France (Rockman 2003b). The orientation of Britain in northwestern Europe, its changing shorelines during the late glacial, current distribution of radiocarbon dates, and comparative studies of late glacial migrations in Europe (Jochim et al. 1999) together suggest a point-and-arrow-type pattern (after Anthony 1997) for the British recolonization. Flint was an important lithic raw material in the toolkits of the late glacial hunter-gatherers of northwestern Europe, and therefore identification and use of new sources of flint during recolonization would have been a crucial component of environmental familiarization (Rockman 2003b). Although flint is widely distributed across the southern and eastern edges of England (Mortimore and Wood 1984), the flint-free north and west as well as changing vegetation cover during the late glacial (Walker et al. 1994; Walker and Harkness 1990; Whittow 1992) suggest that development of knowledge about its specific exposures and qualities would have been necessary.

Field survey and trace element analysis by means of inductively coupled plasma mass spectrometry (ICP-MS) identified five regions within the Cretaceous chalk flint formation of England and France and linked the majority of flint artifacts from five late glacial sites across England to flint sources in southwestern England, particularly the Salisbury Plain region (Rockman 2003b; Rockman et al. 2003). The current body of late glacial radiocarbon dates (see Barton 1997; Charles 1996; Fischer and Tauber 1986; Housley et al. 1997, 2000; Street and Terberger 1999) suggests that this procurement pattern may have persisted for several hundred years.

Although the identified flint exposures have a wide range of physical characteristics and vegetation causes substantial visual interference (even accounting for late glacial vegetation cover), the recolonizers of Britain appear to have successfully used wayfinding schemata that brought them at least into proximity to flint-bearing deposits. The earliest hunter-gatherers did not extensively use the first flint resources they would most likely have encountered along probable reentry routes in southern, southeastern, and eastern Britain (see shoreline reconstructions in Lambeck 1995). Rather, the landform characteristics of the tested British flint regions strongly suggest that southwestern England, particularly the northern Salisbury Plain region, was not only topographically the most "legible" (using a model based on Golledge

2003) of the tested regions, but it was also the most similar in navigational characteristics to one of the most probable colonization population source areas, the Paris Basin (Rockman 2003b).

Additional research is needed to identify the scope and rate of subsequent lithic resource use and change. The research on which this case study is based (Rockman 2003b) was designed to identify the establishment of lithic procurement systems during an initial colonization. The durability of the Salisbury Plain as a source area through the climatic changes of the Younger Dryas and into the Mesolithic has not yet been examined. The contribution this example makes to macroevolutionary investigation is the apparent absence of substantial adaptive change during and following major climatic and location shifts, the apparent success of lithic schemata and the persistence of the northern European Palaeolithic socioeconomic strategy; the thresholds of macroevolutionary change are higher.

South Pass City, Wyoming 1867–1872

The gold rush that centered South Pass City did not mark the first arrival of humans to the area. South Pass City was named for the geographic South Pass, which was the crossing point of the Oregon, California, and Mormon Trails over the Continental Divide, located 10 miles to the southwest. The gold rush miners were, however, the first to exploit the gold resources of the region. Hostility of the local Native American groups limited any information exchange (McDermott 1993), and so the miners had to overcome both knowledge as well as social barriers. The South Pass Gold Rush took place near the middle of a long series of mineral rushes in the American West (Hardesty 2003; Huseas 1991). Therefore, many of the South Pass miners arrived with schemata developed from previous mineral rushes and the capitalist frontier economy (*Bauplan*) had already accommodated setting up of mining ventures, including activities as diverse as bodies of mining law and provision of foodstuffs and lodging (Conlin 1986).

Early use of fauna at South Pass was split almost 50:50 between wild game and domestic animals, with a focus on larger mammals, such as deer, cattle, and pig. Within three years, faunal use had changed to a 25:75 wild to domestic fauna ratio, with notable numbers of wild and domestic fowl (Rockman 1995). Primary documents report several trips in which men from South Pass City attempted to hunt but failed as they did not know how to proceed or follow animals (Chisholm and Homsher 1960). Professional hunters worked in the area (Kingman to Johnnie, letter, 25 November 1869, South Pass City) although available data are not sufficient to determine the extent to which they were able to provision the town; overall patterns indicate a rapid establishment of an urban-style food supply chain.

Gold in the South Pass mining district is found principally in quartz veins. Some placer deposits have developed, but due to a lack of water the majority of effort focused on development of hard rock mining. It is now known that the South Pass area is geologically complex and has undergone at least four periods of deformation.

Gold-bearing rocks are generally confined to a broad shear zone approximately 8 miles long and 4 miles wide, but the current distribution of ore individual veins is very irregular and does not follow the major topographic trends of the area (Hausel 1984, 1989).

The distribution of mining claims in relation to the local gold-bearing geology shows that at least most of the miners did not grasp, or at least respond to, this effectively almost random placement of gold-bearing deposits. Rather, mining claims were organized in lodes, which are long lines of claims centered on gold discovery. The distribution of mappable claims shows that the lodes tend to follow the major topographic trends of the area (Rockman 2001a). One early report (Raymond 1870) likened South Pass ore to the gold-bearing quartz found in California.

Mining claims and census records show that it took approximately four months for miners to develop a functional familiarity with the process of gold claiming at South Pass (Rockman 2001a). It took approximately 30 years more to develop the scientific understanding of the gold-bearing geology of the area (Raymond 1870 in Rockman 2001a; see discussion of Hayden 1872; Knight 1901). This understanding did not come quickly enough for the town of South Pass City, however, as the area went economically by 1872 and as a residential community in 1967 (Huseas 1991; Huseas and Doherty 1984).

With respect to food, the miners and town inhabitants do not appear to have succeeded in developing even functional locational knowledge. With respect to gold, locational knowledge was gathered, and the limitations of the resource were keenly felt, as shown by the general collapse of the mining boom within 4 years. Social knowledge about the nature of the gold was developed only under a very different, non-gold rush, scientific situation. This may be seen as a failure of the landscape learning process. Such a failure, however, does not appear to have affected the behavioral holon of gold rush practice. Miners and others left the South Pass area for Montana, Colorado, and elsewhere in the West, and mineral rushes in this style continued in the West for another 30 years. Collectively, the "profitable" signal of the mineral rushes was so strong that it did not respond substantively to individual "failures" such as that of the Wyoming Gold Rush.

This example demonstrates the absence of appropriate landscape knowledge in multiple resource categories and the capacity of a complex social system to absorb and promote practices and situations that were economically maladaptive at the local and individual level.

Conclusions

Richerson and Boyd (2005) noted that

> Darwinian analysis reveals a mass of largely unexplored questions surrounding the psychology of cultural transmission and the biases that affect what we learn from others. Small, dull effects at the individual level are the stuff of powerful forces of evolution at the level of populations. Understanding rather precisely how individuals deploy their kit of imitation

heuristics is necessary to understand the rates and direction of cultural evolution, and work on the problem has hardly begun.

The same holds true for landscape learning. It is possible to study human knowledge structures in the present and model environments of the past, but an archaeology of environmental learning is necessary to assess the "small, dull" details of what people did when in both new and old places, how they moved from one to the other, and how those details changed over space and time.

Landscape learning is possible because of behavioral flexibility, which in turn evolved in response to unpredictable environmental fluctuations (Potts 1998). The combination of behavioral flexibility and social learning processes (prestige-biased) (Richerson and Boyd 2005) create a viable explanation for the dispersal of hominids out of Africa (after Gamble 1993).

The phenomena of dispersal and colonization have interesting roles in macroevolutionary investigations due to early culture-historical explanations for cultural change. Recent theoretical developments make them a useful starting point for assessing population-environment interactions, as a colonization approach requires examination of assumptions about environmental interaction decisions and what the constraining parameters of such decisions may be. At the micro-level, such parameters may be described as schemata (Golledge 2003). At the macroscale, they are preadaptations or exaptations (Gould and Vrba 1982; after Gamble 1993).

Landscape learning proceeds through the development of locational knowledge, limitational knowledge, and social knowledge (Rockman 2003a, b). Locational knowledge is constrained by *Bauplan* (see Rosenberg 1994, this volume), limitational knowledge responds dynamically to appropriate scales of environmental change (after Hopkinson 2001, 2007a). Social knowledge, the composite of shared locational and limitational knowledge, develops through social learning and is thereby subject to the multiple biases of the transmission process (Boyd and Richerson 1985; Richerson and Boyd 2005).

Compilation of responses to environment interaction may be one source of information that indicates a shift in adaptive peak is necessary or appropriate (Wright 1932) (see also Spencer, this volume). Behavioral holons (after Hopkinson 2001, 2007a) describe the feedback between environmental information and social response, which creates the lag time or punctuation in group-level environmental learning.

Macroevolutionary shifts in the Middle Palaeolithic may have been based in the capacity to incorporate difference and are archaeologically visible in the expansion of range and use of geography (Hopkinson 2007b). In more recent examples, landscape learning can be seen to be decoupled from changes in social complexity. Rather, examples of Upper Palaeolithic colonization show the capacity of hunter-gatherer systems to persist in unfamiliar but similar environments (Rockman 2003b) and for complex economies to absorb, at least initially, expensive economic failures (Rockman 1995).

Taken together, the combination of development of environmental knowledge structures and the processes of holon feedback described by the landscape learning model has great relevance to our efforts to understand the patterns of long-term human–environment interaction and inform studies of macroevolution. By considering the periodicities of relevant aspects of the environment and individual–group information transmission patterns necessary for developing social knowledge and adjustment of practices to local environmental settings, it is possible to think in a new way about stops, starts, and continuity in the archaeological and cultural record. As well, understanding more clearly the circumstances under which hominids and humans appear to have learned or not learned their environments has implications not only for our understanding of the past but may also provide useful information about how we will (or should) adjust our own present-day environmental knowledge and practice to environments we may encounter in the future.

Acknowledgments Preparation of this chapter was supported in part by the Wenner-Gren Foundation Richard Carley Hunt Fellowship Program (Grant No. 7310). Initial research on this topic was supported by the National Science Foundation (Grant No. 003709), the Wenner-Gren Foundation (Dissertation Fieldwork Grant 6776), and the University of Southampton Centre for the Evolutionary Analysis of Culture. Steven L. Kuhn, Katharine MacDonald, David Meltzer, Margaret Beck, and Kara Cooney provided invaluable comments on earlier versions.

References

Ammerman, A. J. and Cavalli-Sforza, L.L. (1984). *The Neolithic Transition and the Genetics of Populations in Europe*. Princeton: Princeton University Press.

Anderson, A. (2003). Entering Uncharted Waters: Models of Initial Colonization in Polynesia. In M. Rockman, and J. Steele (eds.), *Colonization of Unfamiliar Landscapes: The Archaeology of Adaptation* (pp. 169–189). London: Routledge.

Anthony, D. W. (1990). Migration in Archaeology: The Baby and the Bathwater. *American Anthropologist* 92, 895–914.

Anthony, D. W. (1997). Prehistoric Migration as Social Process. In J. C. Chapman and H. Hamerow (eds.), *Migrations and Invasions in Archaeological Explanation* (pp. 21–32). Oxford: Archaeopress.

Ashmore, W. and Knapp, A. B. (eds.) (1999). *Archaeologies of Landscape: Contemporary Perspectives*. Oxford: Blackwell Publishers.

Ashton, N. and Lewis, S. (2002). Deserted Britain: Declining Populations in the British Late Middle Pleistocene. *Antiquity* 76, 388–396.

Barton, R. N. E. (1997). Fifth Interim Report on the Survey and Excavations in the Wye Valley, 1997 and New AMS Radiocarbon Dating Results from Madawg Rockshelter. *Proceedings of the University of Bristol Spelaeological Society* 21, 99–108.

Basso, K. (1996). *Wisdom Sits in Places: Landscape and Language Among the Western Apache*. Albuquerque: University of New Mexico Press.

Beaton, J. M. (1991). Colonizing Continents: Some Problems from Australia and the Americas. In T. D. Dillehay and D. J. Meltzer (eds.), *The First Americans: Search and Research* (pp. 209–230). Boca Raton: CRC Press.

Blanton, D. B. (2003). The Weather is Fine, Wish You Were Here, Because I'm the Last One Alive: 'Learning' the Environment in the English New World Colonies. In M. Rockman and J. Steele (eds.), *Colonization of Unfamiliar Landscapes: The Archaeology of Adaptation* (pp. 190–200). London: Routledge.

Bogucki, P. (1979). Tactical and Strategic Settlements in the Early Neolithic of Lowland Poland. *Journal of Anthropological Research* 35, 238–246.

Boyd, R. and Richerson, P. (1985). *Culture and the Evolutionary Process*. Chicago: University of Chicago Press.
Chapman, J. and Hamerow, H. (1997). On the Move Again: Migrations and Invasions in Archaeological Explanation. In J. C. Chapman and H. Hamerow (eds.), *Migrations and Invasions in Archaeological Explanation* (pp. 1–10). Oxford: Archaeopress.
Charles, R. (1996). Back into the North: The Radiocarbon Evidence for the Human Recolonisation of the North-Western Ardennes after the Last Glacial Maximum. *Proceedings of the Prehistoric Society* 62, 1–17.
Chisholm, J. and Homsher, L. M. (1960). *South Pass 1868: James Chisholm's Journal of the Wyoming God Rush*. Lincoln: University of Nebraska Press.
Conlin, J. R. (1986). *Bacon, Beans, and Galantines*. Reno: University of Nevada Press.
Dean, J. S. (1988). A Model of Anasazi Behavioral Adaptation. In Gumerman, G. J. (ed.), *The Anasazi in a Changing Environment* (pp. 25–44). New York: Cambridge University Press.
Dennell, R. (1997). The World's Oldest Spears. *Nature* 385, 767–768.
Dennell, R. (2003). Dispersal and Colonisation, Long and Short Chronologies: How Continuous is the Early Pleistocene Record for Hominids Outside East Africa? *Journal of Human Evolution* 45, 421–440.
Dincauze, D. F. (2000). *Environmental Archaeology: Principles and Practice*. Cambridge: Cambridge University Press.
Feblot-Augustins, J. (1997). *La Circulation des Matières Premières au Paléolithique*. Liege: ERAUL 75.
Fedele, F. G. (1984). Toward a Human Ecology of Mountains. *Current Anthropology* 25, 688–691.
Fiedel, S. J. and Anthony, D. W. (2003). Deerslayers, Pathfinders, and Icemen: Origins of the European Neolithic as Seen from the Frontier. In M. Rockman and J. Steele (eds.), *Colonization of Unfamiliar Landscapes: The Archaeology of Adaptation* (pp. 144–168). London: Routledge.
Fischer, A. and Tauber, H. (1986). New C-14 Datings of Late Palaeolithic Cultures from Northwestern Europe. *Journal of Danish Archaeology* 5, 7–13.
Gamble, C. (1993). *Timewalkers: The Prehistory of Global Colonization*. Phoenix Mill, UK: Alan Sutton Publishing.
Gamble, C. (1999). *The Palaeolithic Societies of Europe*. Cambridge: Cambridge University Press.
Golledge, R. G. (1987). Environmental Cognition. In D. Stokols and I. Altman (eds.), *Handbook of Environmental Psychology* (pp. 131–174). New York: John Wiley and Sons.
Golledge, R. G. (2003). Human Wayfinding and Cognitive Maps. In M. Rockman and J. Steele (eds.), *Colonization of Unfamiliar Landscapes: The Archaeology of Adaptation* (pp. 25–43). London: Routledge.
Gould, S. J. (1991). Exaptation: A Crucial Tool for Evolutionary Psychology. *Journal of Social Issues* 47, 43–65.
Gould, S. J. and Vrba, E. S. (1982). Exaptation; A Missing Term in the Science of Form. *Paleobiology* 8, 4–15.
Graves, M. W. and Addison, D. J. (1995). The Polynesian Settlement of the Hawaiian Archipelago: Integrating Models and Methods in Archaeological Interpretation. *World Archaeology* 26, 380–399.
Hardesty, D. (2003). Mining Rushes and Landscape Learning in the Modern World. In M. Rockman and J. Steele (eds.), *Colonization of Unfamiliar Landscapes: The Archaeology of Adaptation* (pp. 81–96). London: Routledge.
Hausel, W. D. (1984). *Tour Guide to the Geology and Mining History of the South Pass Gold Mining District Fremont County, Wyoming*. Laramie: Geological Survey of Wyoming.
Hausel, W. D. (1989). *Geology of Wyoming's Precious Metal Lode and Placer Deposits*. Laramie: Geological Survey of Wyoming.
Hayden, F. W. (1872). *Preliminary Report of the United States Geological Survey of Wyoming and Portions of Contiguous Territories*. Washington, DC: Government Printing Office.
Henrich, J. and Boyd, R. (1998). The Evolution of Conformist Transmission and the Emergence of Between-Group Differences. *Evolution and Human Behavior* 19, 215–241.

Hewlett, B. S. and Cavalli-Sforza, L. L. (1986). Cultural Transmission Among Aka Pygmies. *American Anthropologist* 88, 922–934.

Hopkinson, T. (2001). *The Middle Palaeolithic Leaf Points of Europe: An Ecological Geography*. Department of Archaeology. Cambridge: University of Cambridge.

Hopkinson, T. (2007a). *The Middle Palaeolithic Leaf Points of Europe: Ecology, Knowledge and Scale*. Oxford: John and Erica Hedges, Ltd.

Hopkinson, T. (2007b). The Transition from the Lower to the Middle Palaeolithic in Europe and the Incorporation of Difference. *Antiquity* 81, 294–307.

Housley, R. A., Gamble, C. S., and Pettitt, P. (2000). Reply to Blockley, Donahue & Pollard. *Antiquity* 74, 119–121.

Housley, R. A., Gamble, C. S., Street, M., and Pettitt, P. (1997). Radiocarbon Evidence for the Lateglacial Human Recolonisation of Northern Europe. *Proceedings of the Prehistoric Society* 63, 25–54.

Huseas, M. M. (1991). *Sweetwater Gold: Wyoming's Gold Rush 1867–1871*. Cheyenne: Cheyenne Corral of Westerners International Publishers.

Huseas, M. and Doherty, R. (1984). Century of Decline: A Study of the Interpretation and Restoration of South Pass City, 1867–1967. In M. Kornfeld and J. Francis (eds.), *South Pass City: Changing Perspectives on a Nineteenth Century Frontier Town* (pp. 23–28). Cheyenne: Wyoming Recreation Committee.

Jochim, M., Herhahn, C. and Starr, H. (1999). The Magdalenian Colonization of Southern Germany. *American Anthropologist* 101, 129–142.

Kelly, R. L. (2003). Colonization of New Land by Hunter-Gatherers: Expectations and Implications Based on Ethnographic Data. In M. Rockman and J. Steele (eds.), *Colonization of Unfamiliar Landscapes: The Archaeology of Adaptation* (pp. 44–58). London: Routledge.

Kelly, R. L. and Todd, L. C. (1988). Coming into the Country: Early Paleoindian Hunting and Mobility. *American Antiquity* 53, 231–244.

Knight, W. C. (1901). *The Sweetwater Mining District, Fremont County, Wyoming*. Laramie: University of Wyoming.

Koestler, A. and Smythies, J. R. (eds.) (1969). *Beyond Reductionism: New Perspectives in the Life Sciences: Proceedings of the Alpach Symposium 1968*. London: Hutchinson.

Lahr, M. M. and Foley, R. (1994). Multiple Dispersals and Modern Human Origins. *Evolutionary Anthropology* 3, 48–60.

Lambeck, K. (1995). Late Devensian and Holocene Shorelines of the British Isles and North Sea from Models of Glacio-hydro-isostatic Rebound. *Journal of the Geological Society, London* 152, 437–448.

McDermott, J. D. (1993). *Dangerous Duty: A History of Frontier Forts in Fremont County, Wyoming*. Lander: Fremont County Historic Preservation Commission.

Meltzer, D. J. (2004). Modeling the Initial Colonization of the Americas: Issues of Scale, Demography, and Landscape Learning. In C. M. Barton, G. A. Clark, D. R. Yesner, and G. A. Pearson (eds.), *The Settlement of the American Continents: A Multidisciplinary Approach to Human Biogeography* (pp. 123–137). Tucson: University of Arizona Press.

Minc, L. D. (1986). Scarcity and Survival: The Role of Oral Tradition in Mediating Subsistence Crises. *Journal of Anthropological Archaeology* 5, 39–113.

Minc, L. D. and Smith, K. P. (1989). The Spirit of Survival: Cultural Responses to Resource Variability in North Alaska. In P. Halstead and J. O'Shea (eds.), *Bad Year Economics: Cultural Responses to Risk and Uncertainty* (pp. 8–39). Cambridge: Cambridge University Press.

Mortimore, R. N. and Wood, C. J. (1984). The Distribution of Flint in the English Chalk, with Particular Reference to the 'Brandon Flint Series' and the High Turonian Flint Maximum. In G. d. G. Sieveking and M. B. Hart (eds.), *The Scientific Study of Flint and Chert: Proceedings of the Fourth International Flint Symposium Held at Brighton Polytechnic 10–15 April 1983* (pp. 7–20). Cambridge: Cambridge University Press.

Pavlov, P., Roebroeks, W. and Svendsen, J. I. (2004). The Pleistocene Colonization of Northeastern Europe: A Report on Recent Research. *Journal of Human Evolution* 47, 3–17.

Potts, R. (1998). Variability Selection in Hominid Evolution. *Evolutionary Anthropology* 7, 81–96.
Raymond, R. H. (1870). *Section VII- Wyoming Territory. Mining Statistics West of the Rocky Mountains* (pp. 325–338). Washington, DC: Government Printing Office.
Richerson, P. J. and Boyd, R. (2001). Built for Speed, Not for Comfort: Darwinian Theory and Human Culture. *History and Philosophy of the Life Sciences* 23, 425–465.
Richerson, P. J. and Boyd, R. (2005). *Not by Genes Alone: How Culture Transformed Human Evolution.* Chicago: University of Chicago Press.
Rockman, M. (1995). *Investigation of Faunal Remains and Social Perspectives on Natural Resource Use in an 1867 Wyoming Gold Mining Town.* Unpublished Master's Thesis, Department of Anthropology. Tucson: University of Arizona.
Rockman, M. (2001a). The Landscape Learning Process in Historical Perspective. In J. Gillespie, S. Tupakka, and C. de Mille (eds.), *31st Annual Chacmool Conference Proceedings* (pp. 493–509). Calgary, Canada: Archaeological Association of the University of Calgary.
Rockman, M. (2001b). The Role of Topography in Initial Colonizations. Paper presented at the 66th Society for American Archaeology Annual Meeting. New Orleans, Louisiana.
Rockman, M. (2003a). Knowledge and Learning in the Archaeology of Colonization. In M. Rockman and J. Steele (eds.), *Colonization of Unfamiliar Landscapes: The Archaeology of Adaptation* (pp. 3–24). London: Routledge.
Rockman, M. (2003b). *Landscape Learning in the Late Glacial Recolonization of Britain.* Ph.D. Dissertation, Department of Anthropology, Tucson: University of Arizona.
Rockman, M., Glascock, M. D., and Baker, M. (2003). *Learning the Lithic Landscape: Trace Element Characterization of Flint Using ICP-MS and the Recolonization of Great Britain at the End of the Last Ice Age.* Paper presented at the 68th Society for American Archaeology Annual Meeting. Milwaukee, Wisconsin.
Roebroeks, W. (2001). Hominid Behavior and the Earliest Occupation of Europe: An Exploration. *Journal of Human Evolution* 41, 437–461.
Roebroeks, W. (2003). Landscape Learning and the Earliest Peopling of Europe. In Rockman, M. and J. Steele (eds.), *Colonization of Unfamiliar Landscapes: the Archaeology of Adaptation* (pp. 99–115). London: Routledge Press.
Roebroeks, W., Conard, N. J., and Kolfschoten, T. V. (1992). Dense Forests, Cold Steppes, and the Palaeolithic Settlement of Northern Europe. *Current Anthropology* 33, 551–586.
Rosenberg, M. (1994). Pattern, Process, and Hierarchy in the Evolution of Culture. *Journal of Anthropological Archaeology* 13, 307–340.
Shennan, S. (2000). Population, Culture History, and the Dynamics of Culture Change. *Current Anthropology* 41, 811–835.
Smith, K. P. (1988). Ritual and Resource Variability: Mechanisms for Transmission and Storage of Information Regarding Low-Frequency Resource Cycles in Hunter-Gatherer Societies. In B. V. Kennedy and G. M. LeMoine (eds.), *Diet and Subsistence: Current Archaeological Perspectives: Proceedings of the Nineteenth Annual Conference of the Archaeological Association of the University of Calgary* (pp. 86–107). Calgary: University of Calgary Archaeological Association.
Steele, J. and Rockman, M. (2003). Where Do We Go from Here? Modeling the Decision-Making Process During Exploratory Dispersal. In M. Rockman and J. Steele (eds.), *Colonization of Unfamiliar Landscapes: The Archaeology of Adaptation* (pp. 130–143). London: Routledge.
Street, M. and Terberger, T. (1999). The Last Pleniglacial and the Human Settlement of Central Europe: New Information from the Rhineland Site Wiesbaden-Igstadt. *Antiquity* 73, 259–272.
Thieme, H. (1997). Lower Palaeolithic Hunting Spears from Germany. *Nature* 385, 807–810.
Tolan-Smith, C. (1997). Landscape Archaeology. In C. Tolan-Smith (Ed.), *Landscape Archaeology in Tynedale.* Newcastle-upon-Tyne: Department of Archaeology, University of Newcastle-upon-Tyne.
Tolan-Smith, C. (2003). The Social Context of Landscape Learning and the Lateglacial – Early Postglacial Recolonization of the British Isles. In M. Rockman and J. Steele (eds.),

Colonization of Unfamiliar Landscapes: The Archaeology of Colonization (pp. 116–129). London: Routledge.
Trigger, B. G. (1989). *A History of Archaeological Thought.* Cambridge: Cambridge University Press.
van Andel, T. H. and Runnels, C. N. (1995). The Earliest Farmers in Europe. *Antiquity* 69, 481–500.
Vansina, J. (1985). *Oral Tradition as History.* Madison: University of Wisconsin Press.
Walker, M. J. C., Bohncke, S. J. P., Coope, G. R., O'Connell, M., Usinger, H. and Verbruggen, C. (1994). The Devensian/Weichselian Late-glacial in Northwest Europe (Ireland, Britain, North Belgium, the Netherlands, Northwest Germany). *Journal of Quaternary Science* 9, 109–118.
Walker, M. J. C. and Harkness, D. D. (1990). Radiocarbon Dating the Devensian Lateglacial in Britain: New Evidence from Llanilid, South Wales. *Journal of Quaternary Science* 5, 135–144.
Webb, R. E. (1998). Problems with Radiometric "Time": Dating the Initial Human Colonization of Sahul. *Radiocarbon* 40, 749–758.
Whittaker, C. R. (1989). Supplying the System: Frontiers and Beyond. In Barrett, J. C., Fitzpatrick, A. and Macinnes, L. (eds.), *Barbarians and Romans in the North West Europe* (pp. 64–80). Oxford: BAR International Series.
Whittow, J. (1992). *Geology and Scenery in Britain.* London: Chapman & Hall.
Wright, S. (1932). The Roles of Mutation, Inbreeding, Crossbreeding and Selection in Evolution. *Proceedings of the Sixth International Congress on Genetics* 1, 356–366.

Chapter 3
"The Multiplication of Forms:" Bering Strait Harpoon Heads as a Demic and Macroevolutionary Proxy

Owen K. Mason

Introduction

The trajectory of western arctic prehistory is marked by increasing societal complexity (Table 3.1), with hierarchic control of several resource hot spots emerging by the later centuries B.C., when the caloric bounty of bowhead and gray whales (Krupnik 1987, 1993) was reliably secured (Dinesman et al. 1999; Mason 1998, Mason n.d., unpublished manuscript; Mason and Gerlach 1995). Several evolutionary hinge points define the prehistory of the western arctic. Rapid sea level rise prior to 3000 B.C. transformed the Bering and Chukchi continental shelves (Mason and Jordan 1993) and appears to have destroyed any archaeological evidence of maritime-adapted societies north of the Aleutian Islands; the earliest sites derive from the Arctic Small Tool tradition, termed Paleo-Eskimo in the eastern arctic. From the onset, successful cultural strategies relied on sea mammal hunting, predominantly of ringed seal (*Phoca hispida*), supplemented with bearded seal, walrus, and, caribou (Darwent and Darwent 2005; Giddings and Anderson 1986). The toggling harpoon occurs among the earliest sealing communities (Ackerman 1984; Dumond 2008; Giddings and Anderson 1986). For several thousand years, from ca. 3000 B.C. until the last centuries B.C., Arctic Small tool tradition communities remained small and, likely, fairly mobile (Anderson 1984; Dumond 1984; Giddings and Anderson 1986). Sites include only one or two houses, and architecture is minimally documented (Anderson 1984).

Possibly as early as 500 B.C., Bering Strait (Fig. 3.1) witnessed a major transformation, with sedentary, ranked communities developing around Cape Dezhneva (East Cape) and Sivuqaq (Gambell) at Northwest Cape, St. Lawrence Island (Dinesman et al. 1999; Mason 1998). Although the processes associated with its origin cannot be established, several large cemeteries document the elaboration of

O.K. Mason (✉)
Geoarch Alaska, Anchorage, P.O. Box 91554, AK 99509-1554, USA
e-mail: geoarch@ptialaska.net

Acknowledging the surrealist painter Yves Tanguy, whose painting "Multiplication of the Arcs" resides at the Museum of Modern Art, New York.

A.M. Prentiss et al. (eds.), *Macroevolution in Human Prehistory*,
DOI 10.1007/978-1-4419-0682-3_4, © Springer Science+Business Media, LLC 2009

Table 3.1 Cultural Complexes in Northwest Alaska from 1000 B.C. to A.D. 1300. The succession is mostly sequential, one of descent,[1] but has notable crosscutting contemporaneous parallels, although not as profound as Gerlach et al. (1992) argued (cf. Mason 1998, 2000, 2003)

Archaeological culture	Age range calibrated B.C./A.D.	Geographic distribution	Diagnostic tools, motifs	Subsistence base	Social space	Ritual space
Choris (defined by Giddings in the 1950s on Choris Peninsula)	800–500 B.C. from Choris, onion portage.	Kotzebue sound brooks range north Yukon territory	Linear stamp ceramics, dual barbed bone pts. Burinated bifaces parallel flaked lithics. Labrets	Seal hunting, possibly Beluga Caribou hunting. Extent of fishing uncertain	Large oval structures known at Choris Peninsula presumably used for housing. Possibly collective functions	None attested
Norton (defined by Giddings at Iyatayet, in Norton Sound)	Onset uncertain, 500 B.C. at extreme margins (i.e., Pt. Moller, Chukotka), terminal phase ca. A.D. 600 in Norton Sound, grades into Ipiutak. Brooks Range, poss. into Tanana valley.	Southern Alaska Peninsula to western Canada, generally west of treeline, uncertain possible extent in Siberia and Chukotka	Check stamp ceramics, distinctive pentagonal proj. pts. Oval side/end scrapers, toggling harpoon. Minimal linear decorative motives Oval lamps, Medial Labrets	Seal Hunting, "casual" whaling, possibly attested at Point Hope, Caribou Hunting. Villages often at salmon bearing streams, extent of fishing variable.	Square Houses, ca.16 m^2 larger gender or community specific structures occur at some localities in later phases. Communities of several houses	Groupings of burials occur at Battle Rock and at Point Hope (=Near Ipiutak). Otherwise extent of cemeteries undocumented

Table 3.1 (Continued)

Archaeological culture	Age range calibrated B.C./A.D.	Geographic distribution	Diagnostic tools, motifs	Subsistence base	Social space	Ritual space
Ipiutak (defined by Larsen and Rainey 1948 from Point Hope) Considered derivative from Norton by most archaeologists	Possibly as early as 200 B.C., but most ages between A.D. 200 and 800, None younger than A.D. 900	Golovnin lagoon (Norton Sound) to Point Hope; Brooks Range, Seward Peninsula Kobuk and Noatak River drainages. Major sites: Deering, Pt. Hope, Cape Krusenstern. Possible occupation in forested Koyukuk valley—very rarely associated with forest cover.	Birch Bark containers aceramic, and no lamps. Distinctive lithics, bow/arrow pts. Barbed bone pts for warfare.Harpoon heads less complex than OBS but decorated, occas. Barbed, spurred, Complex linear motifs, circle/dot, Rays and dashes. Cosmo-logical symbols in burials. Abstract objects assoc. with shamans. Use of Iron for engraving	Seal, Walrus, Caribou Hunting, Possibly beluga in southern range. Extent of fishing unclear but significant in interior lakes.	Strong evidence of warfare Drift wood log structures—Large structures 3 or 4 times the smaller houses, which show both winter and summer forms (the latter lacks entry). Houses variable sizes.	Extensive cemeteries at limited locations, primarily Point Hope, small number graves at Deering. Status differences in burials. Especially apparent in elaborate children's. Numerous cases of violence with burials.

Table 3.1 (Continued)

Archaeological culture	Age range calibrated B.C./A.D.	Geographic distribution	Diagnostic tools, motifs	Subsistence base	Social space	Ritual space
Old Bering Sea (defined by Jenness and Collins from Diomede and St Lawrence Isl.)	Earliest phase, **Okvik** dated between A.D. 200 and 400, possibly earlier at Ekven between 600–400 B.C.	Eastern Chukotka littoral, northwest of East Cape, Diomede Isl., St. Lawrence Isl. Isolated artifacts at Point Barrow, Golovnin Lagoon, uncertain implications. No inland occurrences.	Ceramics, bowls, cooking containers, lamps, Less well-flaked lithics, elaborate harpoon heads, multiple spurs, and line holes. Motifs incl. circle, dot, winged objects common in burials only at Siberian sties. Diverse figural tradition, human and animal representations. Use of iron for engraving	Seal and Walrus Hunting most prevalent, Whaling Extent of Caribou hunting uncertain, likely trade for skins with Ipiutak	Houses with elaborate slab flooring, whale bone, drift wood. Limited evidence for larger structures	Dedicated cemetery precincts common at Ekven, Uelen, Chiini on Chukotka, St Lawrence Island Sivuqaq; cemetery at Kukulek undocumented outside mound. Status differences in Ekven and Uelen but not at St. Lawrence Isl. burials.

Table 3.1 (Continued)

Archaeological culture	Age range calibrated B.C./A.D.	Geographic distribution	Diagnostic tools, motifs	Subsistence base	Social space	Ritual space
Punuk (defined by Collins in 1929–37 fr. St. Lawrence Island)	A.D. 800–1200, possibly as late as A.D. 1400	St. Lawrence Island, eastern Chukotka, Eastern Norton sound, Kobuk valley Kotzebue Sound. Northwest AK coast, Nunagiak, northern Ellesmere, Greenland. Only sporadic artifacts in Kotz sound and Pt. Hope.	Ceramics, lamps, Distinctive Sicco, Thule 2 open socket harpoon heads, atlatl counter weights. Curvilinear abstract motifs, lacks figural art. Line and dot, circles, ceramics, Slate use for end blades, scrapers, bola weights, drum handles, Bow guards.	Seal and Walrus Hunting, Whaling. Caribou Hunting uncertain contribution. Fishing uncertain within Bering Strait but stronger orientation in Norton Sound and Kobuk River.	Houses with elaborate slab floors and whale bone supports. Larger structures specialized use for collective purposes.	Fewer graves reported, continuity in cemetery use with earlier OBS communities at Ekven, Uelen and Sivuqaq—but not at Point Hope

Table 3.1 (Continued)

Archaeological culture	Age range calibrated B.C./A.D.	Geographic distribution	Diagnostic tools, motifs	Subsistence base	Social space	Ritual space
Birnirk (defined by Mathiassen in 1930, based on Mason's descriptions published in 1930)	Possibly pre A.D. 600, but most A.D. 600–1200; most sites poorly dated, early range uncertain, most cluster A.D. 900–1100	Type site near Point Barrow Mostly restricted to shores of Chukchi Sea, limited southern influence at Ekven, St. Lawrence uncertain, Norton Sound influence problematic—no diagnostics reported at Safety Sound. No interior sites. Limited occupation at Point Hope, Cape Krusesntern, uncertain at Wales, likely significant.	Ceramics, lamps, distinctive open socket harpoon heads: Birnirk (barbed, spurred, side bladed) Natchuk form (lacks blade); Well-flaked lithics,, Decorative restricted to simple linear arrays, Iron use, No figural art.	Seal, Walrus and Whaling. Caribou Hunting in N Alaska. Extent of Fishing uncertain.	Small houses common, few larger structures, except at Cape Baranov (Siberia). Possible large structure at Kugusugruk.	Burials common in structures, one spectacular example at Kugusuguruk, described as charnel house contains skin clothing, diverse tool inventory. Burial goods not linked to status.

Table 3.1 (Continued)

Archaeological culture	Age range calibrated B.C./A.D.	Geographic distribution	Diagnostic tools, motifs	Subsistence base	Social space	Ritual space
Thule (defined by Mathiassen in 1927)	Possibly as early as A.D. 1000, definition at older range controversial, intergrades with Punuk and Birnirk Younger range, A.D. 1700, "Late Prehistoric"	Bering Strait to Greenland; southern extent ambiguous, possibly as far south as Alaska Peninsula. Expansion into Brooks Range. Chukotka.	Wide range of harpoon head types, Thule types 1, 2, 3, Antler bone points, ceramics, lamps, slate use, ownership marks, dolls common, but abstract designs only linear, simple, evidence of dog traction ca. A.D. 1400	Seal, Walrus, Whaling, Caribou Hunting, Use of Dogs leads to intensified salmon harvesting,	Multiroomed houses with long entry way passages, specialized kitchens. Elaborate roofing and side walls of driftwood and/or whale bone. Evidence for Large structures limited, especially at Utqiagvik A.D. 1400, otherwise rare.	Large cemeteries at Point Hope, Barrow region, likely other; St. Lawrence Island, East Cape. Evidence of status differences.

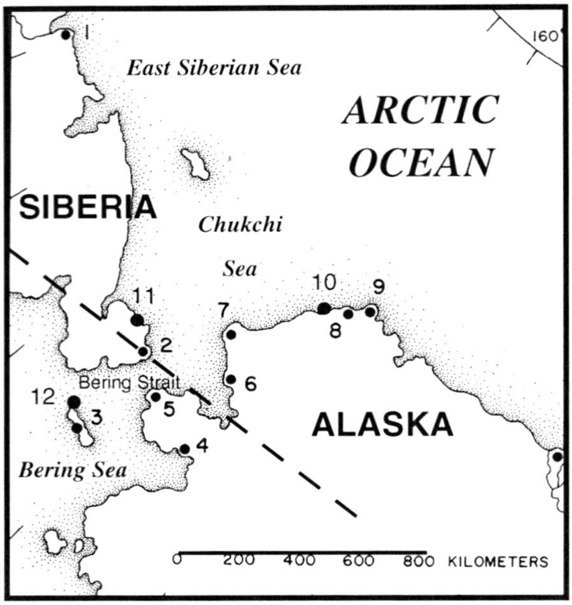

Fig. 3.1 Map of the Bering Strait region with major archaeological sites indicated by number. **1** Cape Baranov; **2** Cape Dezhneva (Uelen and Ekven), **3** Kukulik, St. Lawrence Island; **4** Safety Sound; **5** Kurigitavik (Wales), **6** Cape Krusenstern, **7** Point Hope (Ipiutak and Jabbertown), **8** Walakpa, **9** Birnirk, Nuvuk, **10** Nunagiak, **11** Papelyghak, Chegitun River, **12** Sivuqaq region (Gambell), St. Lawrence Island [Map modified from Stanford 1976:98]

hunting technology and spiritual capital, largely associated with a small fraction of the internees (Table 3.1). For the next millennium from A.D. 1 to A.D. 1000, the Bering Strait experienced a cathartic release of artistic energy that was not matched. The styles evident on many classes of artifact are termed Old Bering Sea (Collins 1937; Jenness 1928) or Ipiutak (Larsen and Rainey 1948), designations that Mason (1998) believes were associated with larger polities organized on a pan-regional basis.

By A.D. 900 the Old Bering Sea/Ipiutak system (Mason 1998) witnessed a cultural transformation (Collins 1937) marked by the development of another series of whaling societies (Whitridge 1999), some that developed in situ within Bering Strait, defined by their aesthetics as the Punuk culture (Collins 1937), while others arose at the margins of the Chukchi Sea, termed Birnirk from a variety of distinctive artifacts (Jenness 1940; Mason 2000; Mason and Bowers 2009).

Climate operated as an external driver, in some measure, on cultures in the Bering and Chukchi Seas (McGhee 1981; Mason and Barber 2003; Mason and Gerlach 1995). Marine productivity shifts in the North Pacific during the first millennium A.D. (Finney et al. 2002) may also have been related to artistic production (a proxy for spiritual capital), which was intensified during warmer and adverse conditions with frequent open water (Mason and Barber 2003). By contrast, innovation in sea mammal hunting technologies co-occurred with cooler-temperature-related, heightened productivity during the last centuries B.C., cooling associated with the Medieval Climatic Anomaly ca. A.D. 900 (Hughes and Diaz 1994) and the Little

Ice Age. The Little Ice Age represented the intensification of storms and upwelling, between A.D. 1500 and 1850 (D'arrigo et al. 2006; Mason and Barber 2003).

The elaboration of harpoon styles in the Bering Strait area offers the opportunity to comprehend a most contentious issue in evolutionary archaeology, the distinction between style and function of artifacts (Bettinger et al. 1996; Carr 1995; Dunnell 1978; Neiman 1995; O'Brien et al. 1994). The implications of this study are significant: From a macroevolutionary vantage point (Prentiss et al., introduction chapter), smaller scale, microevolutionary process (changes in harpoon styles) could be overridden by evolutionary processes acting on more complex levels. Put differently, changes in a single artifact class (harpoon) within a larger package (Ipiutak and OBS sea mammal hunting tool kits) may have had a stronger influence on evolution of cultural systems than simple shifts in relative frequencies of forms within that artifact class alone. The development of an improved harpoon head (Figs. 3.2 and 3.3) conferred an adaptive advantage that altered hunting success to the extent that it affected the fate of the entire community. This macroevolutionary process will be elucidated in this chapter by presenting current knowledge of harpoon history, examining the correlations between harpoon styles (Figs. 3.4a,b), by changes within more complexly integrated cultural entities (Table 3.1), and by tracking the changes through time in relation to climate change. The improved technology is a material manifestation of an altered psychic state that represents a revitalized mentality confronting the weather instabilities of the Medieval Climatic Anomaly (Hughes and Diaz 1994).

The Conundrum of Style and Function in Arctic Archaeology

Either implicitly or explicitly, style has served arctic archaeologists as a "category of ascription" (Barth 1969; cf. Wiessner 1983; Wobst 1977)—a proxy for ethnicity—for nearly three generations. Even Lewis' (1995) effort to reject style on Bering Strait harpoon heads, in order to re-inductively infer cultural relationships from the function of artifacts, resorted in its final analysis to employing the accepted stylistic categories to frame its conclusions. The styles, termed Old Bering Sea, Punuk etc. (Fig. 3.4a) are so engrained in archaeological discourse that possibly we cannot reason about the entities without reference to the constellation of motifs (cf. Collins 1937) that are so easily distinguished. A central problem in all efforts involves the conflation of purely functional attributes with aesthetic attributes. The present effort focuses on several decorated types of artifacts as crucial components in examining the social changes accompanying the Old Bering Sea/Ipiutak interaction sphere.

The purported dichotomy between style and function in technology troubles many macroevolutionary theorists. On the one hand, tools with a specific function have undoubted selective value that powers success and survival among human populations (O'Brien and Lyman 2000:269ff). Stylistic addenda, by contrast, appear to offer no selective value and seem to be merely art for art's sake. Nonetheless, elaborate designs required a substantial investment of time, time better expended on hunting. Ethnographic accounts, if applicable to archaeological cultures, imply that

Fig. 3.2 The Inuit harpoon and throwing board complex, as illustrated by O.T. Mason (1902)

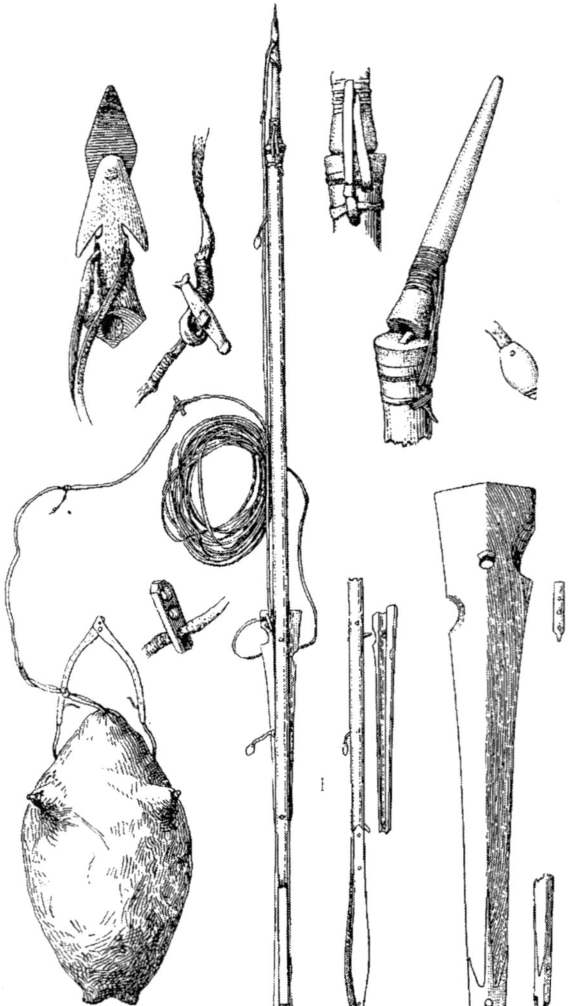

SEAL HARPOON FROM WEST GREENLAND.
Collected by N. P. Scudder.
Cat. No. 35670, U.S.N.M.

artistic addenda served magico-religious purposes and imbued artifacts and their bearers with spiritual force that reflected and conferred status, providing critical display-related information (Carr 1995; Roe 1995).

The investment in items of display arose in diverse areas during the Neolithic as a phenomenon associated with the ascendance of status-striving individuals (Hayden 1998, 2003). The creative energy of aggrandizers is catalyzed by the opportunities

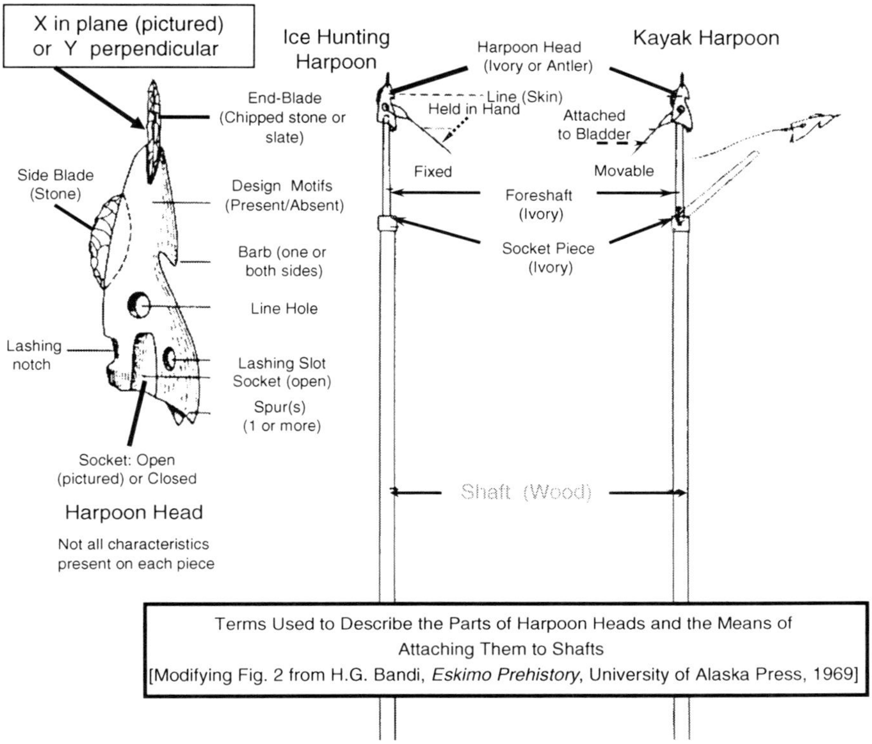

Fig. 3.3 Functional and stylistic constituents of a harpoon head (modifying Bandi 1969)

that arise with the entrance of a new population into regions with previously untapped resources or by the discovery or appearance of a resource that has previously been untapped. Prestige objects are generated that represent the success and thus potential biological fitness of the elites in control. An obvious parallel exists with the biological advertisements of plumage. In the Bering Strait, several categories of object were affixed with emblems of high status and cosmological significance—schemata that are believed to have reflected success in hunting and in guaranteeing survival. The accumulated social capital of individuals inherent in such objects is often reflected in mortuary offerings (O'Shea 1984). To monitor these trends, I will focus primarily on the single object that enabled and conferred success to marine mammal hunters: the harpoon head (Figs. 3.2 and 3.3) and its associated decoration and technology (Fig. 3.2, from Mason 1902).

Regional or Ethnic Differences in Aesthetic Overlays

The various peoples represented in the Bering Strait archaeological record differed in their approaches to decoration; only two or three localities were obsessive in

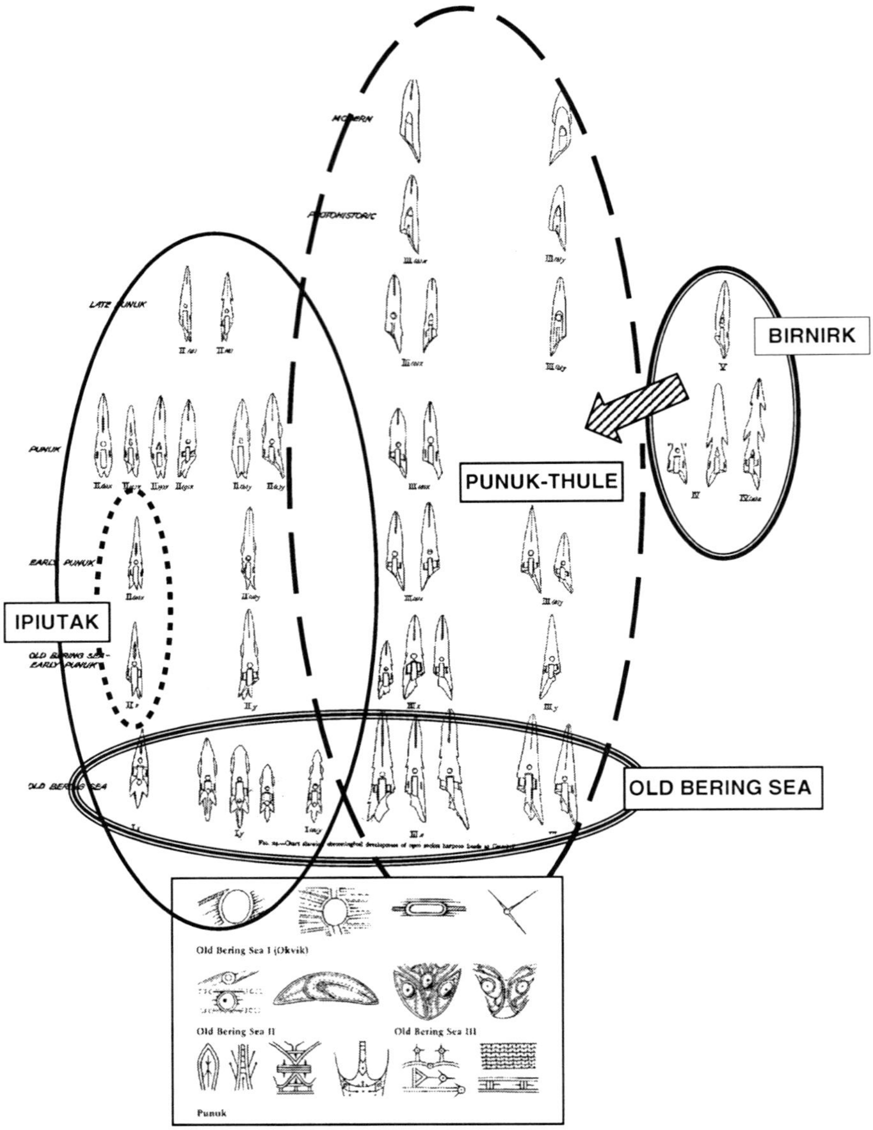

Fig. 3.4 Principal schemes of decorative motifs,(at base). defined by Collins (1937), as simplified by Fitzhugh and Crowell (1988). Although not fully substantiated by radiocarbon dates, the Old Bering Sea styles represent successive styles that are earlier than the Punuk style. Birnirk and Thule (not pictured) decorations involve lines with minimal additions (b) Classification of open socket harpoon heads showing various lineage-like groupings, in *bold*. Old Bering Sea forms, lower, were the progenitor for the Punuk lineages at *left*. The Ipiutak forms are a subset of the Old Bering Sea types. The Birnirk and Thule grouping, at *right*, is intrusive

their aesthetic impulses, applying elaborate motifs to even everyday objects like bag handles and needle cases. Not all groups decorated harpoon heads; most objects associated with warfare were not typically decorated—an exception being the side-bladed daggers (cf. Larsen and Rainey 1948: Figure 40). To some measure, decoration follows from surpluses of walrus bone and ivory, the preferred media for engraving. A bounty of walrus may explain the rise of more complex societies in Bering Strait, as postulated by Collins (1940:549): "it was probably the abundance of walrus, more than any other single factor that made possible the initial concentration of people at numerous places around Bering Strait." Very notably, elaborate art and mortuary practices occur at locations (Cape Dezhneva, Point Hope) with ready access to migratory walrus, which track resource hot spots and the annually receding ice cover of the Bering and Chukchi Seas (Mason and Gerlach 1995).

The intellectual culture of the various peoples, especially its cosmology, can be deciphered to some extent, from the context of the designs (Table 3.1, Fig. 3.4a) (Auger 2005). The aesthetic language that commenced with the Okvik and Old Bering Sea cultures represents a common tradition (Arutiunov and Bronshtein 1985), whose motives were furthest and earliest elaborated on St. Lawrence Island and the adjacent coast of Chukotka (Leskov and Müller-Beck 1993).

Old Bering Sea culture artisans/artists applied designs to a wide variety of objects: fasteners, sealing harpoon heads and foreshafts, snow goggles, containers, handles, and so on (as illustrated in Arutiunov and Fitzhugh 1988). The system of dots, dashes, chevrons, and circles was systematized by Collins (1937) into four stages, OBS 1, 2, 3, and Okvik. In large measure, the enigmatic Ipiutak is only a special subset of OBS 2 (Arutiunov and Bronshtein 1986; Larsen and Rainey 1948). Human and animal representations are common; the human figures portray transformation, occasionally submission, and contain evidence of tattoos. The circumstance of OBS tattooing reminds us that body art likely extended the decorative system (Krutak 1998) and was very likely also paralleled with clothing designs that are not preserved in the archaeological record.

By contrast, Ipiutak artisans rarely decorated harpoon heads and applied decorations to more specialized items, many apparently used exclusively for ritual or in mortuary contexts (Larsen and Rainey 1948). Two classes of Ipiutak objects bear complex animal motifs: Bone tubes (considered shamanic sucking devices) and brow bands for wooden hats (cf. Larsen and Rainey 1948:135–146). The most dramatic Ipiutak pieces are usually termed "masks" ("mask-like" in Larsen and Rainey 1948: Plates 54 and 55); however, the objects were probably clothing overlays or necklaces, considering the attachment holes on their obverse sides. Only three sets are known—two from Pt. Hope and one from Deering. Two were found associated with children's graves (Larsen and Rainey 1948:241; Reanier et al. 1998); while the most elaborate recovered outside, but close to burials (Larsen and Rainey 1948:238, Plate 55). One special class of objects, "open-work carvings" often was decorated; their use remains uncertain, but is assumed to relate to shamanic performance. Most decorated Ipiutak pieces were recovered in mortuary contexts.

With the development of the succeeding Punuk and Thule cultures, fewer graves are known and fewer decorated pieces are found in any context (Ackerman 1984;

Collins 1937; Staley and Mason 2004). Punuk groups continued to favor harpoon heads for engraving when they did decorate something, but fewer elaborate pieces were placed in graves. Graves instead more often contain unworked whale bone and possible food offerings (Bandi and Blumer 2002; Staley and Mason 2004). Punuk-decorated pieces are possibly as common in domestic settings as they are in graves. The largest Birnirk culture mortuary assemblage, from Kugusuguruk, south of Barrow (Dikov 2003; Ford 1959), may represent a profound shift (or alternative) in practice, reflecting a preference for abandoning household inventories in situ in the face of death. This contrasts sharply with the dedicated cemeteries favored by OBS (Arutiunov and Sergeev 2006a, b) or Ipiutak peoples (Larsen and Rainey 1948; Mason 2006). The extent of decoration in the Birnirk culture is underestimated because its largest collection (e.g. that of archaeologist Wilbert Carter (1966) remains unanalyzed (cf. Mason 2000). While a few Birnirk harpoon heads did bear designs resembling Old Bering Sea (Wissler 1916:408), most with decorations were less complex than seen in OBS or Ipiutak cultures.

The Utility of Harpoon Heads in Defining Cultural Blue Prints

The harpoon head (Fig. 3.2) is but a small part of a specialized repertoire for marine mammal hunting (Mason 1902), evidence of a specific, historically based cultural blue print or *Bauplan*. The toggling harpoon system is perhaps the most complex weapons system employed by hunter-gatherers (Oswalt 1973:137). Its development, as ancient as 3000 B.C. (Dumond 2008:278ff), represents a radical improvement in adaptive strategy, a new *Bauplan*, that represents a complex of adaptive strategies[2]. A number of technological features are functionally requisite (cf. Mason 1902:199ff; Porsild 1915:174ff) for the toggling form that penetrates the animal and detaches in one type maintaining a link to a seal skin bladder. The required attributes include at least one line hole, socket attachments (open or closed), end or side blade slots, and lashing slots. Side blades seem optional, occasionally replaced by barbs carved in the bone or walrus tusk. The toggling harpoon system with bladder was propelled by an atlatl from a kayak in open water or in ice conditions with wide leads, although walrus hunting required occasional floating ice for successful recovery—fully ice-free seas presented insurmountable conditions for walrus hunting as in the famine of 1878–1879 off Gambell (Burgess 1974). Other methods of sealing, e.g., netting or breathing hole spearing, do not require the harpoon and/or float (cf. Murdoch 1892:268ff). A float was sewn out of seal skin (Nelson 1899:140ff) and resembles the storage bladders from seal or caribou stomachs used for oil (Nelson 1899:73).

Stylistic addenda were comparatively rare on harpoon heads across the arctic during the nineteenth century, and those first collected by anthropologists rarely had decorations. Some early researchers only noticed the functional attributes; for example, Wissler (1916:410) failed to acknowledge motifs on an elaborate head from the Barrow region (subsequently classifiable as Old Bering Sea).

The earliest classification schemes developed in the eastern arctic lacked temporal control (other than precontact and post-contact) and emphasized functional characteristics, reflecting the absence of decorative and other embellishments (Mason 1902; Porsild (1915:140ff; Thalbitzer 1914:430). The harpoon head is but a small component of a complex propulsion device propelled either by the throwing board or by spear, as recounted by Mason (1902:199ff) and in ethnographic accounts, some from the eighteenth century. The principal distinction among the two harpoon head types lies in the manner in which the head is detained or fastened within the prey. In one, the hold is maintained by the barbs; in the other, the entire head toggles to "place itself across the wound" (Mathiassen 1927b:11). Barbed harpoon heads apparently confer better advantage in prey with thicker skin. Combining barbs with the toggling form yields a third type of harpoon head (Mason 1902).

Further classificatory refinements were offered by Mathiassen (1927b:12–13) who distinguished four basic harpoon head types, contrasting "flat" ("A") from "thin" ("B") and distinguished open ("I") and closed socket ("II") alternatives within the Thule culture. *Thin* harpoon heads, widest at right angles to the line hole, revealed the most variable attributes, occasioning a further 15 subtypes ("a" to "d"), in relation to presence or absence of barbs, and the direction of line holes. *Flat* heads are wider in the direction of the line. The taxonomy was guided by an inferred hierarchical principle from simple to complex (Mathiassen 1927b:13). The inference was not supported by chronological data in 1927, and it remains unsupported at present. Thus, due to its cumbersome complexity, Mathiassen's taxonomy is rarely employed by archaeologists.

Researchers have had highly variable approaches to classifying harpoon heads (i.e., lumpers vs. splitters), not always responding to collection size (cf. Collins 1937; Geist and Rainey 1936). Within the very extensive Kukulek mound on St. Lawrence Island, three types, mostly late prehistoric, predominated (66, $n = 437$), and although Geist and Rainey (1936:88ff, 184, 200) distinguished 10 types and 15 "rare" types, their efforts were not systematic and had limited impact. At Point Hope, only four types (one subtype, IA) sufficed for a sizable collection ($n = 159$), albeit masking considerable variability within the open socket Type I ("Oopik" to Ford 1959) that predominated (45%) (Larsen and Rainey 1948:68ff).

Partially due to large sample size (nearly 400) within the Gambell (Sivuqaq) region of St. Lawrence Island, Collins (1937) offered the most definitive system employed by many researchers (Fig. 3.4b). For open socket harpoon heads, Collins (1937) defined 29 total types within five major classes (I, II. . ..), distinguished as open (a) socket, with a polarity between end blade positions in relation to the line hole (x or y). The earliest type (I) Old Bering Sea, is marked by decorative basal spurs or prongs, often trifurcated, which became smaller, morphing into vestigial "quotations" of the original forms (Fig. 3.4b). The two Punuk types encompass the most stylistic variability in types II and III. All four types undergo a simplification through time, as functional attributes gain ascendance over decorative elements such as vestigial barbs or spurs. The dual barbed, open socket Type IV corresponds to Mathiassen's (1927a) Thule type 2 and is, apparently, an accessory, nonlocal (non-St. Lawrence Island, non-Bering Strait) lineage that includes the types of Ford

(1959), termed, Birnirk, Natchuk, and Nunagiak (close socketed) harpoon heads, marked as a sub-lineage in Fig. 3.5.

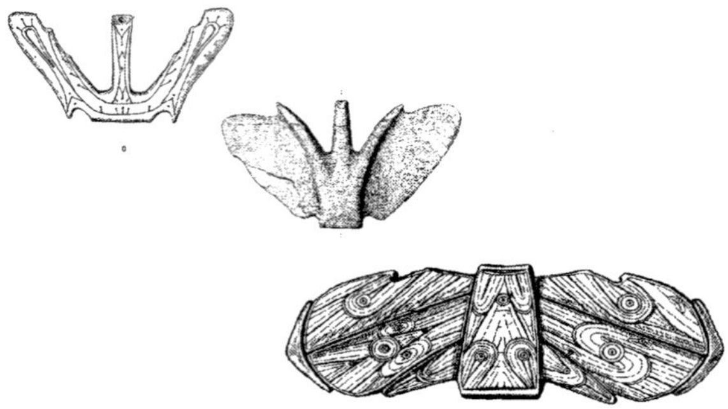

Fig. 3.5 Atlatl stabilizers ("winged objects") of the Old Bering Sea and Punuk cultures (from Fitzhugh and Crowell 1988). *Upper*: Punuk from Nunagiak (Ford 1959:62), measures 15 cm width; *Middle*: Punuk from Grave M4, Meruchta midden, Dovelayaq Bay (Hoffman-Wyss 1987:plate 4) measures 15 cm in width; Bottom, Old Bering Sea from Burial 133, Ekven measures 20 cm in width (Arutiunov and Sergeev 2006b:166)

Ford (1959:75–96) employed his 1930s (purchased and excavated) collections from several northern Alaska sites to elaborate a harpoon head typology for the Birnirk and Thule cultures. Ford's (1959) scheme returned to qualitative classes, verbal tags from local place names (Utqiagvik, Barrow, Nunagiak, etc.), and subsumed variability in possibly minor decorative accoutrements such as spurs and prongs (cf. Ford 1959: Fig. 27). The 1969–1970 excavations by Stanford (1976:29) at Walakpa required several new types, and combined both Collins' and Ford's schemes, and was based on limited use of ^{14}C dates for the development of chronology. It is possibly significant that considerable data are still unanalyzed, especially, the early 1950s Harvard-sponsored excavations at the Birnirk and Nuvuk (Pt. Barrow) mounds by Wilbert Carter (1966). The fullest analysis of a single harpoon head type may be that of the Thule form in the Wales area, as subdivided by Yama'ura (1984).

Contemporaneously, with Ford's synthesis, Russian archaeologists Arutiunov and Sergeev (2006a:77) offered an alternative classification following a decade of excavation at Ekven and Uelen from 1955 to 1965, that produced harpoon heads (especially at Uelen) that did not fit into Collins' scheme (Arutiunov and Sergeev 1972:305). The Russian flow-chart classification incorporated lashing slots, spurs and prongs, number of line holes (1 or 2), open (A) vs. closed socket (B), and presence and position of end and side blades into their system. The resulting harpoon head designation can contain between five and seven letters and numerals (compare Collins, I-a-x with Arutiunov/Sergeev, 2A2x2M3). The Russian classification system aims to include all variations in attribute combination, even types not yet discovered (Dumond 2008:277). Considering that 2/3rd of the Cape Dezhnev harpoon heads fall within only seven of the 44 types (16%); the variability may represent idiosyncratic workmanship or sub-groups (clans, families, etc.) or communities with small populations.

In addition to revising and expanding its classification, Arutiunov and Sergeev (1972) extended harpoon head classification into an explicitly ecological context, one that may have robust evolutionary implications. Based on their hunting expertise, local Chukchi informants told Arutiunov and Sergeev (1972:308) that the various harpoon head types were related to hunting success under different environmental conditions. The critical variable in harpoon head morphology lies in the orientation of the end blade, either in the x or y plane, relative to the line hole: The "x" orientation provides a "better push to the foreshaft," and to heavier harpoon heads. Thus, the "x" type is more advantageous in ice-covered seas (Arutiunov and Sergeev 1972:308). Nineteenth century informants in Barrow offered the same rationale and physics ("the harpoon would have a surer hold since the strain on the line would always draw it at right angles") to Murdoch (1892:221) for using the perpendicular "y" oriented harpoon head. The lighter and less stable "y" form (more prone to end blade breakage) can be employed profitably in open water. The Cape Dezhneva data show the prevalence of "x" harpoon heads in the north-facing Uelen with the "y" in dominance at the more open south-facing Ekven site. Harpoon head types may also have advantages in relation to the skin thickness of the various prey, as observed by Arutiunov and Sergeev (1972:308).

The use of harpoon heads as ethnic diagnosta can be greatly bolstered with ethnographic documentation from the pre-contact period, for the representation of harpoon head type within particular communities. In the Barrow region, Murdoch (1892:225) collected 45 specimens that represented the types in use at ca. 1882; nearly all were close-socket "x" forms. In western Greenland, Porsild (1915:242) similarly reported that several types (out of a wide number of Thule forms) were in use within a single community with 60 kayakers; nonetheless, other nearby communities had alternative preferences.

In the succeeding decades after 1920, the harpoon head served several generations of archaeologists as index fossils and as ethnic markers for specific cultures and temporal periods, especially in the decades preceding widespread use of chronometric dating methods (Collins 1937; Ford 1959; Mathiassen 1927a). Ford's (1959) seriation of harpoon heads for the Barrow region serves as a regional paradigm. Few arctic specialists have distinguished between functional and stylistic attributes, preferring to conflate both in their classifications (Lewis 1995). The history of classification reflects archaeological discovery and, in some sense, the accommodation of researchers to increasing sample size through additional discoveries.

Resolving the Classificatory Mire and Defining Macroevolutionary Trends

At the highest order, as observed above, harpoon heads can be classified within two broad clusters or domains (Fig. 3.4b) as in a venn diagram (cf. Ford 1959). Therefore, the design motifs of Okvik, OBS styles I, II, and III are congruent with and likely antecedent to Punuk types, as well as Ipiutak types—which form one macro lineage (Fig. 3.4b). Another offshoot (Fig. 3.4b) includes the Birnirk forms, which are only loosely bound with OBS/Ipiutak/Punuk forms. The opposition between the OBS-Punuk lineage and a Birnirk Thule lineage remains uncertain, the subject of contention since the 1930s (cf. Mason 2009; Mason and Bowers 2009). The boundaries between the lineages are largely mutually exclusive—although Thule forms reach south, very few OBS objects occur outside of the Chukotka peninsula, fewer than 10 specimens being reported on the mainland of Alaska and virtually all of those resulting from the efforts of collectors. Birnirk types were limited to the Chukchi Sea, and the major florescence in its development occurred in the greater Nunagiak to Barrow region and, paradoxically along the coast of Chukotka (Mason 2009). Despite 75 years of research, much remains uncertain about the chronological limits of the two lineages.

Three major elaborations or expansions in harpoon head types are evident, one centralized in the Anadyr Strait region and the other from Barrow to Nunagiak, with the accessory node from Cape Baranov to Cape Dezhnev (Dikov 2003; Okladnikov and Beregovaia 2008). The timing of these developments can be broadly identified; the first OBS/Okvik series occurs during the late first century B.C. and the first centuries A.D., as late as A.D. 600 (cf. Dumond 2000b, 2008). The second elaboration of Punuk types occurred between A.D. 700 and 1200, largely between Sivuqaq and Wales, where the Sicco form was overwhelming used (Collins 1937; Yama'ura 1984; cf. Ford 1959:73). The final elaboration occurred from Nunagiak to Barrow and apparently postdates A.D. 1000. Other richly decorated objects also prove useful in delineating lineage boundaries, most notably the winged object that was placed at the end of the harpoon head, as a stabilizer on atlatl spears.

The Winged Object as an Ethnic Denominator

While the ubiquitous harpoon head serves archaeologists well as a cultural and temporal diagnosta, the "winged object" provides an alternative conceptual system to define ethnicity. As recognized by nonindigenous (European, American, and Asian) art collectors the winged objects possess a numinous character (Hollowell 2008:249–250) that is apparently far in excess of their everyday use as weight stabilizers during the flight of a throwing stick. Winged objects were only used by societies immediately adjacent to Bering Strait and on St. Lawrence Island. Although two winged object preforms were recovered from Ipiutak houses at Point Hope (Larsen and Rainey 1948), none were found within Ipiutak graves.

The Old Bering Sea winged object (Fig. 3.5) was first purchased by Gordon (1916) in Nome, Alaska in the early twentieth century and was defined as a banner stone or a ceremonial staff. Only 20 years later did Collins (1937:200–201) propose that the objects were atlatl counter weights. Most archaeologists believe that the OBS winged object "evolved" into the streamlined Punuk trident, another type of stabilizer assumed to be younger (Collins 1961:5–7). The highest numbers of winged objects occur only at two general locations: Cape Dezhneva and the greater Sivuqaq (Dovelayaq to Meregta) region, with scattered finds from Chini (Chukotka), Kukulek and Point Hope (Bronshtein 2002; Collins 1961; Dikov 2003). Prior to the Russian excavations of late 1980s and 1990s (Leskov and Müller-Beck 1993), most OBS winged objects were purchased from Native diggers (Hollowell 2008). Of those with secure context, most are from mortuary contexts—few, if any, are from middens or houses. Possibly, OBS winged objects operated as specialized symbolic items and were deposited within graves as trophies or prized offerings. By contrast, Punuk tridents are reported only from the latter contexts; none occurs within the few documented Punuk burials.

Two classification schemes are employed for OBS and Punuk stabilizers, one developed by the American, Henry Collins (1940), the other produced by the Russians, Arutiunov and Sergeev (2006:175; cf. Bronshtein 2002). Collins' (1961:8) typology relies on plan view morphology and minimizes intragroup variability with OBS types. Although an evolutionary sequence is postulated by Bronshtein (2002), the several graves with such objects dated by Russian researchers indicate that either the objects were curated or their elaboration occurred within a single generation or two (Mason 2009). The Russian classification (Bronshtein 2002; Dumond 2008) melds apparently functional characteristics (side holes, possibly for attachment to the spear) with a variety of aesthetic overlays, also vaguely defined.

The most simple, presumably earliest, winged object occurs within House 1 at the Hillside site (Collins 1937:40–42) and the Okvik site (Rainey 1941:26). The dating of House 1 is reasonably good (based on three dates), indicating an occupation from A.D. 125 to 270 (Blumer 2002:71). Otherwise, the OBS winged object, mostly recovered in mortuary contexts, may have continued in use for many centuries, as revealed in comparisons of radiocarbon ages from the Ekven burials with the seriation of OBS winged object types of Bronshtein (2002). The sample size is very low, but 7 dates on well-provenienced winged objects from Ekven indicate use within graves from A.D. 400 to possibly as late as A.D. 1500. No forms or designs can be isolated within specific time ranges; one may hypothesize that formal elaboration was favored. For example, the Russian scheme has two broad classes A and B based on the number of accessory holes, with letters U, M, and E referring to the Main, Experimental, and Unique forms. An implicit evolutionary scheme is added with Arabic numerals 1–4. The tentative conclusion is that winged objects, employed as grave offerings at Ekven (Leskov and Müller-Beck 1993) and Sivuqaq (Blumer 2002), may be quite young, possibly persisting until the fifteenth century A.D. Considering the abundance of winged objects that nonindigenous collectors had obtained during the last 30 years (J. Hollowell 2004, personal communication; Hollowell 2008), the reservoir of OBS objects may have continued well beyond its age of production. No clear temporal priority can be documented for the types distinguished by Bronshtein (2002). Fourth, radiocarbon dating offers support for the florescence of the OBS winged object during the principal Ipiutak occupation at Point Hope, during the latter part of the first millennium A.D. (Gerlach et al. 1992; Mason 2009). Fifth, the dates may overlap with Punuk tridents, providing an explanation for the mutually exclusive character of Punuk and OBS winged objects. Hence, tentatively, these data offer little or no support for the oft-cited evolutionary sequence of OBS to Punuk.

Defining Cultural and Ethnic Units in Bering Strait: Old Bering Sea, Punuk and Birnirk/Thule

Aesthetic analysis of Old Bering Sea designs relies on the ruminations of Arutiunov and Bronshtein (1985), Bronshtein (2006), Collins (1937), and de Laguna (1933). Collins' formulation retains the widest currency, but has limitations, because it mixes purely descriptive criteria (i.e., lines, dashes, chevrons) with more complex interpretative schemata such as rayed eyes or faces[3]. The maximal elaboration of Old Bering Sea and Ipiutak designs co-occurs with status differentiation in graves, evidence of conflict (i.e., warfare) and objects of cosmological import often linked to shamanism (Arutiunov and Sergeev 1969, 1972; Dikov 2004; Larsen and Rainey 1948; Mason 1998).

The co-occurrence of artistic elaboration by OBS with iron supply was postulated from experimental replication studies by Semenov (1964:162–167) who argued that iron gravers were a requisite for working walrus ivory, although Collins (1937:303) thought iron was not used significantly until Punuk times (McCartney 1988). The lineage of OBS/Ipiutak styles is largely limited to the first millennium A.D., although its origin was likely several centuries prior to A.D. 1. Many Ipiutak motifs closely resemble OBS, and both systems are evidence of communities operating across societal boundaries within a common tradition, both in craft and cosmology (Arutiunov and Bronshtein 1985). The character of Old Bering Sea symbolic language decreases in complexity, with only a few linear motifs applied by Thule types (Collins 1937). The disappearance of the Old Bering Sea tradition would imply a similar breakdown or population replacement under the influence of Birnirk or Thule peoples.

The Punuk decorative system represents a drastic shift to a more abstract vocabulary that rarely (if ever) includes human figures or animals (Collins 1937). By Thule, only a limited number of lines and an occasional circle were employed, in addition to ownership marks (Reynolds 1989). As mentioned, Birnirk harpoon heads and other objects bear only linear motifs.

Ethnicity in Burial Assemblages

The cemeteries of the greater Bering Strait, northern Bering and Chukchi Seas (St. Lawrence Island to Point Barrow) have produced 1200 interments that likely date before A.D. 1200; the extensive Ipiutak cemetery (ca. 150 graves) at Point Hope will not be considered here (cf. Mason 2006). The Bering Strait mortuary data base confirms the persistence of distinct, regionally specific lineages, each constrained within its own systemic properties within an adaptive system.

The Cape Dezhneva cemeteries at Ekven and Uelen yielded diagnostic artifacts, mostly of the Old Bering Sea culture, in significantly higher numbers, between 40 and 60%, compared to Sivuqaq on St. Lawrence Island, where less than 5% yielded diagnostic pieces (Bandi and Blumer 2002, Staley and Mason 2004). The Old Bering Sea culture is preeminent within the cemetery precincts of the Dezhneva; OBS in its three phases comprises nearly three-quarters of all graves (Fig. 3.6a). Very few graves can be attributed to Punuk, Birnirk, or Thule. Over 100 Birnirk burials are known from the Barrow vicinity (Hollinger et al. 2009), either within houses at the Birnirk site or within the collapsed "charnel" or mortuary structures at Kugusuguruk (Ford 1959:19ff). Three solid carbon ^{14}C assays on wooden artifacts, when averaged, bracket the Kugusuguruk burials within 1212±84 B.P., calibrated to A.D. 667–980 (Mason 2009).

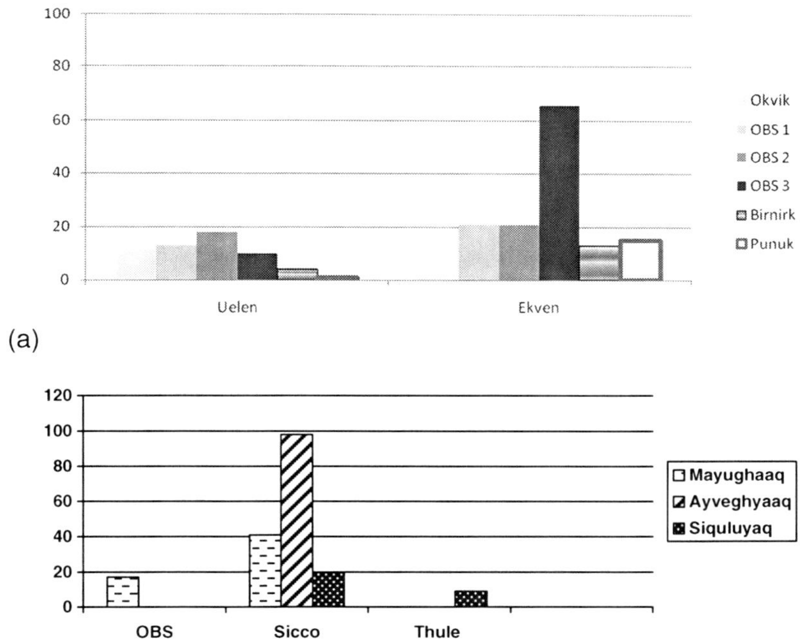

Fig. 3.6 Representation of various styles (Fig. 3.4) within (**a**) Cape Dezhneva and harpoon heads (cf. Fig. 3.5) (**b**) St. Lawrence Island cemeteries. (**a**), *upper*: Graves at Uelen and Ekven with decorated motifs attributed to culture. The Uelen data resemble a normal distribution: the implication is that its use life declined after OBS II. By contrast, Ekven witnessed a temporary increase in use during OBS III and return to low levels of internees buried with socially marked tools (Data from Bronshtein and Plumet 1995:34). (**b**), *lower*: Frequency of Open socket Harpoon Heads at Mounds in the Sivuqaq vicinity, St. Lawrence Island (Data from Collins 1937:Table 2)

The Uelen mortuary data resemble a normal distribution: The implication is that its use life declined after OBS II. By contrast, Ekven witnessed a temporary increase in use during OBS III and a return to low frequencies of internees buried with socially marked tools. A single culture dominated the Cape Dezhneva cemeteries (Fig. 3.6a): Old Bering Sea, divided into its three sub-styles (I, II, and III). At Ekven, 79% of definable graves are OBS (108 of 137 graves), and slightly under half (48%) are from the single style OBS III, while Birnirk comprises only 9.5%. The Uelen cemetery differs, containing a lower proportion of OBS III but has a considerably higher proportion of Okvik graves: nearly 20%. Nonetheless, OBS still comprises nearly three quarters, 72% (41 of 57), while 19% fall within the Okvik (11 of 57). Of the >450 graves, but with only half with identifiable decorated pieces, only 10% in the Cape Dezhneva massif (northern Uelen and southern Ekven) contained objects considered diagnostic of the Birnirk or Punuk cultures: Birnirk (n = 15), early Punuk (n = 14), Birnirk-Punuk (n = 2), and OBS/Birnirk/Punuk (n = 14) (Dneprovsky and Bronshtein 2002).

Because so few graves at Ekven or Uelen reflect the Birnirk or Punuk phenomena, several factors are involved. First, the OBS persisted for a longer period of time than the other cultures—the most likely possibility. Second, Punuk was intrusive in the Ekven area that was predominantly OBS in ethnicity. Third, the area witnessed a population decline so that fewer people produced fewer graves. Present radiocarbon data show that OBS persisted for up to a millennium, in contrast to the apparent several-hundred-year duration of the Punuk phase. The persistence of OBS as an aesthetic tradition, and a socioeconomic strategy for up to a millennium (or more) implicates the role of Cape Dezhneva as a central place that dominated the cultural landscape (Mason and Gerlach 1995).

To the south, at the northwest cape of St. Lawrence Island, across the Anadyr Strait, the Sivuqaq burial grounds extend over a considerable portion of the beach ridge plain east of the nineteenth-century village near Gambell (Bandi 1984; Bandi and Blumer 2002). The notion of dedicated cemetery precincts is less apparent, as late twentieth century construction (Staley and Mason 2004) has nearly doubled the number of graves reported. Very few of the nearly 400 Sivuqaq graves contained any artifacts or offerings; hence, ethnic inferences are tentative (cf. Bandi and Blumer (2002:38; Staley and Mason 2004). Punuk burials are more frequent, though this is a fairly tentative conclusion in view of the low number (n = 16, out of 363 total, or 4.4%) of graves with diagnostic artifacts. Poor preservation accounts for some of the differences—many burials preserved only a "bone paste" and no other organic fragments (Staley and Mason 2004). Burial practices apparently differed considerably across the Anadyr Strait—or the nature of society differed even more dramatically, in view of the dedicated precincts and richly apportioned burials.

Harpoon head frequency may provide a measure of the intensity of occupation in the Sivuqaq region (Fig. 3.6b, from Collins 1937: Table 2). Excluding the "old Gambell" material, for which depth measures are lacking, 302 of the 367 harpoon heads, nearly 82%, were recovered in the upper 1.2 m ("47 in"). Very few harpoon heads were recovered in the lowest 1–2 m of Mayughaaq or Avyeghyaaq, while the upper 1.5 m yielded 151 specimens. The Punuk III-a-x type (related to the Sicco harpoon head) comprises 43% ($n = 161$) of the total, with only 8 of this type in the lower 1 m of Sivuqaq midden. It remains uncertain if the succession Birnirk to Thule at Kurigitivik represents continuity or conquest, i.e., replacement.

Harpoon heads vary spatially across the Chukchi and Bering Sea, with a tendency toward mutually exclusionary relationships across restricted intervals during the first millennium A.D., as reviewed by Mason (2009), which revises the "contemporaneity" model of Gerlach et al. (1992). The mutual exclusivity of harpoon heads implies that the operation of internally imposed cultural boundaries associated with polities ("nations") similar to that observed in the nineteenth century (Burch 1998, 2007). Between the last centuries B.C. and ca. A.D. 700, Old Bering Sea forms, very distinctive with two line holes and multiple barbs, occured only on the Chukotsk Peninsula and St. Lawrence (Fig. 3.4), with only a few outlier occurrences (i.e., Point Barrow, for example, as noted by Wissler [1916]). The integrity of OBS as an archaeological type suggests that a single ethnicity, polity, or interaction sphere provided the mechanism for spreading the use of the type. Stylistic similarities indicate that Ipiutak societies had frequent interactions with OBS. Harpoon heads, termed Oopik, and arrow points with linear designs delineate Ipiutak societies (Larsen and Rainey 1948:108) and are broadly contemporaneous with OBS, between A.D. 400 and 900 (Mason 1998, 2006) from the Alaskan coast, within Norton Sound (Mason et al. 2007) to Point Barrow (Jensen 2009; Wissler 1916).

Following A.D. 900, two harpoon head manufacturing traditions, one marginal and one at the center, are dominant on the opposite coasts of the Chukchi Sea, termed Birnirk and Punuk, both of which contributed to the formation of Thule culture and technology (Collins 1937, 1940; de Laguna 1947; Ford 1959). The marginality of Birnirk was first recognized by Jenness (1940), who postulated its entry from the north. The distribution and origins of Birnirk differ profoundly from its rival, Punuk which arose within a center (if not, *the* center) of the Old Bering Sea culture on St. Lawrence Island.

Mutual Exclusivity of Birnirk and Punuk Across Bering Strait

Punuk developed from the Old Bering Sea culture on St. Lawrence Island at ca. A.D. 900 (Mason 2009). Defined stylistically by Collins (1937:179ff), Punuk also records a shift in social organization, the intensification of subsistence practices (especially bird hunting and whaling), and the adoption of various military technologies, including slat armor, wrist guards, and the composite bow (Bandi 1995; Collins 1937; Mason 2009). The earliest Punuk occupations, reasonably well-dated occur around Sivuqaq between A.D. 700 and 1000 (Blumer 2002; Mason 2009; Staley and Mason 2004,), apparently contemporaneous with OBS at Cape Dezhneva and Ipiutak (Dneprovsky and Bronshtein 2002; Mason 1998, 2006). While the artistic repertoire of Punuk is considerably less intricate than OBS or Okvik

(Collins 1937, 1940, 1964, 1971), the evolutionary significance of this apparent simplification is unclear. On one hand, the Punuk culture seems to represent an expanding culture: Punuk motifs are more widespread on a regional basis, with a dense concentration of sites on St. Lawrence Island, several along the south coast of Chukotka (Chard 1955, 1957; Rudenko 1961) and a number of localities on the Alaska mainland, as far north as Nunagiak (Dumond 2000b; Ford 1959; Harritt 2004; Mason 2003; Mason et al. 2007; Yama'ura 1984).

Punuk objects extend exclusively in a linear and littoral direction along the northwest of Alaska, with several objects serving as index fossils (Fig. 3.5b). The Punuk harpoon head technology is typified by the wide open socket ("III-a-x") "Sicco" head that is used from its core on St. Lawrence Island and the south coast of Chukotka (Ford 1959). The distribution implies that a migration of Punuk peoples occurred during the tenth or eleventh centuries A.D. (Ford 1959:64; Mason 2003:143ff; Mason 2009). This Punuk migration out of Bering Strait is linked with a distinctive atlatl counter weight (Fig. 3.6) (Ford 1959:67; Mason 2003:143; Mason 2009). The occurrence of Punuk harpoon heads may extend the migration even as far as Greenland. For example, a well-appointed grave at Pt. Barrow contains the diagnostic "Sicco" harpoon head (Jensen 2007), evidence of a Punuk passage north (Ford 1959:71). Its final arrival point, marked by Punuk decorations in far northern Canada and Greenland, dates variously between A.D. 800 and 1200 (Schledermann and McCullough 1980), the young age forcefully argued by Gulløv and McGhee (2006).

The distinctive Birnirk culture, by contrast, issued from the margins of the Chukchi Sea and spread toward the center. Two very distinctive single-barbed harpoon heads ("Birnirk" and "Natchuk") serve as type fossils for Birnirk polities (Ford 1959:75ff, 83). The frequency distribution of Birnirk artifacts shows four principal societies on the north coasts of the Chukchi Sea, with high numbers of harpoon heads: Cape Baranov (Okladnikov and Beregovaia 2008, near Wales (Dumond 2002), Pt. Barrow (Ford 1959), and adjacent to the Paipelygak site in north Chukoka (Dneprovsky 2006). No definitively Birnirk sites occur south of Bering Strait—the objects defined as Birnirk by Bockstoce (1979) or Geist and Rainey (1936) are not truly diagnostic (cf. Mason 2000, 2009). A variety of artifacts and practices distinguish Birnirk from its contemporaries; some of the objects are ritually precise, for example, the bark human figures used in shamanic practice (cf. Ford 1959). Some are employed in hunting practices (the wound pin used to secure bleeding seals). However, Birnirk technology is also marked by intensification of subsistence, especially in bird hunting and fishing. Birnirk is marked by a variety of harpoon heads (Ford 1959:75ff), possibly representing a major burst in innovation, population growth, or cross-cultural communication.

The chronology of Birnirk polities remains poorly constrained, but the phenomenon may have arisen as early as A.D. 600 and continued until A.D. 1300 (Mason 2009; Mason and Bowers 2009). Birnirk attained its greatest distribution, extending even into Kotzebue Sound during the centuries after A.D. 1000 and may have supplanted a resident Ipiutak population by conquest or by occupying an empty landscape (Mason 2009; Mason and Bowers 2009).

The Timing of the Shifts in Bering Strait Adaptive Strategies

While chronological issues bear strongly on issues of classification, population pressure may loom as a larger concern to evolutionary theorists, as high populations presumably reflect success in cultural adaptation. To retrodict the chronology of Bering Strait demographic changes, archaeologists can use as a proxy the ^{14}C dates on burials and associated offerings within the two major cemeteries at Sivuqaq and Ekven (Mason 1998). The ^{14}C data base from Sivuqaq reflects a random sample selection process due to the lack of evidence of cultural affinity within the graves and the inclusion of samples on terrestrial materials that are not biased by marine carbon (Bandi 1984). This contrasts with the problematic array of radiocarbon ages from Ekven that were run directly on human bone with a significant fraction of marine carbon due to heavy walrus consumption. The Ekven dates are almost entirely on human bone and often are associated with diagnostic artifacts. Graves from the youngest part (<300 years BP) of the record were not dated. Two major chronological hinge points occur within the nearly 80 radiocarbon dates from the two centers of OBS culture: one prior to A.D. 1, the other from A.D. 1000 to 1300. In the first, the OBS interaction sphere or culture arose during the last centuries B.C. (Dinesman et al. 1999; Dumond 2000a; Mason 1998). At both Ekven and Sivuqaq, the number of burials declined after A.D. 1000–1200. The thirteenth century A.D. marks the ascendance of Thule culture. Furthermore, the Birnirk cultural system may have originated as early as A.D. 600 (Gerlach et al. 1992); however, refinements in dating indicate that the florescence of the Birnirk culture occurred following A.D. 1000 (Dneprovsky 2006; Mason 2003, 2009; Mason and Bowers 2009; Morrison 2001). Next, I shall consider the role of artistic output as evidence for macroevolution in the first millennium A.D. Bering Strait, considering that craft specialization associated with abundant art production is inherently linked with sedentism, *contra* McGhee (1976).

Causation of the Shifts in Bering Strait Adaptive Strategies

Several cultural factors, both internal and external, may have influenced the development of the Bering Strait adaptive strategies (Mason n.d.). Shifts in adaptive strategy may have occurred in the relationship with the greater East Asian world system (i.e., iron supply) (Mason 1998), including its disease pool. Historic factors, such as strong leadership (i.e., agency) may have galvanized communities for a generation or two. The randomness of intercommunity conflict might reflect idiosyncratic patterns of revenge and conflict. Supra-regional conditions of stress might be related to a surging population or sudden shifts in either resource abundance or scarcity. Conditions of scarcity might have catalyzed the development of cosmological ideas that mitigated adversity and rewarded those with success in harnessing spiritual power. A similar view was proposed for the contemporary Dorset phenomenon in the eastern arctic (Taçon 1983). A more nihilistic view is that of McGhee (1976) that suggests art is a random process that is not correlated to sedentism or resources. Adversity also seems to have favored aggregation at the hotspots with the best chances of obtaining marine mammal resources (Mason 1998). Most certainly, the florescence of aesthetic production is concurrent with an abundant walrus resource base, evident in Old Bering Sea and Ipiutak cultures. In the twentieth century, walrus followed the seasonally retreating ice margin far north of Bering Strait (Fay 1982); during previous cold anomalies more extensive summer ice may have concentrated walrus farther south near Point Hope and Chukotka.

The paleo-productivity of the Bering and Chukchi fauna is not known in any detail. Mason and Gerlach (1995) hypothesized that productivity followed intensified upwelling of nutrients. Higher productivity thus resulted during colder and stormier conditions (Mason and Barber 2003) with converse that lowered productivity during warmer and less stormy seas. Significantly, the two hinge points co-occur with shifts in productivity in salmon populations (Finney et al. 2002). The expansion of whaling and walrus-hunting commenced during a more productive cooler period during the last centuries B.C. Then, the OBS adaptive strategy represents the resource challenged period of the early centuries A.D., while the development of Punuk co-occurs with an expansion in resources during the early Medieval glacial expansion ca. A.D. 800 (Hughes and Diaz 1994). The development of Thule, the florescence of a whaling economy, occurred during the Little Ice Age. This co-occurrence implies a correlation between scarcity and artistic embellishment that is emblematic of cosmological needs in stressful circumstances (Taçon 1983). Quite noteworthy is the drop-off in cosmological representation during the relative plenty of Thule times.

The "hotspots" core and periphery model, espoused by Mason and Gerlach (1995), derives from oceanographic data that link the upwelling of nutrients to intensified storminess, circumstances that are linked to colder seas (Mason and Barber 2003, *contra* Krupnik 1993). Support for this model is often indirect: Greater hunting success occurred during the Little Ice Age, as indicated by the faunal remains from the East Cape middens (Dinesman et al. 1999; Savinetsky et al. 2004). Glacial, pollen, ostracodes, tree rings, and beach ridges provide proxy records—each not without its limitations and dating problems (Hu et al. 2001; Mason and Gerlach 1995; Oswald et al. 2001). Data from Northwest Alaska establish parallel global records of an Early Medieval cool period around A.D. 800 (Dansgaard et al. 1975; Hughes and Diaz 1994; Mason 1999, Mayewski et al. 2004). The timing of this cold period may be earlier, commencing around A.D. 400, based on an oxygen isotopic proxy employing northern Alaska Range ostracodes (Hu et al. 2001:10554). The temperature depression in this period might have been twice as deep as the Little Ice Age and may have persisted until A.D. 900, with a very slight warming until A.D. 1400 (Hu et al. 2001:10554).

By contrast, Kobuk River tree rings indicate the period from A.D. 1000 to 1400 was unlike either the present century or the Little Ice Age (D'Arrigo et al. 2005). Decades of exceptionally favorable precipitation alternated with very poor conditions (Juday et al. 2003), a boom and bust that offered daunting challenges to prehistoric societies (Mason and Gerlach 1995). No lake cores at the headwaters of

Bering or Chukchi Sea drainages are available to parallel the isotopic signatures of salmon within North Pacific headwater lakes (Finney et al. 2002). Curve matching shows that the Chukchi and Bering Seas parallel the North Pacific. Lacustrine carbonates in the southern Yukon also record the increased strength of the Aleutian low around A.D. 900 and during the Little Ice Age—correlated with increased marine productivity (Anderson et al. 2005:32).

In summary, most proxy climate records imply that a drastic climatic event, likely a "regime shift" (Juday et al. 2003) occurred around A.D. 900, *from* a less stormy, less productive Ipiutak period *to* a much stormier and more productive Birnirk-Thule period. The inverse relationship exists between the production of symbolic capital meant that the less reliable the seas were, the need for spiritual power intensified, and this accentuated the benefits of those allied with the powerful. At both Ekven and Sivuqaq, the number of dated graves peaks only after A.D. 1000 (Bandi 1984; Blumer 2002; Mason 1998).

Aesthetics, Mobility, and Subsistence Production

The intensity of decoration co-occurs with high walrus exploitation and is inversely related to success in whaling. On the broadest scale, communities with a surplus of whale bone were less likely to engrave elaborate designs. Very likely, relatively porous whale bone offered a less attractive palette for aesthetic design. However, bone supply cannot be the complete explanation since the Birnirk whalers were more interested in engraving (cf. Carter 1966) than their Thule descendents who primarily inscribed linear arrays onto their antler harpoon heads, congruent with contact period ownership marks (Reynolds 1989). Economic differences between societies do not pertain, because as walrus hunting remained at high levels on St. Lawrence Island throughout the Little Ice Age (Collins 1937; Geist and Rainey 1936), the interest in decoration also declined to a negligible amount. The need for decoration would also seem high in view of the expansion in military hardware (e.g., slat armor that occurred in the fourteenth and fifteenth centuries A.D. (Mason 2009).

In macroevolutionary terms, the florescence of harpoon head technology, a specialized "cultural lineage" (Shennan 2002:73) of arctic sea mammal hunters, likely conferred an adaptive advantage, either in affirming ethnicity, status, the production of symbolic capital or as a proxy for hunting success. The greatest decorative elaboration of harpoon heads occurs in Old Bering Sea and Punuk cultures, in association with access to iron and in bountiful provision in mortuary contexts. The production of psychic and political capital is restricted to several villages adjacent to, and husbanding, the critical resource hotspots of the western arctic (Leskov and Müller-Beck 1993; Mason 1998; Mason and Gerlach 1995). The early and sudden development of Old Bering Sea remains an enigma but appears related to a propitious constellation of circumstances: (a) heightened upwelling during cold weather (Mason and Barber 2003); (b) the entry of exotic materials (iron) and technologies from the East Asian world system (Collins 1937); (c) generational or demographic factors such as families endowed with risk-taking behaviors or life histories of successful hunting (Whitridge 1999); (d) an ideology crystallized in stylistic representation, catalyzed by collective experiences in a revitalized shamanic tradition (Mason 2006).

Of equal significance is the co-occurrence of the florescence of the harpoon head with the first firm evidence of success during the last centuries B.C., in whaling, of both gray and bowheads (Dinesman et al. 1999), and in walrus hunting. The reliance of St. Lawrence Islanders on the walrus also co-occurs with a major increase in the size and scale of villages and in cemeteries. The extreme differential in mortuary goods implies that high status individuals akin to the nineteenth century umialit, or whaling captains, were instrumental in this transformation. Heightened aesthetic production was associated with specialization within households. At Point Hope, ivory working occurred within households (Fig. 3.7) that were also producing more clothing and in possession of more military armaments (cf. Mason 2006). Once again, success in hunting was the underpinning that catalyzed craft specialization and with it the military and trading advantages of the very sizable community of Ipiutak people at Point Hope (Mason 2006)

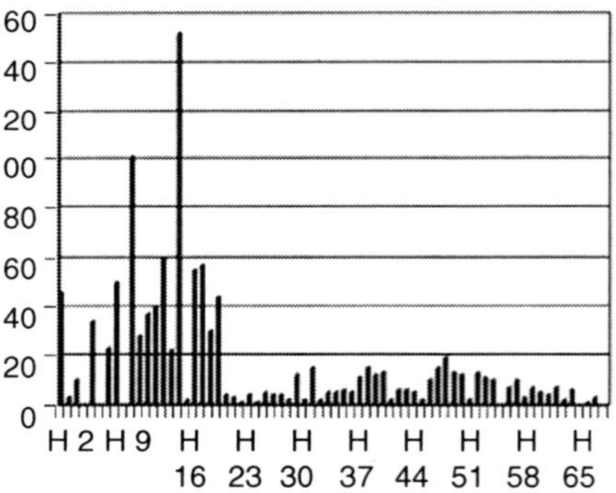

Fig. 3.7 Evidence for craft specialization at the Ipiutak site, near Point Hope, as tabulated from appendices in Larsen and Rainey (1948). The houses with high amounts of ivory working debitage also had high frequencies of *arrow* heads used in warfare and in needles for clothing manufacture (Mason 2006)

Conclusions

The chronology of various harpoon head variants is still known only in outline, although data from St. Lawrence Island are approaching a satisfactory state in a few areas. The earliest occupations associated with the Old Bering Sea culture are known from the Hillside locality (Dumond 1998). Blumer (2002) graphically displays the temporal progression of St. Lawrence Island harpoon heads, substantially confirming the claims of Mason and Gerlach (1992) that many types were used contemporaneously across the western arctic. Old Bering Sea types remained in use as late as A.D. 1200, while Punuk and Thule types arose successively during A.D. 650–900. Punuk designs occur in subsequent centuries, between A.D. 1000 and 1100 at Cape Prince of Wales (Harritt 2004) and Cape Lisburne (Mason 2003). Thule types are first documented on the North Slope only in the early eleventh century A.D. (Mason et al. 2003; Morrison 2001). The formative conditions of Thule culture still elude archaeologists, but emerging data that indicate an initial occupation around A.D. 1200 (Mason and Bowers 2009) do not support Kotzebue Sound and Seward Peninsula as a likely hearth region, much as Collins (1937) originally hypothesized. By examining art traditions, it is possible to define blue prints for culture. The persistence of unique adaptive strategies in the arctic is coterminous with several lengthy mortuary traditions at Ekven, Sivuqaq, and Ipiutak. Major shifts occur in these, correlative with productivity shifts in the North Pacific. The various adaptive strategies embodied in technological systems reflect structural/psychological changes in how stress and status were managed in the face of shifting resource conditions.

In evolutionary terms, change in harpoon styles was dependent upon broader processes of macroevolutionary change, co-occurring with innovation in walrus hunting initially and in whaling as well. This implies that evolutionary archaeologists working on stylistic change elsewhere would be wise to consider broader context in generating explanatory models. Results of this study are anticipated by those of Eldredge (this volume) in that major change in cornets often depended not on simple microevolution but on macro-cultural forces well beyond the performance capacity of the instruments themselves. Similar patterns on different scales are documented by Chatters and Zeder (both this volume) who recognize the necessity of macroevolutionary modeling as fundamental in developing a comprehensive evolutionary understanding of technological evolution.

Notes

1. As per other discussions in the book a *Bauplan* is a structure or blueprint but not an actual evolving entity—it does denote some interesting implications for interpreting change in actual entities, however; Prentiss et al. (intro chapter) and Bettinger (discussion chapter—this volume) (following Rosenberg 1994; as elaborated by Prentiss and Chatters 2003; Chatters and Prentiss 2005; Prentiss et al. this volume).
2. The efforts of Van Pelt (2008) to disentangle Okvik from Old Bering Sea seem rather too inclusive, but this M.A., written in the 1970s was only recently published and the author has had too little time to assess it.

References

Ackerman, R.E. (1984). Prehistory of the Asian Eskimo zone. In D. Damas (Ed.), *Handbook of North American Indians*, volume 5, *Arctic* (pp. 106–118). Washington, DC: Smithsonian Institution.

Anderson, D.D. (1984). Prehistory of North Alaska. In D. Damas (Ed.), *Handbook of North American Indians*, volume 5, Arctic (pp. 80–93). Washington, DC: Smithsonian Institution

Arutiunov, S.A. and Bronshtein, M. (1985). The problem of distinguishing between Old Bering Sea and Okvik ornamental styles. *Jahrbuch des Bernischen Inter-Nord Historische Museum* 63–64, 17–21.

Arutiunov, S.A. and Fitzhugh, W.W. (1988). Prehistory of Siberia and the Bering Sea. In W.W. Fitzhugh and A. Crowell (Eds.), *Crossroads of Continents: Cultures of Siberia and Alaska* (pp. 117–129). Washington, DC: Smithsonian Institution Press.

Arutiunov, S.A. and Sergeev, D.A. (1972). *Ecological Interpretation of Ancient Harpoon Heads* 12, 305–309.

Arutiunov, S.A. and Sergeev, D.A. (2006a) [1969]. *Ancient cultures of the Asiatic Eskimos: The Uelen Cemetery*. Anchorage: Shared Beringian Program, National Park Service [Translation by Richard L. Bland of *Drevnie kul'tury aziatskikh eskimosov (Uelenskii mogil'nik)*, Moscow: Akademia Nauk SSSR.

Arutiunov, S.A. and Sergeev, D.A. (2006b) [1975]. *Problems in the ethnic history of the Bering Sea: Ekven cemetery*. Anchorage: Shared Beringian Program, National Park Service [Translation by Richard L. Bland of *Problemy etnishiskoi istorii Beringomoria: Ekven mogil'nik]*. Akademiia Nauk, SSSR, Institut Etografi Imeni, N.N. Moscow: Milkukho-Maklaia.

Auger, E. (2005). *The Way of Inuit Art*. Jefferson, NC and London: McFarland and Company.

Bandi, H.G. (1984). St. Lorenz Insel-studien. Vol. 1. *Berner beitärge zur archäologischen und ethnologischen erforschung des Beringstrassengebeites. Band I. Allegemeine einführung und gräberfunde bei Gambell am Nordwestkap der St. Lorenz Insel,* Alaska. Bern and Stuttgart: Academica Helvetica 5,I.

Bandi, H.G.(1995). Siberian Eskimos as Whalers and Warriors. In A.P. McCartney (Ed.), *Hunting the Largest Animals: Native Whaling in the Western Arctic and Subarctic* (pp. 165–184), Studies in Whaling No. 3, Occasional Paper 36, Edmonton: Circumpolar Institute, University of Alberta.

Bandi, H.G. and Blumer, R. (2002). Investigations by Swiss archaeologists on St. Lawrence Island.InD.E. Dumond and R.L. Bland (Eds.),*Archaeology in the Bering Strait Region: Research on Two Continents*. University of Oregon Anthropological Paper 59, 25–60.

Bettinger, R.L., Boyd, R. and Richerson, P.J. (1996). Style, function, and cultural evolutionary processes. In H.D.G. Maschner (Ed.), *Darwinian Archaeologies* (pp. 133–164). New York: Plenum Press.

Blumer, R. (2002). Radiochronological Assessment of Neo-Eskimo Occupations on St. Lawrence Island. In D.E. Dumond and R.L. Bland (Eds.), *Archaeology in the Bering Strait Region: Research on Two Continents*. University of Oregon Anthropological Papers 59:61–106.

Bockstoce, J.R. (1979). *The Archaeology of Cape Nome, Alaska*. University Museum Monograph 38, Philadelphia: University of Pennsylvania.

Bronshtein, M. (2002). Structural and artistic features of 'winged objects': The discussion continues. In D.E. Dumond and R.L. Bland (Eds.), *Archaeology in the Bering Strait Region: Research on Two Continents*, University of Oregon Anthropological Paper 59, 127–138.

Bronshtein, M. (2006). Variability in ancient Eskimo graphic designs: On the problem of the ethnic and cultural history of the Bering Sea form the 1st millennium B.C. to the 1st millennium A.D. *Alaska Journal of Anthropology* 4(1–2), 162–175.

Burch, E.(t.)S. (1998). *The Inupiaq Eskimo Nations of Northwest Alaska*. Fairbanks: University of Alaska Press

Burgess, S. (1974). *The St. Lawrence Islanders of Northwest Cape: Patterns of Resource Utilization*. Unpublished Ph.D. Departments of Geography and Anthropology, Dissertation, Fairbanks, AK: University of Alaska Fairbanks.

Carr, C. (1995). A Unified Middle Range Theory of Artifact Design. In C. Carr and J.E. Neitzel (Eds.), *Style, Society and Person: Archaeological and Ethnological Perspectives* (pp. 171–258). New York: Plenum Press.

Carter, W.K. (1966). *Archaeological Survey of Eskimo or Earlier Material in the Vicinity of Point Barrow, Alaska*. Final Report to the Office of Naval Research and Arctic Institute of North America. Cambridge: Harvard University.

Chard, C.S. (1955). Eskimo archaeology in Siberia. *Southwestern Journal of Anthropology* 11, 150–177.

Chard, C.S. (1957). The southwestern frontier of Eskimo culture. *American Antiquity* 22(3), 304–305.
Chatters, J.C. and. Prentiss, W. (2005). A Darwinian macro-evolutionary perspective on the development of hunter-gatherer systems in northwestern North America. *World Archaeology* 37(1), 46–65.
Collins, H.B. (1929). Prehistoric art of the Alaskan Eskimo. *Smithsonian Miscellaneous Collections* 81(14).
Collins, H.B. (1937). Archaeology of St. Lawrence Island, Alaska. *Smithsonian Miscellaneous Collections* 96(1).
Collins, H.B. (1940). *Outline of Eskimo Prehistory. In Essays in Historical Anthropology of North America.* Smithsonian Miscellaneous Collections 100. Washington, DC: Smithsonian Institution.
Collins, H.B. (1961). Eskimo cultures. In *Encyclopedia of World Art*, volume 5 (pp. 3–28). New York: McGraw-Hill Book Co.
Collins, H.B. (1964). The Arctic and Subarctic. In J.D. Jennings and E. Norbeck (Eds.), *Prehistoric Man in the New World* (pp. 85–114). Chicago: University of Chicago Press.
Collins, H.B. Jr. (1971). Composite Masks: Chinese and Eskimo. *Anthropologica n.s.* 13(1-2): 271–278.
Dansgaard, W., Johnsen, S.J., Reeh, N., Gundestrup, N., Clausen, H.B. and Hammer, C.U. (1975). Climatic changes, Norsemen and modern man. *Nature* 255, 24–28.
D'Arrigo, R., Wilson, R., and Jacoby, G. (2006). On the Long Term Context for Twentieth Century Warming. *Journal of Geophysical Research,* 111:1-12
Darwent, J. and Darwent, C.(2005).Occupational history of the old whaling site at Cape Krusenstern, Alaska. *Alaska Journal of Anthropology* 3(2), 135–154.
de Laguna, F. (1933). A Comparison of Eskimo and Paleolithic Art. Reprinted in single folio from *American Journal of Archaeology* 36(4), 477–511, plates 19–23, and *American Journal of Archaeology* 37(1), 77–107, 18–24.
De Laguna, F. (1947). The Prehistory of Northern North America as Seen from the Yukon. *Memoirs of the Society for American Archaeology* 3.
Dikov, N. (1977) [2003]. *Archaeological Sites of Kamchatka, Chukotka and the Upper Kolyma*, Anchorage Shared Beringian Project, National Park Service [Translation by Richard L. Bland of *Arkheologischeskie pamiatniki Kamchatki, Chukotki, I Verkhnei Kolymy*, Moscow: Nauka].
Dikov, N. (1979) [2004]. Early Cultures of Northeast Asia. Anchorage: Shared Beringian Project, National Park Service [Translation by Richard L. Bland of *Drevnie kul'tury severovostochnoi Azii: Aziia na styke s Amerikoi v drevnosti*. Moscow: Izdatel'stvo. Nauka].
Dinesman, L.G., Kiseleva, N.K., Savinetsky, A.B. and Khassanov, B.P. (1999). *Secular Dynamics of Coastal Zone Ecosystems of northeastern Chukchi Peninsula, Chukotka: Cultural Layers and Natural Depositions from the Last Millennia*. Tübingen MoVince Verlag.
Dneprovsky, K. (2006). New Early Eskimo SIte of Paipelghak in Chukotka: Preliminary Publication Based on materials from 2002-2004. *Alaska Journal of Anthropology* 4:30–40.
Dneprovsky, K. and Bronshtein, M.(2002). Ekven H-18: A Birnirk and early Punuk-period site in Chukotka. In D.E. Dumond and R.L. Bland (Eds.), *Archaeology in the Bering Strait Region: Research on Two Continents.* University of Oregon Anthropological Papers 59, 166–206.
Dumond, D.E. (1984). Prehistory of the Bering Sea region. In D. Damas (Ed.), *Handbook of North American Indians*. Vol. 5, *Arctic* (pp. 94–105). Washington, DC:Smithsonian Institution.
Dumond, D.E. (1998). *The Hillside Site on St. Lawrence Island: An Examination of Collections from the 1930s.* University of Oregon Anthropological Papers 55.
Dumond, D.E. (2000a). *Henry B. Collins at Wales 1936: A Partial Description of Collections.* University of Oregon Anthropological Papers 56.
Dumond, D.E. (2000b). The Norton tradition. *Arctic Anthropology* 37(2), 1–22.
Dumond, D.E. and Griffin, D. (2002). Measurements of the marine reservoir effect on radiocarbon ages in the Eastern Bering Sea. *Arctic* 55(1), 77–86.

Dumond, D.E. (2008). The story of Okvik. In D.E. Dumond (Ed.), *Aspects of Okvik: Four Essays on Things of Bering Strait*. University of Oregon Anthropological Papers 68, 261–308.

Dunnell, R.C. (1978). Style and function: a fundamental dichotomy. *American Antiquity* 43, 191–202.

Fay, F. (1982). *Ecology and Biology of the Pacific walrus*, Odobenus rosmarus divergens Illger. North American Fauna No. 74, Washington.DC: U.S. Fish and Wildlife Service.

Finney, B.P., Gregory-Eaves, I., Douglas, M.S.V. and Smol, J.P. (2002). Fisheries productivity in the northeastern Pacific Ocean over the past 2,200 years. *Nature* 416, 729–733.

Fitzhugh, W.W. and Crowell A. (Eds.) (1988). *Crossroads of Continents: Cultures of Siberia and Alaska*. Washington, DC: Smithsonian Institution Press.

Ford, J.A. (1959). Eskimo prehistory in the vicinity of Point Barrow, Alaska. *Anthropological Papers of the Museum of Natural History* 47(1).

Geist, O.and Rainey, F. (1936). *Archaeological Excavations at Kukulik, St. Lawrence Island, Alaska*. Volume II, Fairbanks, Miscellaneous Publications University of Alaska and Washington, DC: Government Printing Office.

Gerlach, S.C., Mason, O.K. and Mason, O.T. (1992). Calibrated radiocarbon dates and cultural interaction in the western Arctic. *Arctic Anthropology* 29(1), 54–81.

Giddings, J.L. and Anderson, D.D. (1986). *Beach ridge archaeology of Cape Krusenstern: Eskimo and pre-Eskimo settlements around Kotzebue Sound, Alaska*. Publications in Archeology 20.Washington, DC: National Park Service, U.S. Department of the Interior.

Gordon, G.B. (1916). The double axe and some other symbols. *The* [University of Pennsylvania] *Museum Journal* 7(1), 46–68.

Gullov, H.C. and McGhee, R. (2006). Did Bering Strait People Initiate the Thule Migration. *Alaska Journal of Anthropology*4:254–263.

Harritt, R.K. (2004). A preliminary reevaluation of the Punuk-Thule interface at Wales, Alaska. *Arctic Anthropology* 41(2), 163–176.

Hayden, B. (1998). Practical and prestige technologies: The evolution of material systems. *Journal of Anthropological Method and Theory* 5(1), 1–55.

Hayden, B. (2003). *Shamans, Sorcerers and Saints: A Prehistory of Religion*. Washington, DC: Smithsonian Institution Press.

Hoffmann-Wyss, A.B. (1987). Prähistorische eskimogräber an der Dovelavik Bay und bei Kitnepaluk im westen der St. Lorenz Insel, Alaska. St. Lorenz Insel-studien. *Berner beitärge zur archäologischen und ethnologischen erforschung des Beringstrassengebeites. Band* II. Bern and Stuttgart: Academica Helvetica 5,II.

Hollowell, J. (2008).A history of "Alaska Eskimo Ivories" as art. In D.E. Dumond (Ed.), *Aspects of Okvik: Four Essays on Things of Bering Strait*, University of Oregon Anthropological Papers 68, 221–260.

Hu, F.S., Ito, E., Brown, T.A., Curry, B.B. and Engstrom, D.R. (2001). Pronounced climatic variations in Alaska during the last two millennia. *Proceedings of the National Academy of Sciences of the United States of America* 98:10552–10556.

Hughes, M.K. and Diaz, H.F. (1994). Was there a "Medieval warm period," and if so, where and when? *Climate Change* 26, 109–142.

Jenness, D. (1928). *Archaeological Investigations in Bering Strait*, Annual Report for 1926, Bulletin 50, Ottawa: National Museum of Canada.

Jenness, D. (1940). Prehistoric culture waves from Asia to America. *Journal of the Washington Academy of Sciences* 30(1), 1–15.

Jensen, A.M. (2007). Nuvuk Burial: An Early Thule Hunter of High Status. *Alaska Journal of Anthropology* 5, 119–122.

Juday, G., Barber, V., Rupp, S., Zasada, J. and Wilmking, M. (2003). A 200-year perspective of climate variability and the response of white spruce in interior Alaska. In D. Greenland, D.G. Goodin, and R.C. Smith (Eds.), *Climate Variability and Ecosystem Response at Long-Term Ecological Research Sites* (pp. 226–250). Oxford: Oxford University Press.

Krupnik, I.(1987). The bowhead vs. the gray whale in Chukotkan aboriginal whaling. *Arctic* 40(1), 16–32.
Krupnik, I. (1993). *Arctic Adaptations: Native Whalers and Reindeer Herders of Northern Eurasia.* Hanover: University Press of New England.
Krutak, L. (1998). *One Stitch at a Time: Ivalu and Sivuqaq tattoo.* Unpublished M.A. Thesis, Dept. of Anthropology, Fairbanks: University of Alaska.
Larsen, H. and Rainey, F. (1948). *Ipiutak and the Arctic Whale Hunting Culture.* Anthropological Papers No. 42. New York: American Museum Natural History.
Leskov, A.M., and MüllerBeck, H. (1993). *Arktische Waljäer vor 3000 Jahren: Unbekannte Sibirische Kunst.* Mainz: V. Hase and Koehler Verlag.
Lewis, M. (1995). *Technological Development and Culture Change on St. Lawrence Island: A Funcitonal Typology of Toggle Harpoon Heads.* Unpublished Ph.D., Department of Anthropology, University of Alaska, Fairbanks.
Mason, O.T. (1902). *Aboriginal American Harpoons: A Study in Ethnic Distribution and Invention.* Reprinted from the United States Museum for 1900 (pp. 189–304). Washington, DC: Smithsonian Institution.
Mason, J.A. (1930). Excavations of Eskimo Thule culture sites at Point Barrow, Alaska. In Proceedings, 23rd International Congress of Americanists, New York, pp. 383–394.
Mason, O.K. and Jordan, J.W. (1993). Heightened North Pacific storminess and synchronous late Holocene erosion of northwest Alaska beach ridge complexes. *Quaternary Research* 40(1), 55–69.
Mason, O.K. (1998). The contest between Ipiutak, Old Bering Sea and Birnirk polities and the origin of whaling during the First Millennium A.D. along Bering Strait. *Journal of Anthropological Archaeology* 17(3), 240–325.
Mason, O.K. (2000). Archaeological Rorshach in delineating Ipiutak, Punuk and Birnirk in NW Alaska: Masters, Slaves or Partners in Trade? In M. Appelt, J. Berglund and H.C. Gulløv (Eds.), *Identities and Cultural Contacts in the Arctic* (pp. 229–251), Publication No. 8. Copenhagen: Danish Polar Center.
Mason, O.K. (Ed.) (2003). *Uivvaq Heritage Project*: Field Season 2002. Final Report to Aglaq/CONAM\ J.V. No. 2 in Fulfillment of Contract #2103-004, Cape Lisburne Clean Sweep Remedial Action Anchorage: U.S. Air Force, 615th Wing, Elmendorf Air Force Base.
Mason, O.K. (2006). Ipiutak Remains Mysterious: A Focal Place Still Out of Focus. In B. Grønnow (Ed.), *Dynamics of Northern Societies. Proceedings of a Symposium* (pp. 106–120). Copenhagen: Danish National Museum and Danish Polar Center.
Mason, O.K. (2009). Flight from Bering Strait: Did Siberian Punuk/Thule military cadres conquer Alaska between A.D. 1150 and 1350? In H.D.G. Maschner, R. McGhee and O.K. Mason (Eds.), *The Northern World A.D. 1100–1350: The Dynamics of Climate, Economy and Politics*, Salt Lake City: University of Utah Press, in press.
Mason, O.K. and Barber, V. (2003). A paleogeographic preface to the origins of whaling: cold is better. In A.P. McCartney (Ed.), *Indigenous Ways to the Present: Native Whaling in the Western Arctic* (pp. 69–108). Edmonton: Circumpolar Institute, University of Alberta, and Salt Lake City: University of Utah Press.
Mason, O.K. and Bowers, P.M. (2009). The Origin of Thule is Always Elsewhere: Early Thule within Kotzebue Sound, Cul de sac or nursery? In B. Gronnow (Ed.), *On the Track of the Thule Culture: New Perspectives in Inuit Prehistory: An International Symposium in Honor of Research Professor H.C. Gulløv* (pp. 25–44). Copenhagen: SILA, Danish National Museum.
Mason, O.K., Ganley, M.L., Sweeney, M., Alix, C. and Barber, V. (2007). *An Ipiutak Outlier: A Late 1st millennium AD* Qarigi *in Golovnin Bay.* Final Report on the Qitchauvik Field School, 1998–2000. NPS Technical Report NPS/AR/CRR/2007/67. Anchorage: National Park Service.
Mason, O.K. and Gerlach, S.C. (1995). Chukchi sea hot spots, paleo-polynyas and caribou crashes: Climatic and ecological constraints on northern Alaska prehistory. *Arctic Anthropology* 32(1), 101–130.

Mathiassen, T. (1930). Archaeological collections from the western Eskimos. Reports of the Fifth Thule Expedition 1921–1924, volume 10(1), 1–98.
Mathiassen, T. (1927). (a) and (b) Archaeology of the Central Eskimos, Report of the Fifth Thule Expedition, Vol. IV (Parts i and ii, Copenhagen Glendendalske Boghandel Nordsik Forlag.
Mayewski, P.A., Rohling, E.E., Stager, J.C., Karlen, W., Maasch, A., Meeker, L., Meyerson, E.A., F. Gasse, S. van Kreveld, S. Holmgren, K., Lee-Thorp, J., Rosqvist, G., Rack, F., Staubwasser, M., Schenider, R.R., and Steig. E. (2004). Holocene climatic variability. *Quaternary Research* 62(3), 243–255.
McCartney, A.P. (1988). Late prehistoric metal use in the New World Arctic. In R.D. Shaw, R.K. Harritt, and D.E. Dumond (Eds.), *The Late Prehistoric Development off Alaska's Native Peoples* (pp. 57–80), Aurora Monograph 4. Anchorage: Alaska Anthropological Association.
McGhee, R. (1976). Differential artistic productivity in the Eskimo cultural tradition. *Current Anthropology* 17(2), 203–220.
McGhee, R. (1981). Archaeological evidence for climate change during the last 5,000 years In T.M.L. Wigley, M.J. Ingram, and G. Farmer (Eds.), *Climate and History: Studies in Past Climates and Their Impact on Man* (pp. 162–179). Cambridge: Cambridge University Press.
Morrison, D. (2001). Radiocarbon dating the Birnirk Thule transition. *Anthropological Papers, University of Alaska* N.S. 1(1), 73–85.
Murdoch, J. (1892). Ethnological results of the Point Barrow Expedition. *Ninth Annual Report*, Bureau of American Ethnology, Government Printing Office, Washington D.C.
Neiman, F.D. (1995). Stylistic Variation in Evolutionary Perspective: Inferences from Decorative Diversity and Interassemblage Distance in Illinois Woodland Ceramic Assemblages. *American Antiquity* 60:7–36.
Nelson, E.W. (1899). *Eskimo about Bering Strait*. 18th Annual Report of the Bureau of American Ethnology for the Years 1896–1897. Washington, DC: Government Printing Office.
Okladnikov, A.P. and Beregovaia, N.A. (2008) [1971]. Early sites of Cape Baranov. Anchorage: Shared Beringian Heritage Program, National Park Service. [Translation by Richard Bland of *Drevnie poseleniia Baranova Izdatel'stvo* Novosibirisk: "Nauka," Sibirskoe Ordelenie]
Oswald, W., Brubaker, L B., Anderson, P.M. and Gerlach, S.C. (2001). Late Holocene environmental and cultural changes at Tukuto Lake, northwestern Alaska. In S.C. Gerlach and M.S. Murray (Eds.), People and Wildlife in Northern North America: Essays in Honor of R. Dale Guthrie (pp. 102–111), BAR International Series 944, Oxford: Archaeopress.
Oswalt, W.H. (1973). *Habitat and Technology: The Evolution of Hunting*. New York: Holt, Reinhardt and Winston.
O'Brien, M.J., Holland, T.D., Hoard, R.J., and Fox, G.L. (1994). Evolutionary implications of design and performance characteristics of prehistoric pottery. *Journal of Archaeological Method and Theory* 1, 259–304.
O'Brien, M.J. and Lyman, R.L. (2000). *Applying Evolutionary Archaeology: A Systematic Approach*. New York: Kluwer Academic/Plenum.
O'Shea, J. (1984). *Mortuary Variability: An Archaeological Investigation*. New York: Academic Press.
Porsild, M.P. (1915). The Material Culture of the Eskimos in West Greenland. *Meddelelser om Grønland. Bind* LI (pp. 111–250). Copenhagen: C.A. Reitzel.
Prentiss, W.C. and Chatters, J.C. (2003). Cultural diversification and decimation in the prehistoric record. *Current Anthropology* 44(1), 33–58.
Rainey, F. (1941). *Eskimo Prehistory: The Okvik site on the Punuk Islands*. Anthropological Paper No. 37, part 4. New York: American Museum Natural History.
Reanier, R.E., Sheehan, G.W., and Jensen, A.M. (1998). Report of 1997 Field discoveries City of Deering Village safe water cultural resources project. Report to Ukpeagvik Inupiaq Corporation Real Estate Science Division, Barrow.

Reynolds, G. (1989). Ownership marks and social affinity in northwest Alaska during late prehistoric times. *Arctic Anthropology* 26(1), 88–107.

Roe, P.G. (1995). Style, Society, Myth and Structure. In C. Carr and J.E. Neitzel (Eds.), *Style, Society and Person: Archaeological and Ethnological Perspectives* (pp. 27–76). New York: Plenum Press.

Rosenberg, M. (1994). Pattern, Process, and Hierarchy in the Evolution of Culture. *Journal of Anthropological Archaeology* 13:307–340.

Rudenko, S.I. (1961). *Ancient Culture of the Bering Sea and the Eskimo problem*. Arctic Institute of North America, Translations from Russian Sources No. 1, Toronto: University of Toronto Press.

Savinetsky, A., Kiseleva, N.K. and Khassanov, B.F. (2004). Dynamics of sea mammal and bird populations of the Bering Sea region over the past several millennia. *Palaeogeography, Palaeoclimatology, Palaeoecology* 209, 335–352.

Semenov, S.A. (1964). Prehistoric Technology. New York: Barnes and Noble.

Shennan, S. (2002). *Genes, Memes and Human History: Darwinian Archaeology and Cultural Evolution*. London: Thames and Hudson.

Staley, D.P. and Mason, O.K. (2004). A Punuk Whale Bone Grave from Sivuqaq, St. Lawrence Island: Evidence of high social standing, AD 775–1020. *Alaska Journal of Anthropology* 2 (1–2), 126–136.

Stanford, D.J. (1976). *Walakpa: Its Place in the Birnirk and Thule Cultures*. Smithsonian Contributions to Anthropology 20, Washington, DC: Smithsonian Institution.

Taçon, P. (1983). An analysis of Dorset art in relation to prehistoric culture. *Études Inuit Studies* 7(1), 41–65.

Thalbitzer, W. (1914). Ethnographical Collections from East Greenland (Angmagsalik and Nualik) Made by G. Holm, G. Amdrup and J Petersen and described by W. Thalbitzer. *The Ammasalik Eskimo*, 1(7). Meddelser on Gronland 39:319–755.

Van Pelt, V.A. (2008). A Survey of Okvik art style. In D.E. Dumond (Ed.), *The Story of Okvik. In Aspects of Okvik: Four Essays on Things of Bering Strait*, University of Oregon Anthropological Papers 68 (pp. 173–220).

Whitridge, P. (1999). The prehistory of Inuit and Yupik whale use. *Revista de Arqueología Americana* 16, 99–154.

Wiessner, P. (1983). Style and social information in Kalahari San projectile points. *American Antiquity* 48, 253–276.

Wissler, C. (1916). Harpoons and darts in the Stefansson Collection. *Anthropological Papers* of *the American Museum of Natural History* 14, part II.

Wobst, H.M. (1977).Stylistic behavior and information exchange. In C. Cleland (Ed.), *For the Director: Research Essays in Honor of James B. Griffin* (pp. 317–342), University of Michigan Museum Anthropological Paper No. 61, Ann Arbor.

Yama'ura, K. (1984). *Toggle Harpoon Heads from Kurigitavik, Alaska*. Bulletin of the Department of Archaeology No. 3:213–262, Tokyo: University of Tokyo.

Part II
Macroevolutionary Approaches to Cultural Change

Chapter 4
The Emergence of New Socioeconomic Strategies in the Middle and Late Holocene Pacific Northwest Region of North America

Anna Marie Prentiss

Introduction

At its most fundamental level, neo-Darwinism asserts that evolution is a product of sorting mechanisms (e.g. selection) acting on populations of variants over successive generations. Taken to its logical extreme, this model implies a process of selection and drift acting exclusively on genes and their particular phenotypes. Eldredge (1995; see also Dover 2000) refers to this program as "hardened NeoDarwinism" and argues that it does not fully encompass the full range of processes and scales by which evolution operates. If selection only targets the smallest biological entities (genes) then we would expect to see change that only comes piecemeal in form.

Anthropologists have offered similar critiques of the neo-Darwinian paradigm. As noted by Ingold (1990), the hardened Darwinian view implies a process that is akin to assembling a machine with independently designed and manufactured parts. But in order to create an adequately functioning "whole" the parts must somehow fit together regardless of origin, thus requiring significant tinkering to find that fit. The ultimate implication is that not all parts will have evolved through a targeted adaptive process but may be more realistically identified as accommodations (Gould and Lewontin 1979).

Some of the most interesting phenomena for anthropological archaeologists are exactly those characters held by large populations that are made up of myriads of integrated "parts" and utterly irreducible to piecemeal evolutionary process. If evolution produces such cultural entities recognizable only in their manifestation on population-level organizational scales (subsistence strategies, inheritance systems, and the like), then we need an evolutionary theory that can help us understand that process. Paleobiologists (Vrba and Eldredge 1984) have explicated a theory of emergence that can help us grasp the complexities of evolution in these contexts.

In this chapter, I address issues of emergence as critical components in understanding the evolution of such complex cultural entities (see Prentiss et al., this

A.M. Prentiss (✉)
Department of Anthropology, The University of Montana, Missoula, MT 59812, USA
e-mail: anna.prentiss@umontana.edu

A.M. Prentiss et al. (eds.), *Macroevolution in Human Prehistory*,
DOI 10.1007/978-1-4419-0682-3_5,

volume). I explore the utility of these ideas with several examples from the prehistory of the Pacific Northwest region of North America. More specifically I describe two periods in Pacific Northwest prehistory where there were rapid changes resulting in the emergence of new socioeconomic strategies. First, as detailed in Prentiss and Chatters (2003a, b; Chatters and Prentiss 2005) the collector strategy emerged on the northern Northwest Coast and subsequently spread through the region. Second, a "complex collector" strategy appears to have emerged in the Lower Fraser Valley on the central Northwest Coast leading to a somewhat similar expansion. In both cases, groups possessing the strategy appear to have benefited as indicated by proliferation of the cultural package (acquisitive fitness per Chatters, this volume) but also as marked by rising numbers of villages potentially indicating enhanced reproductive or biological fitness.

Emergence in Paleobiology

As evolutionary biologists from morphologists to taxonomists (e.g. Gould 2002; Gould and Lewontin 1979; McGhee 1999) tell us, building an entire system in the evolutionary sense requires a complex process of selection-targeted change in some characters and consequent systemic modifications in others. The result of that process may be new organisms containing characters that are the consequent result of that complex process of evolutionary reorganization among multiple parts. Some characters can be traced to the basic neo-Darwinian process whereby selection favors directional change over time (e.g. longer legs on African antelope evolve as forests give way to savannahs). But others like social organization are only visible on population scales and cannot be traced as easily to simple trait-oriented evolutionary process. The emergence of new forms (visible only on population scales) and their potential for alterations in fitness levels have been the subject of much discussion and debate in paleobiology (Arnold and Fristrup 1982; Gould 2002; Lloyd 1988; Lloyd and Gould 2003; Vrba 1983, 1989; Vrba and Eldredge 1984).

Two models of emergence at the macroevolutionary scale developed during the 1980s and are termed "emergent character" and "emergent fitness." Emergent characters are the result of a process that cannot be reduced to a linear sequence of evolutionary action at lower population levels or as an identifiable combination of lower level parts (Vrba 1983, 1989; Vrba and Eldredge 1984). Examples include a species' adaptive range and various population characteristics (e.g. social arrangements), since they cannot be held by individuals yet undoubtedly affect potential for further evolution at the species scale (Gould 2002).

Some population level characters do come about through a more standard set of Darwinian processes whereby selective action on specific lower level characters eventually gives rise to a population level or aggregate character. These characters (e.g. reproductive designs, larval strategies, etc. [Arnold and Fristrup 1982]) do not fit the more strict criteria for emergent characters developed by Vrba (1983, 1989). However, they may still confer advantages at the population level. In order not to exclude this equally important process, researchers developed the emergent fitness view (Arnold and Fristrup 1982; Damuth 1985; Lloyd 1988), suggesting that however a character emerges (whether reducible or irreducible to lower level Darwinian processes), it is still viewed as emergent if it has fitness implications across entire populations. Emergent fitness can be understood mathematically as the residual after individual fitness is factored out.

Viewed from the perspective of fitness landscapes (e.g. Wright 1931, 1932), emergent characters can be viewed as marking local adaptive "peaks." Peaks represent adaptations shaped by selection; valleys or troughs in between reflect more maladaptive contexts. Severe troughs or chasms threaten extinction. The theory assumes that evolution proceeds in three phases: (1) local populations or demes move to the vicinity of a new peak; (2) they ascend that peak under selection; and (3) ultimately they spread into the broader population. While details of this model have been debated for decades (Coyne et al. 1997; Fisher 1958; Wade and Goodnight 1998) it is still one of the most influential conceptions of evolutionary processes ever developed (Eldredge 1989, 1999; Joshi 1999). It has clear connotations for understanding macroevolutionary change in biological and cultural systems.

In Wright's (1932) conception, demes drift around their adaptive peaks into adjacent maladaptive troughs or chasms and are periodically drawn close enough to new optimum whereby they ascend under selection. The initial drift process can be triggered either by random genetic changes in local populations or by shifts in associated habitat, thereby drawing groups from their peaks into less-well- adapted troughs or valleys. While many researchers agree with Wright that drift could play a key role in drawing characters from their peaks, it is now also clear that adaptive processes can play the same role (Coyne et al. 1997).

Emergence in Archaeology

We cannot examine the role of emergence in cultural evolution without defining the basic cultural units that undergo evolution. If cultures are no more than sets of independent traits then it is no use persisting further with this discussion. However, there is good reason to believe that culture is more complexly integrated than what is implied by the reductionist view (Boyd et al. 1997; Chatters, this volume; Prentiss et al., this volume; Rosenberg, this volume; Holden and Shennan 2005; O'Brien et al. 2008; Zeder, this volume). One productive approach is to view culture as organized hierarchically in a nested series of units (Boyd et al. 1997; Chatters and Prentiss 2005; Eldredge 2000, this volume; Prentiss and Chatters 2003a; Rosenberg 1994; Shennan 2002). At the most basic level are traits that define fundamental actions such as manufacture and maintenance of artifacts. Most cultural characters are linked as integrated sets of information that can be called packages. Complex packages like languages have an internal logic or cultural glue that binds together the disparate memes and packages into definable wholes. These "core" characteristics likely act as evolutionary entities in and of themselves, albeit in complex ways (Bettinger, this volume; Boyd et al. 1997; Chatters, this volume).

Chatters and Prentiss (2005) offer the term Resource Management Strategy (RMS) to describe integrated predation, labor organization, and technology as evolutionary units. These core-like entities are expressed variably by human behavior called tactics that is reflected in the archaeological record much like the way phenotypes of past life are reflected in fossils. RMS vary almost endlessly, there are certain fundamental structures (or *Baupläne*) that show up repeatedly in the archaeological record (Chatters, this volume). Thus, for example, Chatters and Prentiss (2005) define mobile forager, sedentary forager, serial forager, and collector RMS (many other kinds are possible—e.g., high tech [Clovis] foragers). It is notable that the behavioral and material (or phenotypic) expression of each can vary widely. Recognition of variability in RMS depends upon now standard inferential procedures in archaeology as explicated in a broad literature generally classified as "middle range" (Chatters 1987, this volume; Chatters and Prentiss 2005). Substantial additional work will be necessary to sort out many details concerning transmission and inheritance on these complex scales (Prentiss et al., this volume; Shott 2008). However, following from Darwin (1859), this should not prohibit us from thinking about broader processes.

As in biological systems, lower level packages or characters may evolve within and spread throughout a population, and yet may never achieve the level of full population integration required to be considered truly emergent under either the character or fitness models. Many technologies would fall into this category. For example, Paleoindian projectile point technology evolved into a wide variety of designs and local expressions beginning with a basic triangular design in pre-Clovis times (Adovasio and Pedler 2004) and eventually shifting to a variety of concave-based lanceolates, square-based lanceolates (Agate Basin and Hell Gap), and stemmed forms by later Paleoindian periods (O'Brien and Lyman 2000, 2003). While possession of different projectile point styles could have offered some perhaps subtle advantages to individuals (e.g. Frison 2004), measurable changes in human reproductive fitness were probably more dependant on participation in broader systems.

An entire technological repertoire, viewed as an integral component of a larger RMS (Prentiss and Clarke 2008), could be possessed and used equally across a human group with possible population-wide fitness implications (but see Bamforth 2002). Although Bettinger (2003) correctly argues that no single individual transmits the RMS in its entirety, it is still possible for Darwinian microevolutionary processes to give rise to new technological packages with significant implications for fitness at those group levels (see also Boyd and Richerson 1992). For example, between 1000 and 2000 B.P., Eskimo groups of the Bering Strait developed technological packages that included large open boats (umiaks), sealskin floats, and complex harpoon systems that enabled them to achieve success at whaling (Mason 1998, this volume). Groups who possessed the full expression of these had economic and undoubtedly reproductive advantage over their neighbors, particularly under conditions conducive to open water hunting. Consequently, we might view them as emergent under the fitness criterion.

The technology example still fails under the more strict character criterion. However, RMS taken in their entirety do have such emergent characteristics. Once an RMS crystallizes within a population it will define a complex cycle of variation in group actions that might include residential and logistical mobility, resource harvesting and processing, labor management, and economic elements of social organization. While a group character is not visible at the level of the individual person, that character may have significant implications for the competitive viability of all group members. For example, a very successful strategy might permit one group to grow larger and hold more territory more effectively than its neighbors (Bettinger and Baumhoff 1982; Kosse 1994). Therefore we could argue that strategy would impart an emergent fitness and fit the definition of an emergent character. This does not mean that high population density is necessary for the evolutionary process that gives rise to new forms. As Chatters and I have argued elsewhere (Prentiss and Chatters 2003a; Chatters and Prentiss 2005), small, more isolated groups have greater chances of deviating enough from old patterns to form something new. But an emergent characteristic of that new entity, even if it is emergent within a small group, might include the potential to develop a restructured population able to work in a new integrated fashion to operate a redesigned resource collection, processing, and distribution strategy. This new group might now exist within a reorganized cultural niche (e.g. Laland et al. 2000) providing new advantages over other groups without this character.

Following from Spencer (this volume; see also Bettinger, this volume) we can envision this process from the standpoint of Wright's (1931, 1932) adaptive landscapes. Emergence of a new strategy could require a risky crossing between adaptive peaks. Consequently it is entirely possible that the transition between strategies could occur either through conscious adaptive decision making (as in the Monte Alban case) or it could be a consequence of nonadaptive and historically contingent factors, which for hunter-gatherers might include unintentional rearrangements of mobility schedules or harvesting and processing priorities (Chatters and Prentiss 2005), introduction of technologies from adjacent regions (Bettinger 1999; this volume), or altered social environments (Prentiss 2010; Rosenberg, this volume). Whether such crossings are generated by adaptive strategizing or accidental "slides" it is possible that risky trough-crossings could be visible archaeologically in short-lived transition periods (Spencer, this volume) potentially as demographic low points (Goodale et al. 2008).

In summary, we could expect emergent fitness under either the Darwinian microevolutionary pathway to a commonly held cultural character (e.g. a technology) or through a macroevolutionary event occurring on the higher scale resulting in a true emergent character. Either way, cultural chronologies may illustrate brief transition periods associated with such peak crossing events perhaps as indicated by fluctuations in demographic profiles. The prehistoric record from North America's Pacific Northwest region provides examples of these processes.

Pacific Northwest Prehistory

North America's Pacific Northwest (Figs. 4.1 and 4.2) cultures evolved in a punctuated process that included periods of high and low cultural diversity (Prentiss and Chatters 2003a, b). As documented elsewhere, mobile foragers (forager-collector

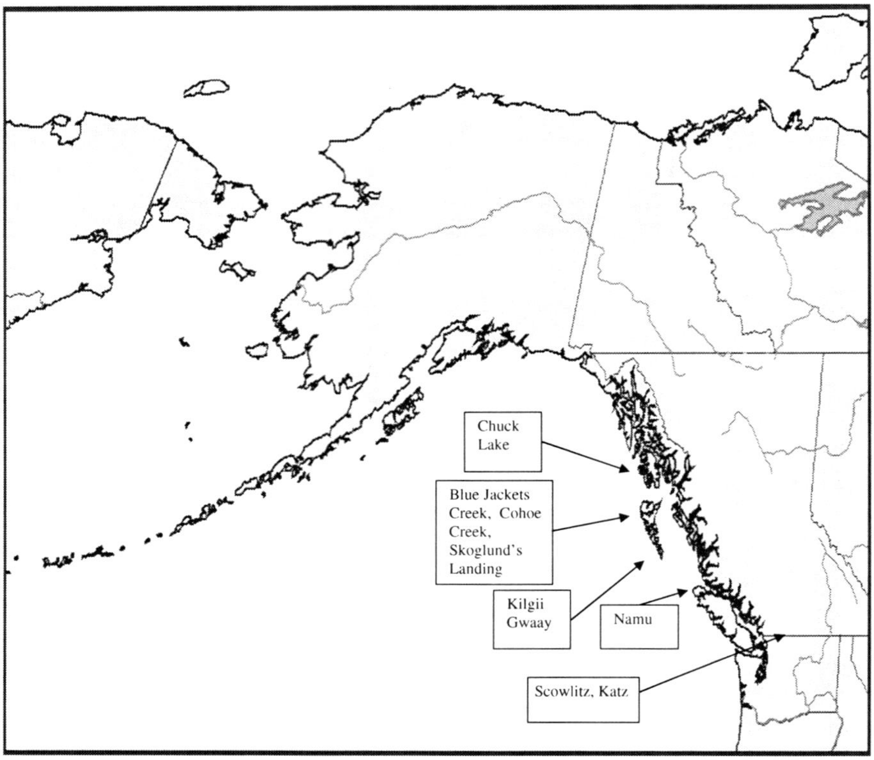

Fig. 4.1 Northwest region of North America showing key sites discussed in the chapter

defined per Binford 1980) of the Early Holocene evolved rapidly into a variety of forms that included serial forager, sedentary forager, and collector strategies during the period of approximately 4000–5300 cal. B.P. (Chatters and Prentiss 2005; Prentiss and Chatters 2003a, b). Later, collectors came to dominate the region through a regional process of cultural decimation (Prentiss and Chatters 2003a). Collectors also evolved further during the Late Holocene resulting in the appearance of the large and complex villages so well known on the Northwest Coast (Matson and Coupland 1995). This "Complex Collector" pattern subsequently spread into portions of the interior plateau region giving rise to the complex hunter-gatherers of the Mid-Fraser Canyon and various segments of the Columbia river system (Prentiss et al. 2005a). Because the evolutionary history of the collector strategy is relatively well understood in the Pacific Northwest, it can be very effectively used as a means of demonstrating that the evolution of cultural systems can include both emergent characters and emergent fitness.

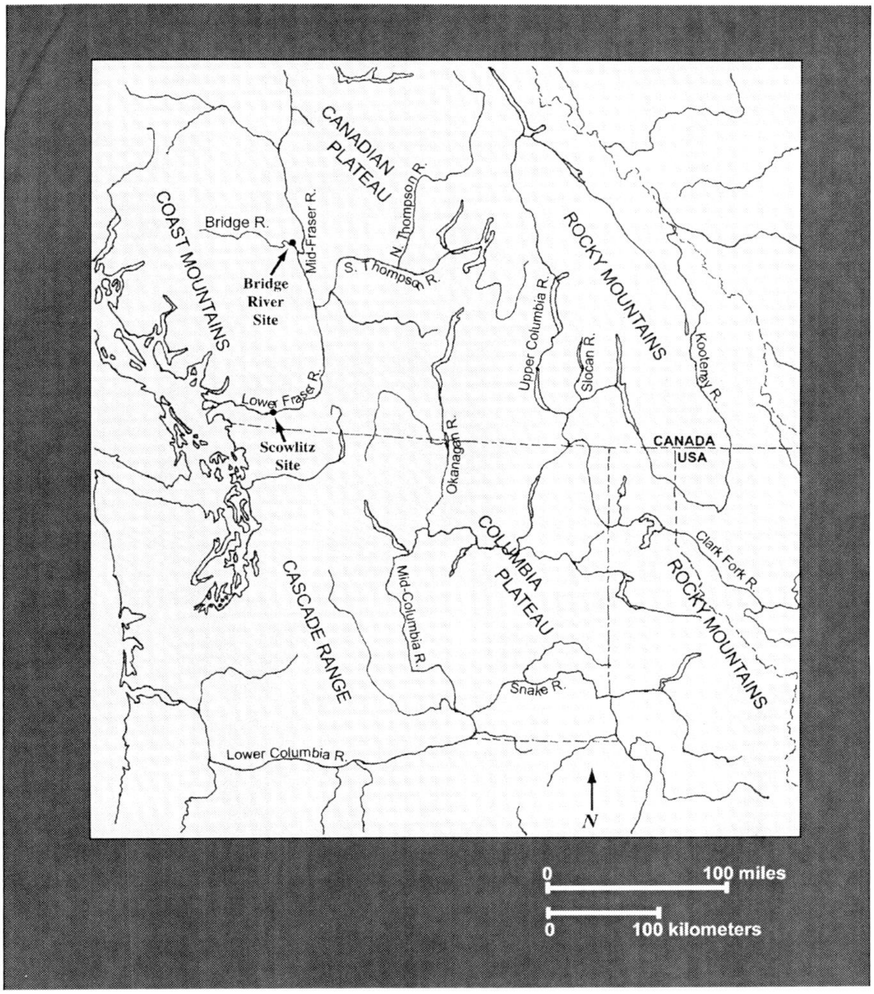

Fig. 4.2 Pacific Northwest region detail showing portions of the central Northwest Coast and Plateau regions as well as locations of Scowlitz and Bridge River villages

Emergent Collectors

Evidence for earliest collector behavior (ca. 5000 cal. B.P.) is extremely sparse and found only in relatively isolated contexts such as the Namu site, located on a particularly remote section of the British Columbia Central Coast, and at the Blue Jackets Creek, Cohoe Creek, and Skoglund's Landing sites of Haida Gwaii (Queen Charlotte Islands). Prior to 5000 cal. B.P., hunter-gatherers of the central and northern Northwest Coast had been highly mobile from a residential standpoint, leaving very little evidence of temporary house structures, which were probably nothing more than temporary shelters similar to many other recent hunter-gatherers of northern and far southern regions around the world. There is no evidence that any group possessed a logistical strategy that engaged in resource targeting and intensification. Ubiquitous thin shell midden deposits reflecting short-term residential occupations indicate a fairly high degree of intersite variability in faunal remains. Elk, deer, and sea mammals dominate the early Glenrose Cannery strata (Imamoto 1976). Salmon and deer are frequent in early Namu deposits (Cannon 1991). Inshore fish and shellfish are typical of Cohoe Creek (Ham 1990) and Chuck Lake (Ackerman et al. 1989), while a wider range of food sources (birds, fish, and mammals) co-occur in the midden deposits at Kilgii Gwaay (Fedje et al. 2004). There is no unambiguous signature for resource-processing strategies (either in bone assemblages or site features) that could be interpreted as oriented toward storage prior to about 5000–5500 cal. B.P. when salmon numbers spike upward at Namu. Lithic technologies between about 9000 and 5000 cal. B.P. are consistent throughout the region (Prentiss and Clarke 2008). Three major technologies are recognized, including flake tool production from cobbles and pebbles (Ackerman 1992; Carlson 1996; Haley 1995; Mitchell and Pokotylo 1996), bifacial technology (leaf-shaped bifaces) (Ackerman 1992; Matson 1996; Matson and Coupland 1995), and microblade production from a variety of core forms (Ackerman 1992; Magne 1996). A significant degree of stylistic similarity in tool production (Washington to Alaska) suggests a pattern of high mobility and probable frequent interaction between bands (Prentiss and Chatters 2003a, b). All in all, the data are consistent with a conclusion that all the pre-5000–5500 cal. B.P. hunter-gatherers operated a mobile, immediate-return forager strategy (Prentiss and Chatters 2003a)

Current evidence suggests that collecting appeared abruptly in no more than two places on the northern Northwest Coast after about 5000 cal. B.P. The Middle Holocene strata at the Namu site feature massive accumulations of shell and herring, salmon, and deer bones. Cannon (2002; Cannon and Yang 2006) interprets the post-7000 B.P. midden as the consequence of occupations by delayed-return, logistically organized collectors occupying a permanent village, many characteristics of the fully developed Northwest Coast Pattern. This is in contrast to Ames' (1998) interpretation of Namu as an aggregation site where multiple mobile bands met and made intensive use of local resource abundance. Chatters and Prentiss (2005) suggest that a collector pattern could be evident in the materials dating ca. 5600–3600 cal. B.P. on the basis of the rapid jump in salmon numbers that appears to be dominated by post-cranial elements (implying storage). However, no other evidence (house features for example) has been found to otherwise support Namu as a permanent village occupied by collectors (Cannon 1991; 2002). Namu will thus have to remain somewhat ambiguous as an early marker of collector behavior.

Potentially better data sets come from Blue Jackets Creek, Cohoe Creek, and Skoglund's Landing on Haida Gwaii. Blue Jackets Creek is another large midden, similar to Namu, with dates marking the beginning of intensive occupations around 4500 cal. B.P. (Fladmark et al. 1990; Mackie and Acheson 2007). Most critically, site features suggest the presence of house floors implicating the possibility of residential sedentism. Human burials indicate some degree of status differentiation (Severs 1974) which could reflect the possibility that individuals used surplus resource accumulations to demonstrate status differentiation (e.g. Hayden 1995). Faunal remains have not been well described but include various terrestrial mammals, marine mammals, salmon, and halibut likely reflecting warm and cold season resource-procurement tactics (Severs 1974). The Cohoe Creek site provides additional evidence for Early Graham tradition residential behavior similar to that of Blue Jackets Creek (Christensen and Stafford 2004). These data lead us to suspect the degree of residential permanence typical of collectors. However, fully sedentary foragers are known elsewhere in the region (Chatters 1995; Chatters and Prentiss 2005), and it is possible that the strategy could have appeared here as well. The nearby Skoglund's Landing site (Fladmark 1970) suggests the possibility of a specialized field camp used by residents of local residential bases (Mackie and Acheson 2007). Blue Jackets Creek, Cohoe Creek, and Skoglund's Landing combined are suggestive of settlement patterns typical of collectors (Binford 1980).

I think that it is realistic to assume that some form of collector strategy emerged on Haida Gwaii by ca. 4500–5000 cal. B.P. and possibly in the Namu area slightly earlier. While details of the transition from foraging to collecting remain to be worked out, there are several factors that we can be clear on. First, collecting clearly emerged first in places that were relatively isolated from major developments elsewhere in the region. Second, it emerged in places that did offer significant resource opportunities. Finally, when it did appear, the transition appears to have been abrupt, especially on Haida Gwaii. From a selective standpoint, it may have offered advantage to those living in an isolated environment that offered high seasonal abundance. Storage could replace frequent social contacts and aid in getting past lean seasons. To those who made the transition, it meant fundamental changes in resource scheduling, settlement-positioning, technologies, and labor management. Regular movement to place small groups in optimal positions for harvesting limited quantities of a broad range of resources shifted to establishment of more permanent residential bases from which task groups could be sent to collect large numbers of specific resources (Ames 2002). Cold season survival was achieved not by frequent moves to new foraging locales but by collecting surpluses for storage. Resource processing became a critical labor element, most likely favoring women's activities (assuming gendered labor division in the past was similar to more recent times), similar to Bettinger and Baumhoff's (1982) processors in the Great Basin. Consequently, collectors could survive more effectively in the more seasonal and periodically resource restricted context of the early Neo-glacial period (Chatters and Prentiss 2005). The strategy could probably also support higher populations in better defined territories. Those groups who successfully developed or otherwise came to possess this strategy could have potential advantage over their neighbors who lacked it. It should be no surprise, therefore, that the collector strategy expanded while all others waned after 4100 cal. B.P. By 3000 cal. B.P. it was the only recognizable strategy in use across the entire region.

The collector RMS of the Middle Holocene Pacific Northwest may have featured emergent characters and fitness. The reorganization of mobility, technology, and labor probably came about through a rapid and complex Darwinian process that probably produced both adaptations and exaptations (or co-opted traits). The emergent strategy, however, appears to have featured characters that could not be reduced to the microevolutionary process that included patterns of labor organization associated with co-residential populations capable of harvesting more resources within larger and certainly better defined (and probably at least partially defended) territory. The collector strategy clearly offered significant reproductive advantages (and thus emergent fitness) to its possessor groups.

Data are insufficient to evaluate fitness changes on Haida Gwaii, though it is significant that the collector strategy crossed the boundary into the Neo-glacial while all other strategies in the region declined. The best data possibly reflecting on changing fitness levels occur in the Gulf of Georgia and Lower Fraser Valley areas of the Central Coast and on the interior Plateau (Fig. 4.3). Following results of research by Lepofsky et al. (2005), we recognize a change in population trends at around 3500–3700 cal. B.P. on the Central Coast, which is when the collector strategy began to manifest itself in this area as the Locarno Beach phase. Although the pattern of change is not highly dramatic it is clear that it does reverse the downward trend associated with the previous, more forager-like, cultural pattern. Indeed the apparent ca. 3500–4000 B.P. demographic trough could reflect the period of lowered fitness associated with a major socioeconomic transition as predicted by the shifting balance theory.

A potentially more dramatic example comes from the Columbia Basin within the Southern Plateau region (Fig. 4.3). Here, Chatters (1995) demonstrates that after a nearly complete hiatus between 3500 and 4100 cal. B.P., entrance of groups possessing the collector strategy (called Pithouse II) resulted in potentially rapid population growth (Prentiss et al. 2005b). A similar though less dramatic pattern is evident in the Mid-Fraser Canyon of the Canadian Plateau. In the Plateau cases, demographic troughs probably reflect local population collapses under socioeconomic stress followed by population replacements rather than in situ cultural transitions (Chatters and Prentiss 2005; Goodale et al. 2008). From an adaptive landscape perspective we could argue that the original groups slid into maladaptive troughs that led to cultural extinctions rather than transitions to newly emergent strategies.

Emergent Complex Collectors

An important problem for archaeologists who work on the Northwest Coast is that of explaining the origin of the complex hunter-gatherer pattern that characterizes its late period (e.g. Ames 1985, 1995; Burley 1980; Lepofsky et al. 2005; Matson 1983; Matson and Coupland 1995). The ethnographic Northwest Coast is famous for its large coastal and riverine context villages each consisting of one or more rows of large wooden houses occupied by ranked corporate groups organized by lineages, clans, and/or moieties. The ethnographic villages of the Northwest Coast could be described as enhanced or complex collectors (e.g. Prentiss et al. 2005b) as they

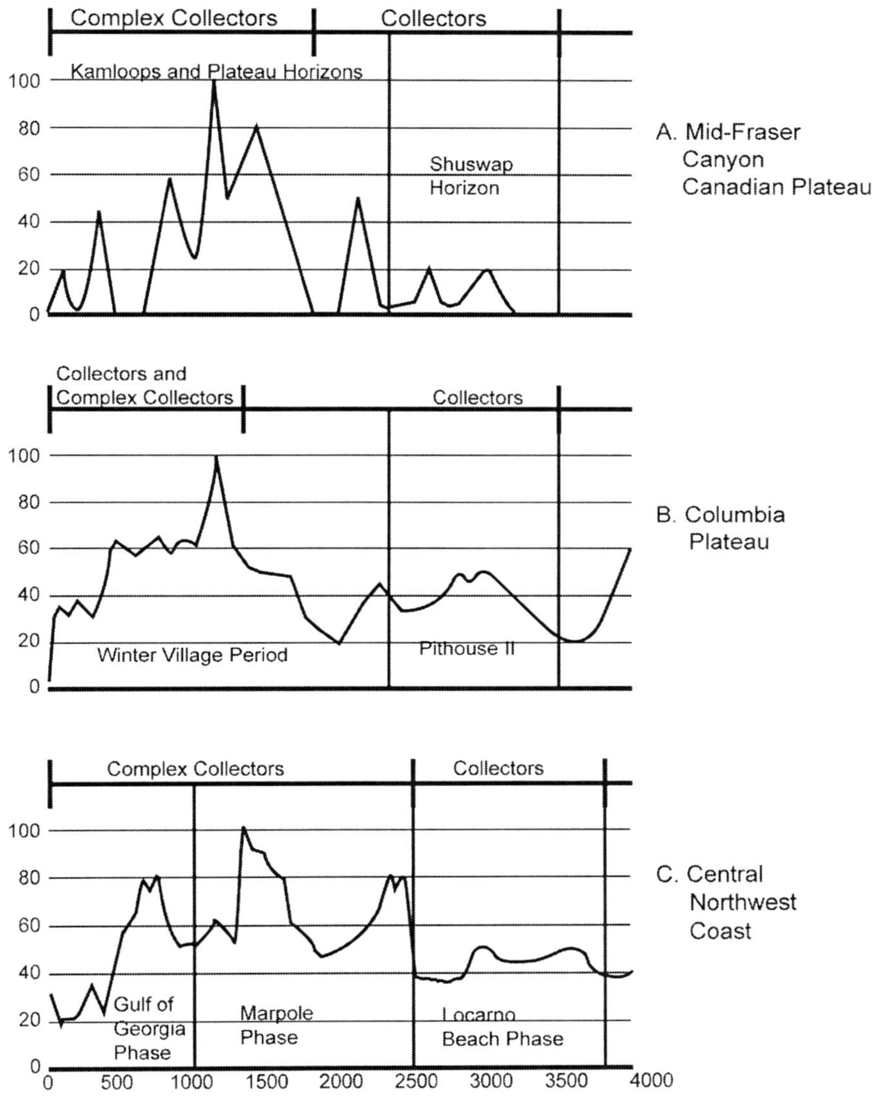

Fig. 4.3 Population proxies of Mid-Fraser Canyon (**A**), Columbia Plateau (**B**), and Central Northwest Coast (**C**). Each chart provides a rescaled depiction of data patterns compiled and originally quantified by researchers using a range of methods (**A**: Prentiss et al. 2005b; **B**: Chatters 1995; Prentiss et al. 2005b; **C**: Lepofsky et al. 2005)

retained all aspects of earlier collector strategies but added new aspects including multifamily house groups acting as economic units controlling aspects of resource harvesting, goods production, land use, and various ritual and spiritual elements. The intensified economies permitted clusters of settlement far more dense than that known for early collectors (e.g. Locarno Beach phase). This in turn permitted the establishment of larger and more powerful units for raiding and warfare (Ames and Maschner 1999; Maschner and Reedy-Maschner 1998).

Early evidence for the appearance of the complex collector strategy comes from the Lower Fraser Valley of southwestern British Columbia[1] (Fig. 4.2). Row villages appear abruptly at approximately 2500 cal. B.P. in two places. The Katz site has two rows of semi-subterranean housepits along the Fraser River west of Hope, B.C. with house sizes large enough for possibly two families to co-reside (Lenert 2007). Located to the west of Katz, the Scowlitz site provides the best evidence for major change to the collector strategy. Scowlitz is a village that includes a variety of house structures and numerous burial features (Lepofsky et al. 2000). House 3 is a 17-m-in-length rectangular structure very similar to those of the ethnographic Coast Salish that dates to about 2500 cal. B.P. Features in House 3 include a variety of large postholes, benches, hearths and rock-piles. House 3 is part of a row of large houses fronting the Harrison River near its entry into the Fraser. Paleoethnobotanical studies suggest that the houses were occupied year round (Lepofsky et al. 2000). Lithic raw materials suggest logistical movement and/or exchange relations spanning central Oregon to the Central Coast of British Columbia (Blake 2004). Lithic tools include well-developed ground slate (knives in particular) and microblade industries. Also present are a wide variety of wood and bone/antler working tools (adzes, wedges, abraders) and hunting/warfare gear (bifacial projectile points and knives) (Lepofsky et al. 2000). Faunal preservation at the site is poor, but location, residential permanence and technologies suggest a likely emphasis on salmon, combined with various upland plant and animal resources. The major implication is that House 3 at Scowlitz reflects a system of multifamily coresidence. Even if we assume a high ratio of space per person, say 4 m^2 (Hayden et al. 1996), over 45 people could have lived in this structure. If there were three large houses simultaneously occupied, the group at Scowlitz would have been well over 100 and maybe over 150 persons. Residential permanence and, possibly, year-round occupation would have required extensive logistical movement and exchange (e.g. Binford 2001) as is indicated by the wide range of lithic raw materials at the site.

The pattern at Scowlitz stands in stark contrast to earlier collectors on the Northwest Coast particularly in the realm of residential permanence, logistical range, and group size. The Locarno Beach phase (ca. 3700–2400 cal. B.P.) is typified by extensive midden development and indicators of processing of salmon for storage. Structures are rare and indicated by small semi-subterranean depressions and occasional hearths and post-holes in middens (Matson and Coupland 1995). While long distance exchange relations existed throughout the region after about 3000 cal. B.P., no single site seems to contain the impressive numbers of exotics seen in villages after 2500 cal. B.P. (Blake 2004). Although Locarno Beach groups were clearly collectors, they appear not to have made use of the same pattern of residential permanence seen at Scowlitz and clearly did not live in large groups (certainly less than 40 per total residential group). Therefore, while they were undoubtedly more competitive under Neoglacial conditions than the earlier more forager-like strategies, their collector strategy was to become less competitive against the new strategy coming from the Fraser Valley under warming conditions after 2500 cal. B.P.

Lepofsky et al. (2005) demonstrate that warming conditions may have played a critical role in the emergence of the complex collectors of the post 2500 cal. B.P. Marpole phase on the Central Coast. Droughts in this region may have been quite frequent between ca. 2500 and 1400 cal. B.P. (see also Hallett et al. 2003) with several important implications. First, frequent fires in higher elevation contexts would have opened forests to easier travel and foraging. Increased production of various shrubs would have produced seasonal bumper-crops of berries and eased access to medium and large ungulates (e.g. deer, elk, sheep) and bears. Second, there were also significant changes in the structure of salmon resources. While marine resources were peaking in productivity in the eastern Pacific (e.g. Tunnicliffe et al. 2001), droughts may have had severe affects on access to coastal and interior peoples. Lepofsky et al. (2005) predict that second-order streams in all areas would have lost their runs or at least endured significant reductions in numbers. Increased sediment and warmer interior water temperatures would have lowered production levels to some degree (e.g. Chatters et al. 1995) but, more significantly, delayed freshets which would have caused salmon to bunch at mouths of major trunks (e.g. the mouth of the Harrison). Consequentially, this meant a far more patchy resource landscape than had been present prior to the post 2500 cal. B.P. period. This may have been particularly true of places like the eastern portion of the Lower Fraser Valley, many kilometers from the coast, where other marine resources were not available.

It is perhaps not surprising that complex collectors emerged first at the mouth of the Harrison and possibly at or near the mouth of the Fraser Canyon. These were places that probably offered the most productive patches in the form of clustered anadromous fish and easy access to upland habitats in the hemlock forests of the Coast Range. Coresidential corporate groups permitted multifamily economic units to simultaneously guard their position while accessing resources at extended ranges, a strategy that would not have been possible in earlier times. The simultaneous expansion of logistical range, refinement of lithic technologies, aggregation of disparate family units into corporate-group households, development of more effective strategies for mass harvest and processing of food resources, and expanding capacity to mark and protect territories happened as a rapid macroevolutionary event at places like Scowlitz. Many elements of the transformation could be reduced to Darwinian processes (e.g. refinement of technologies and foraging strategies). However, like the emergent collector system centuries before, elements of this change appear irreducible to lower level phenomena or parts. Integrated high-density populations organized within house groups with the ability to control key segments of the landscape may represent emergent characters within the new complex collector strategy.

Various data sets suggest a pattern of emergent fitness associated with complex collectors (Fig. 4.3). Lepofsky et al. (2005; see also Burley 1980) demonstrate that the complex collectors (or Marpole phase) fluoresced in the Lower Fraser Valley after 2500 cal. B.P. Lepofsky et al.'s (2005) data suggest a rapid rise in evidence for human populations around 2400 cal. B.P. that may have effectively doubled the size of the human group across the region. Coincident with a likely decline in eastern Pacific fisheries production levels (e.g. Tunnicliffe et al. 2001), Marpole populations appear to have temporarily declined, as villages were abandoned and others consolidated between 2000 and 1800 cal. B.P. As numbers climbed again after this point, we recognize the appearance of larger villages with ranked households in the Gulf of Georgia region (Matson and Coupland 1995).

Population troughs clearly preceded the archaeological manifestations of large house groups and, later, ranked households of the Marpole phase on the central Northwest Coast (Fig. 4.3). While additional research will be necessary to confirm and expand upon these conclusions, I suggest that in each case, the emergence of new patterns was likely preceded by a short stressful period as expected under the peak-crossing model. Subsequently, it also appears likely that once the new strategy had emerged it spread rapidly into adjacent areas of the coast and on to the interior.

The most dramatic example of the expanding complex collector pattern comes from the Mid-Fraser Canyon of the western interior Plateau of British Columbia (Fig. 4.3). The Mid-Fraser area is located several 100 km north of the Lower Fraser Valley and can be easily reached via the Fraser Canyon itself or via a slightly less direct route up the Lillooet River valley. The Mid-Fraser's fishing sites are known as the best on the Canadian Plateau (Kew 1992; Romanoff 1992) and in that sense replicate the patchiness that was valuable to early complex collectors of the Lower Fraser Valley. As documented by Prentiss et al. (2003, 2005a, b, 2007), extremely large villages (up to 115 houses visible on the surface) with unusually large housepits (e.g. 18 m and larger in diameter) appeared abruptly shortly after 1900 cal. B.P. New dates on housepit features at the Bridge River site provide earliest indicators of this phenomenon (ca. 1700–1900 cal. B.P.). Earliest dating houses are large (14–18 m diameter) with multiple hearth features around edges of floors, potentially reflecting multiple family units. They are also characterized by a well-developed ground slate and nephrite tool industry, highly reminiscent of that found in the Lower Fraser Valley. Earliest occupations of large villages on the east side of the Fraser Canyon such as Bell (Stryd 1973) and Keatley Creek (Hayden 1997; Prentiss et al. 2003) postdate 1700 cal. B.P. and tend to, more realistically, cluster around 1500–1600 cal. B.P. The best dated sites (Bridge River and Keatley Creek) appear to have highest densities of occupied housepits at ca. 1100–1400 cal. B.P. before declining and final abandonment by about 800 cal. B.P. A similar pattern is recognizable on the Columbia Plateau, particularly in portions of the Central and Upper Columbia drainage after about 1200 cal. B.P. (Goodale et al. 2008; Prentiss et al. 2005b).

This occupation pattern correlates strongly with eastern Pacific fisheries records (e.g., Tunnicliffe et al. 2001) and regional climatic patterns suggesting cooling temperatures with abundant fish at ca. 1200–1300 cal. B.P. followed by warmer conditions and reduced numbers of fish after 1000 cal. B.P. (Prentiss et al. 2005b). Prentiss et al.'s (2005b) population proxy (reproduced in Fig. 4.3) clearly shows that human numbers increased only with the advent of the complex collector strategy, which subsequently expanded and persisted as long as optimal fishing conditions persisted. This is nearly identical to the later history of the Marpole culture, which was apparently adversely affected by the decline in salmon after 1100 cal. B.P.

Emergent hereditary status inequality appears to be associated with a population trough in the Mid-Fraser after ca. 1200 cal. B.P. (Fig. 4.3) (Prentiss 2010; Prentiss et al. 2007). This could reflect the emergence of the socioeconomic and political structures associated with the historic St'át'imc culture of the Fraser Canyon associated environs in British Columbia (Teit 1906). If so, it reflects an additional peak-crossing event in the late prehistoric period of this region.

Discussion

New forms of organization within human societies can clearly include emergent properties whether viewed as fitness or characters. The Pacific Northwest examples suggest that new strategies (sometimes characterized also by new *Baupläne*) do develop through standard microevolutionary processes, whereby technologies and other adaptations are refined and reorganized (albeit rapidly in some cases). This process may be complex as multiple members of a population participate in development, transmission, and practice of new tactics. When key changes are rewarded (in essence, selected for) they become adaptations (e.g. O'Brien and Holland 1992). However, to work effectively, the new tactics must find a fit within the broader adaptive structure. Consequently, modifications to other traits may be necessary to create a functioning whole (Chatters and Prentiss 2005; Prentiss et al. 2007).

While no single individual holds an entire package at this scale, commonly held ideologies may form the glue that binds its disparate elements (Bettinger 2003). A new strategy thus may carry its own emergent properties that cannot be reduced to those lower level microevolutionary processes (see also Eldredge, this volume; Mason, this volume; Zeder, this volume). It may prescribe new patterns of labor integration, social arrangements, population structure and density, a land-use pattern and range, and ultimately, a new ecological relationship between the human group and its habitat.

Data from this study imply that, as predicted by adaptive landscape theory, transitions to new organizational forms can come with short-lived periods of stress and potentially reduced organic fitness (see also Spencer, this volume). Likewise, once present, group possession of such complex organizational strategies may impart a significant degree of competitive advantage (and thus, potential fitness) over other groups holding less adaptive variants.

Potential examples from the North American record are numerous. Bettinger and Baumhoff (1982) explained the Numic expansion across the Great Basin by virtue of the new processor strategy's ability to make more efficient use of desert resources and support more dense populations. With whaling technology, complex social organization and an Asian-derived war complex, Thule populations exploded across the North American Arctic at the expense of remnant Dorset Paleoeskimo groups

(Mason, this volume; Prentiss and Lenert, this volume). In an earlier, but, perhaps, even more dramatic dispersal, Clovis peoples took advantage of a technologically and organizationally unique strategy designed for surviving in contexts of high isolation, in often poorly known terrain (e.g. Haynes 2002; Kelly and Todd 1988). The early agricultural complex of the American Southwest (e.g. Roney and Hard 2002) clearly offered unique advantages over earlier hunting and gathering strategies in the Tucson Basin and elsewhere. Late prehistoric polities from the upper Mississippi Valley to the Valley of Oaxaca emerged and spread at the expense of their neighbors, sometimes using radically new forms of socioeconomic and political organization (Chatters, this volume; Spencer, this volume).

However, as Bettinger (this volume) points out, we are only at the beginning in our study of cultural macroevolution. We have more to learn about the actual processes associated with the emergence of complex cultural variants. We also require better techniques for defining structure and variation in such entities. As noted in the introduction to this volume, quantitative techniques drawing from network theory may ultimately be useful in this regard. Once we cross this threshold, it may be possible to more rigorously study the phylogenetic and ethnogenetic histories of such complex systems, revealing insights into patterns of ancient human history not yet available today.

Notes

1. Complex collectors may have developed at about the same time on British Columbia's North Coast. However, the archaeological record is less well known in this context precluding much further theorizing.

Acknowledgments Figure 4.1 was partially drafted by Nathan Goodale. Figure 4.3 was drafted by Mark Fritch. I thank Mike Lenert for his comments on the paper and Jim Chatters for years of stimulating and insightful discussions. Archaeological research at the Bridge River and Keatley Creek sites was funded by the National Science Foundation, the Wenner-Gren Foundation for Anthropological Research Inc., and The University of Montana.

References

Ackerman, R. (1992). Earliest Stone Industries on the North Pacific Coast of North America. *Arctic Anthropology*, 29:18–27.

Ackerman, R., Reid, K., and Gallison, J. (1989). Heceta Island: An Early Maritime Adaptation. In B. Onat (Ed.), *Development of Hunting-Fishing-Gathering Maritime Societies along the West Coast of North America*. Pullman: Reprint Proceedings, Vol. IIIC, The Circum-Pacific Prehistory Conference, Seattle, Washington, Washington State University Press.

Adovasio, J.M. and Pedler, D.R. (2004). Pre-Clovis Sites and Their Implications for Human Occupation before the Last Glacial Maximum. In D.B. Madsen (Ed.), *Entering America: Northeast Asia and Beringia Before the Last Glacial Maximum* (pp. 139–158). Salt lake City: University of Utah Press.

Ames, K. (1985). Hierarchies, Stress, and Logistical Strategies among Hunter-Gatherers in Northwestern North America. In T.D. Price and J.A. Brown (Eds.), *Prehistoric Hunter-Gatherers: The Emergence of Cultural Complexity* (pp.155–180). Orlando: Academic Press.
Ames, K. (1995). Chiefly Power and Household Production on the Northwest Coast. In T.D. Price and G.M. Feinman (Eds.), *Foundations of Social Inequality* (pp. 155–188). New York: Plenum Press.
Ames, K. (1998). Economic Prehistory of the Northern British Columbia Coast. *Arctic Anthropology* 35, 68–87.
Ames, K. (2002). Going by Boat: The Forager-Collector Continuum at Sea. In B. Fitzhugh and J. Habu (Eds.), *Beyond Foraging and Collecting: Evolutionary Change in Hunter-Gatherer Settlement Systems* (pp. 19–52). New York: Kluwer Academic/Plenum Publishers.
Ames, K. and Maschner, H.D.G. (1999). *Peoples of the Northwest Coast: Their Archaeology and History*. London: Thames and Hudson.
Arnold, A.J. and Fristrup, K. (1982). The Theory of Evolution by Natural Selection: A Hierarchical Expansion. *Paleobiology* 8, 113–129.
Bamforth, D.B. (2002). Evidence and Metaphor in Evolutionary Archaeology. *American Antiquity* 67, 435–452.
Bettinger, R.L. (1999). What happened in the Medithermal? In C. Beck (Ed.), *Models for the Millenium: Great Basin Anthropology Today* (pp. 62–74). Salt Lake City: University of Utah Press.
Bettinger, R.L. (2003). Comment on "Cultural Diversification and Decimation in the Prehistoric Record" by W.C. Prentiss and J.C. Chatters. *Current Anthropology* 44, 33–58.
Bettinger, R.L. and Baumhoff, M.A. (1982). The Numic Spread: Great Basin Cultures in Competition. *American Antiquity* 47, 485–503.
Binford, L.R. (1980). Willow Smoke and Dog's Tails: Hunter-Gatherer Settlement Systems and Archaeological Site Formation. *American Antiquity* 45, 4–20.
Binford, L.R. (2001). *Constructing Frames of Reference*. Berkeley: University of California Press.
Blake, M. (2004). Fraser Valley Trade and Prestige as Seen at Scowlitz. In W.C. Prentiss and I. Kuijt (Eds.), *Complex Hunter-Gatherers: Evolution and Organization of Prehistoric Communities on the Plateau of Northwestern North America* (pp. 103–114). Salt Lake City: University of Utah Press.
Boyd, R., Mulder, M.B., Durham, W.H., and Richerson, P.J. (1997). Are Cultural Phylogenies Possible? In P. Weingart, S.D. Mitchell, P.J. Richerson, and S. Maasen (Eds.), *Human by Nature: Between Biology and the Social Sciences* (pp. 355–384). Mahwah: Lawrence Erlbaum Associates, Publishers.
Boyd, R. and Richerson, P.J. (1992). How Microevolutionary Processes Give Rise to History. In M. Nitecki and D.V. Nitecki (Eds.), *History and Evolution* (pp. 179–209). New York: State University of New York Press.
Burley, D.V. (1980). *Marpole: Anthropological Reconstructions of a Prehistoric Northwest Coast Culture Type*. Burnaby: Department of Archaeology Simon Fraser University Publication Number 8.
Cannon, A. (1991). *The Economic Prehistory of Namu*. Burnaby: Archaeology Press.
Cannon, A. (2002). Sacred Power and Seasonal Settlement on the Central Northwest Coast. In B. Fitzhugh and J. Habu (Eds.), *Beyond Foraging and Collecting: Evolutionary Change in Hunter-Gatherer Settlement Systems* (pp. 311–338). New York: Kluwer Academic/Plenum Publishers.
Cannon, A. and Yang, D.Y. (2006). Early Storage and Sedentism on the Pacific Northwest Coast: Ancient DNA Analysis of Salmon Remains from Namu, British Columbia. *American Antiquity* 71, 123–140.
Carlson, R.L. (1996). Introduction to Early Human Occupation in British Columbia. In R.L. Carlson and L. Dalla Bona (Eds.), *Early Human Occupation in British Columbia* (pp. 3–10). Vancouver: University of British Columbia Press.
Chatters, J.C. (1987). Hunter-Gatherer Adaptations and Assemblage Structure. *Journal of Anthropological Archaeology* 6, 336–375.

Chatters, J.C. (1995). Population Growth, Climatic Cooling, and the Development of Collector Strategies On the Southern Plateau, Western North America. *Journal of World Prehistory* 9, 341–400.
Chatters, J.C., Butler, V.L., Scott, M.J., Anderson, D.M., and Neitzel, D.A. (1995). A Paleoscience Approach to Estimating the Effects of Climatic Warming on Salmonid Fisheries of the Columbia Basin. *Canadian Special Publication in Fisheries and Aquatic Sciences* 21, 489–496.
Chatters, J.C., Hackenberger, S. and Lenz, B. (2006). *From PaleoIndian to Archaic in the Pacific Northwest: Transition or Replacement?* Paper presented at the 2006 Northwest Anthropological Conference, Seattle, WA.
Chatters, J.C. and Prentiss, W.C. (2005). A Darwinian Macroevolutionary Perspective on the Development of Hunter-Gatherer Systems in Northwestern North America. *World Archaeology* 37, 46–65.
Christensen, T. and Stafford, J. (2004). Raised Beach Archaeology in Northern Haida Gwaii: Preliminary Results from the Cohoe Creek Site. In D.W. Fedje and R.W. Mathewes (Eds.), *Haida Gwaii: Human History and Environment from the Time of the Loon to the Time of the Iron People* (pp. 245–273). Vancouver: University of British Columbia Press.
Coyne, J.A., Barton, N.H. and Turrelli, M. (1997). A Critique of Sewall Wright's Shifting Balance Theory of Evolution. *Evolution* 51, 643–993.
Damuth, J. (1985). Selection Among Species: A Formulation in Terms of Natural Functional Units. *Evolution* 39, 1132–1146.
Darwin, C. (1859). *On the Origin of Species by Means of Natural Selection or Preservation of Favored Races in the Struggle for Life*. London: Murray.
Dawkins, R. (1990) *The Extended Phenotype: The Long Reach of the Gene* (new ed.). Oxford: Oxford University Press.
Dover, G. (2000). Anti-Dawkins. In H. Rose and S. Rose (Eds.), *Alas Poor Darwin* (pp. 55–77). New York: Harmony Books.
Eldredge, N. (1995). *Reinventing Darwin: The Great Debate at the High Table of Evolutionary Theory*. New York: John Wiley.
Eldredge, N. (1989). *Macroevolutionary Dynamics: Species, Niches, and Adaptive Peaks*. New York: McGraw Hill Publishing Company.
Eldredge, N. (1999). *The Pattern of Evolution*. New York: W.H. Freeman and Company.
Eldredge, N. (2000). Biological and Material Cultural Evolution: Are There any True Parallels? *Perspectives in Ethology* 13, 113–153.
Fedje, D.W., Mackie, Q., Dixon, E.J., and Heaton, T. (2004). Late Wisconsin Environments and Archaeological Visibility on the Northern Northwest Coast. In D.B. Madsen (Ed.), *Entering America: Northeast Asia and Beringia Before the Last Glacial Maximum* (pp. 97–138). Salt Lake City: University of Utah Press.
Fisher, R.A. (1958). *The Genetical Theory of Natural Selection*, Revised Edition. New York: Dover.
Fladmark, K.R. (1970). Preliminary Report on the Archaeology of the Queen Charlotte Islands. *B.C. Studies* Special Issue 6–7, 18–45.
Fladmark, K.R., Ames, K. and Sutherland, P. (1990). Prehistory of the Northern Coast of British Columbia. In W. Suttles (Ed.), *Handbook of North American Indians, Volume 7: Northwest Coast* (pp. 229–239). Washington: Smithsonian Institution.
Frison, G.C. (2004). *Survival By Hunting: Prehistoric Human Predators and Animal Prey*. Berkeley: University of California Press.
Goodale, N.B., Kuijt, I. and Prentiss, A.M. (2008). Demography of Prehistoric Fishing-Hunting People: A Case Study of the Upper Columbia Area. In J.-P. Bocquet-Apel (Ed.), *Recent Advances in Paleodemography: Data, Techniques, and Patterns* (pp. 179–207). New York: Springer-Verlag.
Gould, S.J. (2002). *The Structure of Evolutionary Theory*. Cambridge: Belknap Press.
Gould, S.J. and Lewontin, R.C. (1979). The Spandrels of San Marco and the Panglossian Paradigm: A Critique of the Panglossian Programme. *Proceedings of the Royal Society of London B* 205, 581–598.

Haley, S. (1995). The Pasika Complex Revisited. In R.L. Carlson and L. Dalla Bona (Eds.), *Early Human Occupation in British Columbia* (pp. 51–64). Vancouver: University of British Columbia Press.

Hallett, D.J., Lepofsky, D.S., Mathewes, R.W., and Lertzman, K.P. (2003). 11,000 years of Fire History and Climate in the Mountain Hemlock Rainforests of Southwestern British Columbia based on Sedimentary Charcoal. *Canadian Journal of Forest Research* 33, 292–312.

Ham, L.C. (1990). The Cohoe Creek Site: A Late Moresby Tradition Shell Midden. *Canadian Journal of Archaeology* 14, 199–221.

Hayden, B. (1995). Pathways to Power: Principles for Creating Socioeconomic Inequalities. In T.D. Price and G.M. Feinman (Eds.), *Foundations of Social Inequality* (pp. 15–78). New York: Plenum Press.

Hayden, B. (1997). *The Pithouses of Keatley Creek*. New York: Harcourt-Brace.

Hayden, B., Bakewell, E. and Gargett, R. (1996). The World's Longest Lived Corporate Group: Lithic Analysis Reveals Prehistoric Social Organization Near Lillooet, British Columbia. *American Antiquity* 61, 341–356.

Haynes, G. (2002). *The Early Settlement of North America*. Cambridge: Cambridge University Press.

Holden, C.J. and Shennan, S. (2005). Introduction to Part 1: How Tree-like is Cultural Evolution? In R. Mace, C.J. Holden, and S. Shennan (Eds.), *TheEvolution of Cultural Diversity: A Phylogenetic Approach* (pp. 11–29). London: UCL Press

Imamoto, S. (1976). Analysis of the Glenrose Faunal Remains. In R.G. Matson (Ed.), *The Glenrose Cannery Site* (pp. 21–41). Ottawa: National Museum of Man, Archaeological Survey of Canada, Mercury Series No. 52.

Ingold, T. (1990). An Anthropologist Looks at Biology. *Man* 25, 208–229.

Joshi, A. (1999). The Shifting Balance Theory of Evolution. *Resonance* December, 4(12), 66–75.

Kelly, R.L. and Todd, L.C. (1988). Coming into the Country: Early Paleoindian Hunting and Mobility. *American Antiquity* 53, 231–245.

Kew, M. (1992). Salmon Availability, Technology, and Cultural Adaptation in the Fraser River Watershed. In B. Hayden (Ed.), *A Complex Culture of the British Columbia Plateau* (pp. 177–221). Vancouver: University of British Columbia Press.

Kosse, K. (1994). The Evolution of Large Complex Groups: A Hypothesis. *Journal of Anthropological Archaeology* 13, 35–50.

Laland, K.N., Odling-Smee, J., and Feldman, M.W., (2000). Niche Construction, Biological Evolution and Cultural Change. *Behavioral and Brain Sciences* 23, 131–175.

Lenert, M.P. (2007). Coast Salish Household and Community Organization at *Sxwoxwiymelh*: An Ancient Sto:lo Village in the Upper Fraser Valley, British Columbia. Ph.D. dissertation, Department of Anthropology, University of California, Los Angeles.

Lepofsky, D., Blake, M., Brown, D., Morrison, S., Oakes, N., and Lyons, N. (2000). The Archaeology of the Scowlitz Site, sw British Columbia. *Journal of Field Archaeology* 27, 391–416.

Lepofsky, D., Lertzman, K., Hallett, D., and Mathewes, R. (2005). Climate Change and Culture Change on the Southern Coast of British Columbia 2400–1200 B.P. *American Antiquity* 70, 267–294.

Lloyd, E.A. (1988). *The Structure and Confirmation of Evolutionary Theory*. New York: Greenwood Press.

Lloyd, E.A. and Gould, S.J. (2003). Species Selection on Variability. *Proceedings of the National Academy of Science USA* 90, 595–599.

Mackie, Q. and Acheson, S. (2007). The Graham Tradition. In D.W. Fedje and R.W. Mathewes (Eds.), *Haida Gwaii: Human History and Environment from the Time of the Loon to the Time of the Iron People* (pp. 274–302). Vancouver: University of British Columbia Press.

Magne, M.P.R. (1996). Comparative Analysis of Microblade Cores from Haida Gwaii. In R.L. Carlson and L. Dalla Bona (Eds.), *Early Human Occupation in British Columbia* (pp. 151–158). Vancouver: University of British Columbia Press.

Maschner, H.D.G. and Reedy-Maschner, K.L. (1998). Raid, Retreat, Defend (Repeat): The Archaeology and Ethnohistory of Warfare on the North Pacific Rim. *Journal of Anthropological Archaeology* 17, 19–51.

Mason, O.K. (1998). The Contest Between the Ipiutak, Old Bering Sea, and Birnirk Polities and the Origin of Whaling During the First Millennium A.D. Along Bering Strait. *Journal of Anthropological Archaeology* 17, 240-325.

Matson, R.G. (1983). Intensification and the Development of Cultural Complexity: The Northwest versus Northeast Coast. In R. Nash (Ed.), *The Evolution of Maritime Cultures on the Northeast and Northwest Coasts* (pp. 125–148). Burnaby: Simon Fraser University Department of Archaeology Publication No. 11.

Matson, R.G. (1996). The Old Cordilleran Component at the Glenrose Cannery Site. In R.L. Carlson and L. Dalla Bona (Eds.), *Early Human Occupation in British Columbia* (pp. 111–122). Vancouver: University of British Columbia Press.

Matson, R.G. and Coupland, G. (1995). *The Prehistory of the Northwest Coast*. San Diego: Academic Press.

Mitchell, D. and Pokotylo, D.L. (1996). Early Period Components at the Milliken Site. In R.L. Carlson and L. Dalla Bona (Eds.), *Early Human Occupation in British Columbia* (pp. 65–82). Vancouver: University of British Columbia Press.

McGhee, G.R. Jr. (1999). *Theoretical Morphology: The Concept and Its Applications*. New York: Columbia University Press.

O'Brien, M.J. and Holland, T.D. (1992). Variation, Selection, and the Archaeological record. In M.B. Schiffer (Ed.), *Archaeological Method and Theory*, Vol. 2 (pp. 31–79). Tucson: University of Arizona Press.

O'Brien, M.J. and Lyman, R.L. (2000). *Applying Evolutionary Archaeology*. New York: Kluwer Academic/Plenum.

O'Brien, M.J. and Lyman, R.L. (2003). *Cladistics and Archaeology*. Salt lake City: University of Utah Press.

O'Brien, M.J., Lyman, R.L., Collard, M., Holden, C.J., Gray, R.D. and Shennan, S.J. (2008). Transmission, Phylogenetics, and the Evolution of Cultural Diversity. In M.J. O'Brien (Ed.), *Cultural Transmission and Archaeology: Issues and Case Studies* (pp. 39–58). Washington, DC: The SAA Press.

Prentiss, A.M. (2010). Social Histories of Complex Hunter-Gatherers: Pacific Northwest Prehistory in a Macroevolutionary Framework. In K.E. Sassaman and D.H. Holly Jr. (Eds.), *Hunter-Gatherer Archaeology as Historical Process*. Arizona: University of Arizona Press (in review).

Prentiss, A.M., Lyons, N., Harris, L.E., Burns, M.R.P., and Godin, T.M. (2007). The Emergence of Status Inequality in Intermediate Scale Societies: A Demographic and Socio-Economic History of the Keatley Creek Site, British Columbia. *Journal of Anthropological Archaeology* 26, 299–327.

Prentiss, W.C. and Chatters, J.C. (2003a). Cultural Diversification and Decimation in the Prehistoric Record. *Current Anthropology* 44, 33–58.

Prentiss, W.C. and Chatters, J.C. (2003b). The Evolution of Collector Systems in the Pacific Northwest Region of North America. *Senri Ethnological Studies* 63, 49–82.

Prentiss, A.M. and Clarke, D.S. (2008). Lithic technological organization in an evolutionary framework: Examples from North America's Pacific Northwest region. In W. Andrefsky (Ed.), *Lithic Technology: Measures of Production, Use and Curation* (pp. 257–285). Cambridge: Cambridge University Press.

Prentiss, W.C., Chatters, J.C., Lenert, M., Clarke, D. and O'Boyle, R. (2005b). The Archaeology of the Plateau of Northwestern North America During the Late Prehistoric Period (3500–200 B.P.): Evolution of Hunting and Gathering Societies. *Journal of World Prehistory* 19, 47–118.

Prentiss, W.C., Lenert, M., Foor, T.A., Goodale, N.B. and Schlegel, T. (2003). Calibrated Radiocarbon Dating at Keatley Creek: The Chronology of Occupation at a Complex Hunter-Gatherer Village. *American Antiquity* 68, 719–736.

Prentiss, W.C., Lenert, M., Foor, T.A., and Goodale, N.B. (2005a). The Emergence of Complex Hunter-Gatherers on the Canadian Plateau: A Response to Hayden. *American Antiquity* 70, 175–180.

Romanoff, S (1992). The Cultural Ecology of Hunting and Potlatches among the Lillooet Indians. In B. Hayden (Ed.), *A Complex Culture of the British Columbia Plateau* (pp. 470–505). Vancouver: University of British Columbia Press.

Roney, J.R. and Hard, R.J. (2002). Early Agriculture in Northwestern Chihuahua. In S. Schlanger (ed.), *Traditions, Transitions and Technologies: Themes in Southwestern Archaeology; Proceedings of the 2000 Southwest Symposium* (pp. 160–177). Boulder: University Press of Colorado.

Severs, P.D.S. (1974). Archaeological Investigations at Blue Jackets Creek, FlUa-4, Queen Charlotte Islands, British Columbia, 1973. *Canadian Archaeological Association Bulletin* 6:163–205.

Shennan, S. (2002). *Genes, Memes and Human History: Darwinian Archaeology and Cultural Evolution*. New York: Thames and Hudson.

Shott, M. (2008). Darwinian Evolutionary Theory and Lithic Analysis. In M.J. O'Brien (ed.), *Cultural Transmission and Archaeology: Issues and Case Studies* (pp. 146–157). Washington D.C.: SAA Press.

Stryd, A.R. (1973). The Later Prehistory of the Lillooet Area, British Columbia. Ph.D dissertation, Department of Archaeology, University of Calgary, Calgary.

Teit, J. (1906). The Lillooet Indians. *Memoirs of the American Museum of Natural History, Jesup North Pacific Expedition* 2, 193–300.

Tunnicliffe, V., O'Connell, J.M., McQuoid, M.R. (2001). A Holocene Record of Marine Fish Remains from the Northeastern Pacific. *Marine Geology* 174, 197–210.

Vrba, E.S. (1983). Macroevolutionary Trends: New Perspectives on the Roles of Adaptation and Incidental Effect. *Science* 221, 387–389.

Vrba, E.S. (1989). Levels of Selection and Sorting with Special Reference to the Species Level. *Oxford Surveys in Evolutionary Biology* 6, 111–168.

Vrba, E.S. and Eldredge, N. (1984). Individuals, Hierarchies, and Processes: Towards a More Complete Evolutionary Theory. *Paleobiology* 10, 146–171.

Wade, M.J. and Goodnight, C.J. (1998). Genetics and Adaptation in Metapopulations: When Nature Does Many Small Experiments. *Evolution* 52, 1537–1553.

Wright, S. (1931). Evolution in Mendelian Populations. *Genetics* 16, 97–159.

Wright, S. (1932). The Roles of Mutation, Inbreeding, Crossbreeding and Selection in Evolution. *Proceedings of the VI International Congress of Genetics* 1, 356–366.

Chapter 5
Testing the Morphogenesist Model of Primary State Formation: The Zapotec Case

Charles S. Spencer

Introduction

Archaeologists working in the Mexican state of Oaxaca have documented nearly ten millennia of prehistoric cultural evolution, a rich panorama comprising periods of stability or gradual change, as well as episodes of major cultural transformation, such as the shift from nomadic hunter-gathering bands to sedentary agricultural villages around 1700–1500 B.C., the transition from egalitarian to ranked societies or chiefdoms around 1200 B.C., and the emergence of the first state society around 300 B.C. (Blanton 1978; Blanton et al. 1999; Flannery and Marcus 1983a; Marcus and Flannery 1996). The focus of my efforts here will be on the last of these, namely, the rise of the early Zapotec state with its capital at Monte Albán in the Valley of Oaxaca (Fig. 5.1). Current evidence indicates that the Monte Albán state was the first to evolve in ancient Mesoamerica and thus can be considered a first-generation or primary state (Spencer and Redmond 2004a). A primary state is one that evolves in a context of nonstate polities, without precedent and with no connections to other preexisting state systems (Service 1975; Spencer 1998; Wright 1986). A primary state emerges through innovation, not emulation. By contrast, secondary states are those that arise in response to the appearance of a primary state; they adopt state institutions by emulating the primary state, and they sometimes manage to maintain their political independence from the primary state, at least for a while (Price 1978). When the Monte Albán state arose, it quickly annexed several surrounding regions, but other areas remained autonomous for varying lengths of time, often developing into secondary states (Balkansky 1998; Spencer and Redmond 2004a). There have been relatively few cases of true primary state formation in our planet's entire history—perhaps no more than six, as suggested by Service (1975). Each of these cases, once it is well documented, becomes a valuable source of data that we can use to assess theoretical perspectives and models concerning the cultural evolution of the state. In this chapter, I use the empirical record of the rise of the

C.S. Spencer (✉)
Division of Anthropology, American Museum of Natural History, New York, NY 10024, USA
e-mail: cspencer@amnh.org

A.M. Prentiss et al. (eds.), *Macroevolution in Human Prehistory*,
DOI 10.1007/978-1-4419-0682-3_6, © Springer Science+Business Media, LLC 2009

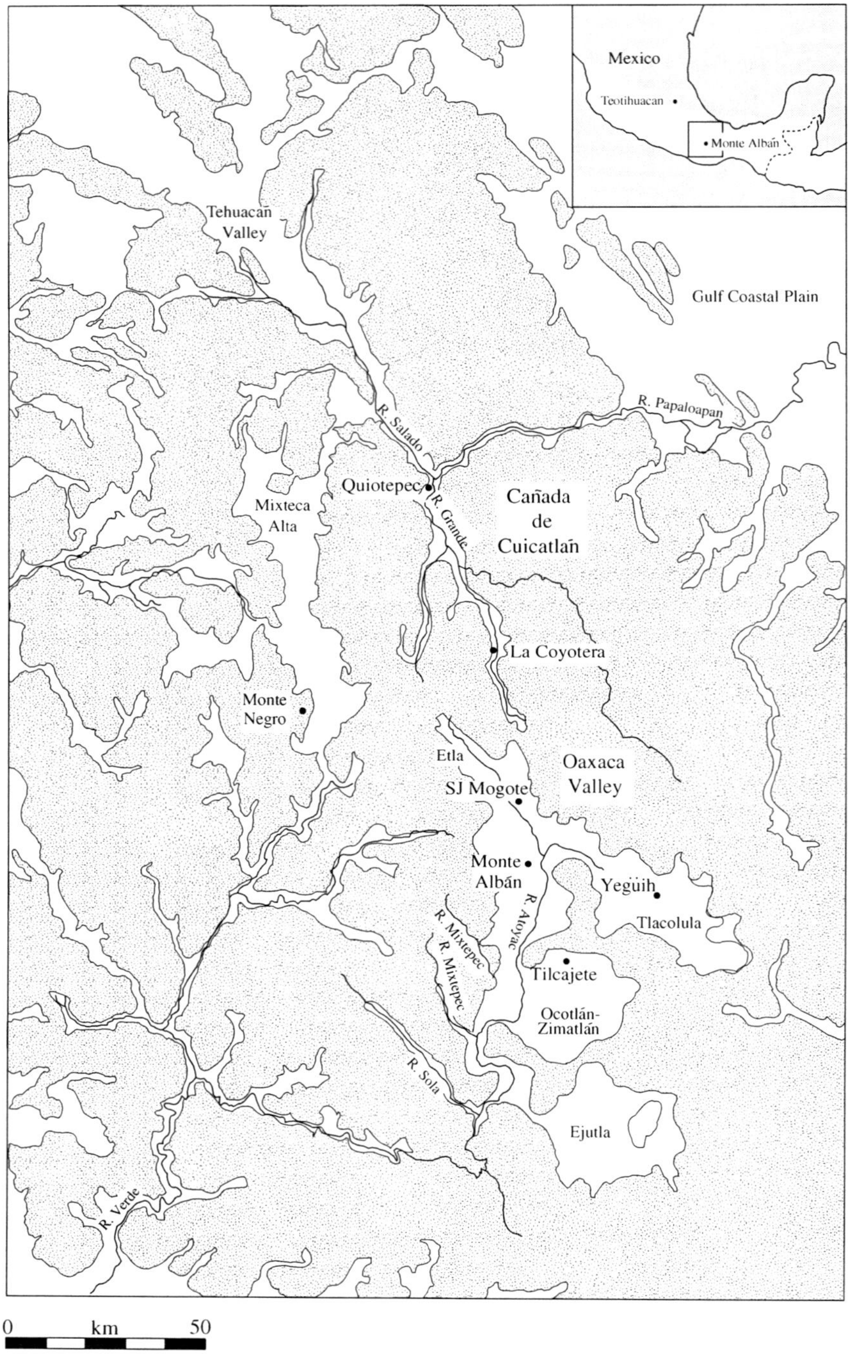

Fig. 5.1 The Valley of Oaxaca, its major subregions, and other surrounding regions, with key archaeological sites indicated

Monte Albán state to examine the validity of what I shall call the *morphogenesist* model of primary state formation.

A Model of Primary State Formation

Anthropological views of state formation are numerous and diverse, but a fundamental distinction among them is whether they see the development of states from simpler societies as a continuous process of quantitative, incremental change or as a transformational process that involves both quantitative and qualitative change. Renfrew (1979) labeled the latter kind of change *morphogenesis*; sociopolitical morphogenesis entails not just an increase in what was already present in precursor societies (i.e., more people, resources, power) but also qualitatively different, *emergent* properties in the form of new rules, new principles, new strategies, and new applications (see also Rosen 1982). This concern with emergent properties makes it reasonable to view the morphogenesist model as an expression of macroevolutionary culture theory (see Prentiss et al., this volume). When applied to the research problem of primary state formation, the two key postulates of a morphogenesist model are as follows: (1) states and the precursor societies known as chiefdoms are both quantitatively and qualitatively distinct cultural developments, with states being not only larger than chiefdoms but also characterized by different administrative principles and regulatory strategies; and (2) the transition from chiefdom to primary state involves an organizational change that is not just incremental and gradual but rather precedent-breaking and transformational (Flannery 1972, 1995; Spencer 1990, 1998; Wright 1977).

In an influential review, Wright (1977) defined the chiefdom as a society with a centralized but not internally specialized administration: "A chiefdom can be recognized as a cultural development whose central decision-making activity is differentiated from, though it ultimately regulates, decision-making regarding local production and local social processes; but it is not itself internally differentiated. It is thus externally but not internally specialized" (Wright 1977:381). He drew a sharp contrast between the chiefdom and the state, which he defined as a society with a centralized and also internally specialized administration: "A state can be recognized as a cultural development with a decision-making process which is both externally specialized with regard to the local processes which it regulates, and internally specialized in that the central process is divisible into separate activities which can be performed at different places at different times" (Wright 1977:383). Not all chiefdoms evolve into states, but it has been argued that all primary states evolved from preexisting chiefdoms (Carneiro 1981; Earle 1987).

A state administration, from this perspective, is inherently bureaucratic. Wright (1977) proposed that states have minimally four tiers of decision-making, three of them above the village level. Because internal administrative specialization is compatible with the effective delegation of partial authority, subsidiary centers of administration can be established at strategic locations within the state's domain, enabling the state to intervene directly in local affairs. Such subsidiary outposts may be distributed in a nested lattice of secondary, tertiary, and even quaternary centers. Population distribution tends to track this administrative lattice, with the result that well-functioning states usually exhibit a four-level hierarchy of settlements according to both administrative functions and population size, while chiefdoms tend to exhibit no more than three levels. Such hierarchies can be detected in the archaeological record through the analysis of site-size distributions (Wright and Johnson 1975). Flannery (1998), Flannery and Marcus (1983b), and Sanders (1974), among others, have suggested that palaces and standardized multiroom temples are also features of state organization that can be detected archaeologically. Palaces in states often have a residential component, where the ruler and his family live, as well as a governmental component, where the ruler carries out various official duties. Standardized, multiroom temples are thought to be manifestations of a state religion staffed by a full-time specialized priesthood.

The morphogenesist model views chiefdoms and states as dynamic entities. They are both capable of exhibiting considerable variation in population size and political centralization (i.e., the concentration of power and resources in elite hands), while nonetheless remaining within the administrative framework characteristic of each. Yet, such variation is not unfettered, because the different administrative principles that define chiefdoms and states are consistent with different optimal regulatory strategies (Wright 1977). Lacking an internally differentiated administration, a chief cannot effectively delegate partial authority; any such delegation amounts to total delegation and leaves the chief vulnerable to insurrection or fissioning off (Wright 1977:381). The optimal strategy for a chief is to avoid the delegation of authority, which means that he has to administer his domain from the first-order center. Consequently, there is a spatial limit to the territory size that a chief can effectively rule; as a rough estimate, I have suggested that this limit may lie in the vicinity of one-half day of travel from the chiefly center, a radius of about 28 km assuming a walking speed of 5.6 km per hour for a 10-hour day of travel (Spencer 1990:7). A territory of this size would allow the chief to get from his capital to the edge of his territory and back in one day. If it becomes necessary to send an assistant out to some part of his realm (to collect mobilized surplus, for instance), the assistant could make a round trip in one day. Such a strategy would be an effective way of temporarily delegating authority in a system that has minimal internal differentiation of central leadership roles.

The optimal regulatory strategy of a state leader is nearly the opposite of a chief's (Wright 1977). A clever state ruler will divide and segment central authority as much as possible, so that his administrative underlings possess only narrowly defined parcels of central authority, weakening their ability to foment insurrection. The capacity to delegate partial authority allows a state ruler to intervene directly into local affairs by dispatching specialized administrative underlings throughout the domain. The state's ability to delegate partial authority is compatible with ambitious strategies of territorial expansion, including the conquest and long-term holding of distant territories, which may be evidenced by burned and abandoned villages, specialized forts and administrative outposts established by the conquering state, and enforced changes in the economic, social, and religious behaviors of subjugated peoples. Some researchers (e.g., Algaze 1993; Carneiro 1970; Marcus 1992; Webster 1975) have argued that predatory expansion plays a central role in the rise of primary states in the first place.

The state's ability to delegate partial authority means that a state can potentially handle local stresses and imbalances more effectively than a chiefdom can. At the same time, a state administration is more expensive to operate, and these extra costs are surely a major factor inhibiting the transformation of chiefdoms into states. The morphogenesist model therefore expects that a successful transition from chiefdom to state will require a substantial increase in the evolving system's capacity to mobilize resources. Previously uncontrolled or exogenous resources can be especially helpful in this regard and might be mobilized through aggressive appropriation (Spencer 1998; Webster 1975; Wright 1977).

The morphogenesist model is not embraced by all researchers and has come under sharp attack by those who prefer what we might call the *continualist* or *gradualist* model of state formation. These critics insist that the morphogenesist model is invalid because it seeks to impose a stepwise, typological scheme on a process of cultural change that is so continuous and gradual that the distinction between a chiefdom and a primary state is arbitrary, little more than a creation of the analyst (Blanton et al. 1996; Feinman and Neitzel 1984; Kottak 1977; Lewis 1968; McGuire 1983; Plog 1974; Upham 1987; Yoffee 1979, 1993). The gradualist assumption underlying this critique has been clearly expressed by Lewis: "evolution generally proceeds through a continuous process of development rather than through a succession of discontinuous movements from stage to stage" (Lewis 1968:103). In a similar vein, Blanton et al. (1996:1) criticize the "static societal types" of a "flawed neoevolutionism," arguing that "researchers should abandon its static ideal-type stages and instead investigate the varying strategies used by political actors to construct and maintain polities and other sociocultural institutions." They go on to highlight two such strategies, the "network strategy" (individual-centered) and the "corporate strategy" (group-oriented), which they believe represent two basic patterns of political action that occur in societies of widely varying complexity (Blanton et al. 1996:2–6). McGuire (1983) also rejects the use of political types such as chiefdoms and states, calling instead for the documentation of continuous variability across the analytical dimensions of "inequality" and "heterogeneity."

I suggest that the continualist critique can be most effectively addressed by *testing* the morphogenesist model with data on the transition from precursor chiefdom to primary state. Such data will inevitably be archaeological in character, since all known cases of primary state formation occurred before detailed written records were kept (including the Zapotec case used here). Some features of state organization that can be identified in the archaeological record include the following: (1) a four-tier settlement hierarchy according to site size, (2) the presence of specialized palaces and temples, and (3) the aggressive annexation of distant territories, i.e., those lying well beyond a half-day of travel from the capital (Flannery 1998; Spencer and Redmond 2004a; Wright and Johnson 1975). A key expectation of the morphogenesist model would be dramatic change in all these features over a relatively short period of time. For example, we might expect to observe a precipitous increase in the territory dominated by the political capital as state organization emerges, and such territorial expansion should be concurrent with the abrupt appearance of a four-tier settlement hierarchy and specialized palaces and temples.

Another key expectation relates to the morphogenesist model's postulate that chiefdoms and states are characterized by different optimal regulatory strategies that derive from the different administrative principles that define them, so that the transition from chiefdom to state involves a shift from one strategic optimum to another. To construct a test of this proposition, let us invoke an analogy with macroevolutionary biology and in particular with Sewall Wright's (1931) "shifting balance" theory, according to which major evolutionary transformations entail a shift from one peak to another on an adaptive landscape—in Eldredge's (1989:18) words, "the most important metaphor in macroevolutionary theory of the past 50 years." Although Wright's original formulation had a genetic focus and was concerned with how more favorable allelic combinations could be seen as peaks (and less favorable combinations as valleys) on such a landscape, later authors expanded his model to include phenotypic variants, embracing populations and even whole species (Dobzhansky 1937; Simpson 1944; Lande 1976). Especially notable was George G. Simpson's (1944) "quantum evolution" theory, which used a two-dimensional adaptive landscape metaphor to present potential combinations of two phenotypic characters, with the third dimension (elevation contours) representing population fitness or adaptedness (Arnold et al. 2001). This phenotypic version of the adaptive landscape model is thought to be consistent with the punctuated equilibria view of macroevolution: relative stasis results from variation and selection toward existing adaptive peaks, while macroevolutionary transformations come about through dramatic peak shifts (Eldredge 1989; Eldredge and Gould 1972; Lande 1986; Newman et al. 1985). Eldredge (1989:21) notes that the adaptive landscape model "afforded a means of handling the apparent suddenness of the appearance of taxa of relatively high rank in the fossil record. The process of switching between two adaptive peaks forms the very heart of George Gaylord Simpson's theory of quantum evolution." Others have emphasized that emergent properties can alter the shape of the adaptive landscape itself—for example, when an unprecedented adaptive strategy turns out to be highly successful, creating an entirely new adaptive peak (Arnold et al. 2001; Gould 1997).

Figure 5.2 presents a simplified model in which a dual-peaked adaptive landscape is given a sociopolitical interpretation, that of chiefdom (on the left) vs. state (on the right). In this tentative application, I suggest we interpret dimension X as polity size, expressed in days of round-trip travel outward from the capital, and dimension Y as administrative complexity, the proxy measure for which is the number of tiers in the settlement hierarchy of the core polity. The model recognizes that political strategies are potentially variable, but they are not all equally optimal and those strategies that adhere closely to the adaptive optima are more likely to persist than those that do not.

I make this application bearing in mind that Arnold et al. (2001) have argued that the adaptive landscape model provides a "conceptual bridge" between microevolution and macroevolution. Accordingly, the process of variation and selection around an existing adaptive peak would be viewed as microevolution, while a peak shift

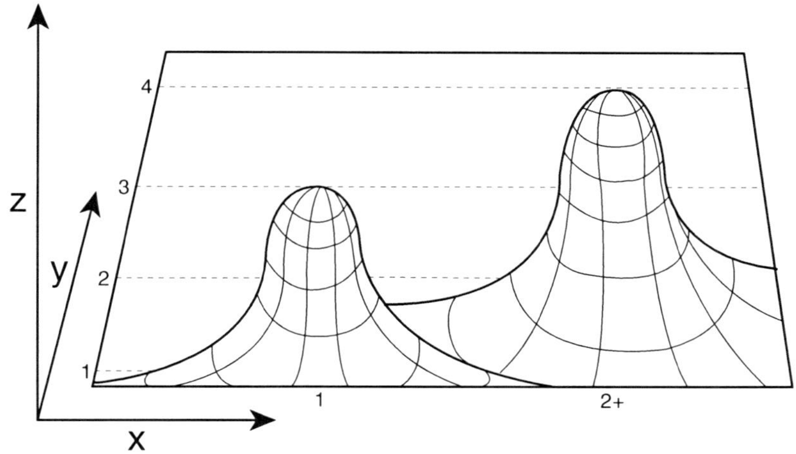

Fig. 5.2 A simple model of the evolutionary transition from chiefdom (*on the left*) to primary state (*on the right*) as depicted by a dual-peaked adaptive landscape. X = polity size, here expressed in days of round-trip pedestrian travel from the capital to the farthest edge; Y = administrative complexity, here expressed in terms of tiers in the polity's regional settlement hierarchy; Z = relative adaptedness

is more analogous to divergence among species, genera, or higher taxa and thus would be considered macroevolution (see Zeder, Chap. 6). From this perspective, a macroevolutionary change (in our sociopolitical application, the shift from chiefdom to primary state) should entail a move through an intermediate phase of relatively lower adaptedness. This condition will tend to be short-lived, since selection favors the peaks, but the adaptive valley must nevertheless be crossed. Empirically, if we can detect this intermediate phase, we might expect to find evidence of such reduced adaptedness in the society undergoing the shift. We might also expect to find evidence of some *novel* or *unusual* strategies that aim to counteract the effects of reduced adaptedness and help speed up the sociopolitical formation onward to the new adaptive peak. Indeed, we might expect such new or emergent strategies themselves to be implicated in the creation of the new peak (Arnold et al. 2001).

My invocation of shifting balance theory inevitably brings with it an obligation to state my assumptions about the relationship between biological and cultural evolution. At base, I assume that biological and cultural evolution can both be usefully seen as "descent with modification" mediated by selection. In both kinds of evolution, variability is generated by individuals and groups of individuals. Evolution entails the selective preservation of this variability over time and may involve increases or decreases in the size and complexity of the entity undergoing change; there is no necessary direction to the evolutionary process. I do not, however, favor a strict Darwinian or selectionist approach to evolutionary archaeology (Dunnell 1980; O'Brien and Lyman 2000; Teltser 1995). I support instead a processual evolutionism, one of whose key assumptions is that biological and cultural evolution are analogous, but not identical, processes (Spencer 1997). I agree with Eldredge (this volume) that we need to make significant adjustments when we attempt to apply evolutionary concepts to the study of culture change. For example, in cultural evolution major roles are played by directed variation, which is an expression of human agency or intentionality, as well as by the inheritance of acquired cultural traits, a consequence of the human capacity for enculturation (Boyd and Richerson 1985). It is through agency that humans launch war parties against other villages and regions, compete for wealth and power in their own political arenas, try to introduce irrigated agriculture into arid lands, and so forth. Whether such strategies succeed, however, is ultimately determined by selection, or "what worked better than what" (Eldredge 1995:34). Furthermore, in cultural evolution, selection is likely to operate at multiple levels—not only at the level of the individual but also at that of the family, the village, and even the state (Wilson and Sober 1994). In some situations (such as during wartime), selection at the level of a larger sociopolitical formation may overpower selection at the individual (and family) level, as people sacrifice their lives for the sake of their village, their chiefdom, or their state. The evolution of complex sociopolitical formations involves not only the emergence of more inclusive strategies of integration but also a corresponding expansion of the scale on which selection operates, determining whether the emergent properties survive to become permanent features in the new landscape.

At the same time, I must say that I see little point in pursuing the analogy between biological and cultural evolution (or, indeed, any analogy) for its own sake. Analogy, as a procedure, is justifiable to the extent that it has heuristic value: e.g., finding new ways of thinking about old problems, finding linkages between what had previously looked like unconnected factors or lines of evidence, finding novel ways to bring data to bear on ideas. In the present application, we can use the analogy with Wright's shifting balance theory and Simpson's quantum evolution theory to generate a specific test implication for the morphogenesist model of primary state formation that we otherwise might not have entertained: that the polity which evolved from precursor chiefdom to primary state had to move through an intervening period of relatively lower adaptedness, a valley between peaks on an adaptive landscape. Moreover, we can use the general connection between the morphogenesist model and macroevolutionary culture theory to view the emergent primary state as a newly created peak on an adaptive landscape and explore the risks, costs, and potential benefits entailed in the construction of that peak. With such considerations in mind, let us now turn to the archaeological record of Oaxaca.

Empirical Application

Just prior to the founding of Monte Albán, during the Rosario phase (700–500 B.C.), the three branches (or subvalleys) of the Oaxaca Valley were occupied by three independent chiefdoms, separated by a sparsely occupied buffer zone in the central area (Marcus and Flannery 1996:123–126, Fig. 128). San José Mogote had a population of about 1,000 and was the chief center of a polity that encompassed the Etla

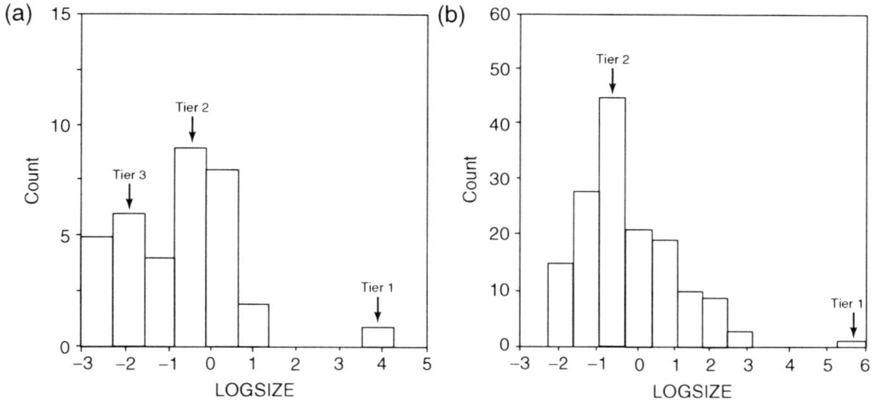

Fig. 5.3 (**a**) Rosario phase (700–500 B.C.) site size in the Etla subvalley, showing three modes of site size, indicating a three-tier settlement hierarchy; (**b**) Early Monte Albán I phase (500–300 B.C.) site size in the Etla-Central subvalley, showing two modes of site size, indicating a two-tier settlement hierarchy. LOGSIZE is the natural logarithm of site size in hectares; original data from Kowalewski et al. (1989:Appendix I)

subvalley (Marcus and Flannery 1996:125). Its rival in the Ocotlán-Zimatlán subvalley was the Rosario phase occupation at San Martín Tilcajete, while the site of Yegüih was the probable center of the contemporaneous chiefdom in the Tlacolula subvalley (Marcus and Flannery 1996:125–126). Each of these chiefdoms had no more than a three-tiered settlement hierarchy, as exemplified by the site-size histogram of the Etla subvalley, which shows at least two, and probably three, modes in its distribution (Fig. 5.3a). The political territory of the Etla chiefdom was limited by the aforementioned buffer zone to the south and the mountainous zone to the north, where Drennan (1989:374) found no evidence of Rosario phase occupation. Thus, the maximum territory size of this chiefdom could not have exceeded a distance of 20 km from the capital of San José Mogote (Fig. 5.4).

The center of Monte Albán was founded around 500 B.C., probably by the leaders of San José Mogote and their followers; San José Mogote was largely abandoned at this time (Marcus and Flannery 1996:139). An estimated 5,000 people lived at Monte Albán during the Early Monte Albán I phase (500–300 B.C.) (Kowalewski et al. 1989: Table 5.9). Only two institutional buildings can be securely dated to this phase. One of them, the Danzantes building, is associated with numerous carvings of sacrificial victims, many of which are similar to the earlier (Rosario phase) Monument 3 at San José Mogote (Flannery and Marcus 1983b, c). It appears that the great increase in population between Rosario-phase San José Mogote and Monte Albán during the Early Monte Albán I phase was not accompanied by an increase in the number of public buildings at the latter site. Nor is there any evidence at either site of a palace or specialized multiroom temple at this time (Flannery and Marcus 1990). It appears that there was substantial growth in population between the Rosario phase and the Early Monte Albán I phase—as well as a shift in the

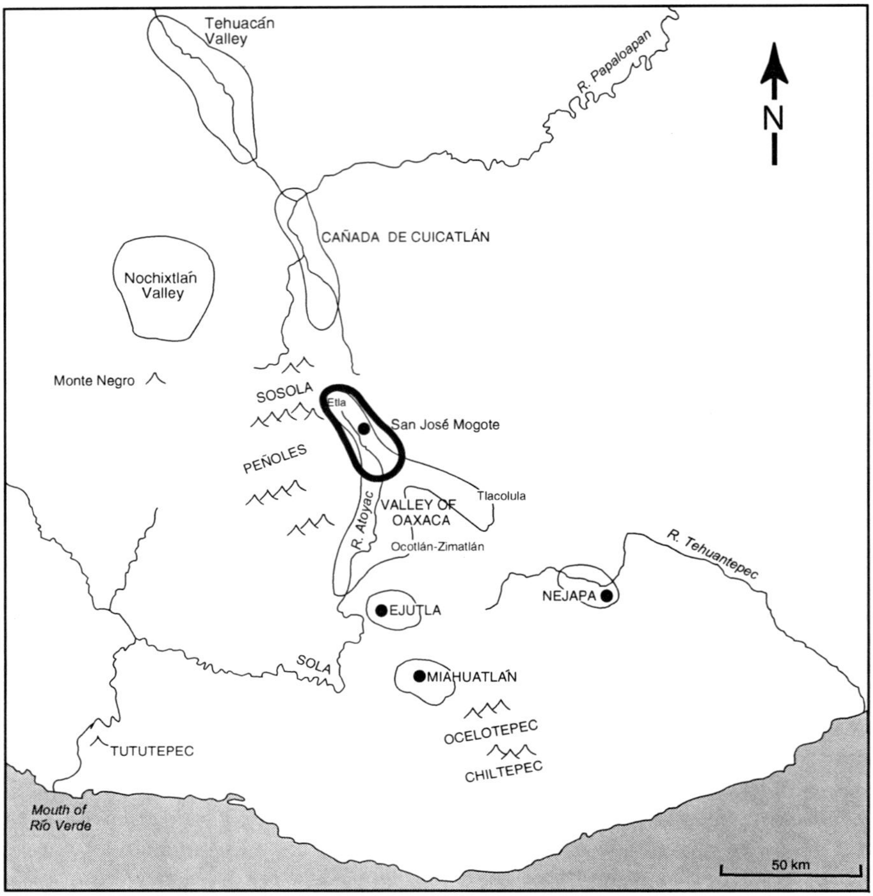

Fig. 5.4 Proposed maximum territory size of the Etla polity during the Rosario phase (700–500 B.C.)

location of the first-order center—but within the context of a chiefly political organization (Marcus and Flannery 1996:160–165; Spencer 1982:17–19).

How much territory fell within the Monte Albán polity during its first phase of occupation? Redmond and I have presented data in support of what we call the "Rival Polity Model" view of Early Monte Albán I political organization in the Oaxaca Valley, which sees Monte Albán's political domain *not* extending into the Tlacolula and Ocotlán-Zimatlán subvalleys (Spencer and Redmond 2001a, 2003). The territory controlled by Monte Albán at this time (Fig. 5.5) likely included the Central zone, the Etla subvalley, and a small portion of the mountainous area north of the Etla subvalley, where Drennan (1989:376) reported some Early Monte Albán I occupations that he interpreted as an "expansion of Elta Valley population northward

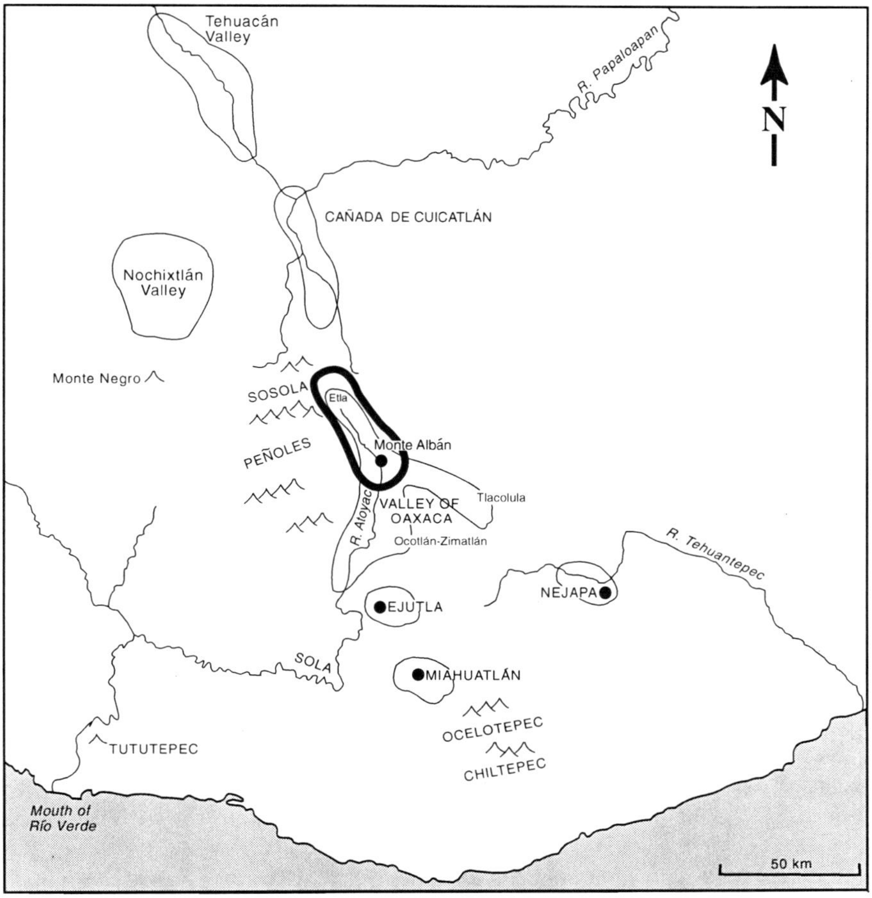

Fig. 5.5 Proposed maximum territory size of the Monte Albán polity during the Early Monte Albán I phase (500–300 B.C.)

into the mountains." Thus, the maximum political "reach" of Monte Albán may have extended some 40 km to the north, roughly a 14-hour round trip from the capital (at a walking speed of 5.6 km per hour); administering over such a distance may well have been difficult for a chiefly form of administration. On the other hand, the southern and eastern edges of Monte Albán's domain lay just 10–15 km from the capital, well within a day's round trip.

Chiefdom organization is indicated by the settlement patterns of Early Monte Albán I. A histogram of site size for sites in the Etla-Central area at this time shows two modes, reflecting the extreme concentration of population at Monte Albán itself at this time (Fig. 5.3b). The two-tiered settlement hierarchy revealed in this graph may represent a decline from the three tiers observed for the Etla subvalley in the

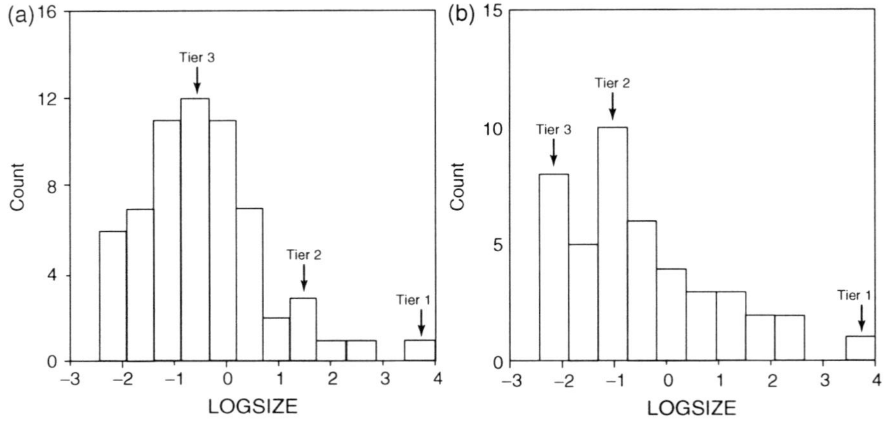

Fig. 5.6 Early Monte Albán I phase (500–300 B.C.) site size for the (**a**) Tlacolula subvalley and (**b**) Ocotlán-Zimatlán subvalley, both showing three modes of site size, indicating a three-tier settlement hierarchy. LOGSIZE is the natural logarithm of site size in hectares; original data from Kowalewski et al. (1989:Appendix I)

Rosario phase (Fig. 5.3a), although a measure of interpretive caution is certainly warranted by the inherent imprecision of site size estimates. That caveat notwithstanding, it appears that both Tlacolula and Ocotlán-Zimatlán show three-tier settlement hierarchies for the Early Monte Albán I phase (Fig. 5.6).

During the following phase, the Late Monte Albán I phase (300–100 B.C.), Monte Albán's estimated population more than tripled to just over 17,000 (Kowalewski et al. 1989: Table 6.4). The site-size histogram for the Etla-Central subvalley shows four modes, which probably reflects a four-tiered settlement hierarchy consistent with state organization (Fig. 5.7a). Although Classic Period construction activities have made it difficult for archaeologists to recover Formative Period buildings at Monte Albán, there are signs that the Late Monte Albán I phase saw an increase in the amount and variety of public buildings, perhaps evidence of the internally differentiated administration of the state. The Danzantes building apparently continued in use, and substantial institutional construction occurred at the North Platform (Flannery and Marcus 1983c). Evidence of a Monte Albán I public building was also recovered within System IV (Acosta 1965). A Late Monte Albán I public building (perhaps a palace) was found on a platform just north of System IV, according to Winter (1994). It is also possible that the first stage of Building J was built in the Late Monte Albán I phase (Caso 1938:11).

Building J has a series of some 40 inscribed slabs that Caso (1947) and Marcus (1976) have interpreted as referring to the conquests of outlying lands. Archaeological projects in the Cañada de Cuicatlán (Spencer and Redmond 1997, 2001b) and Sola (Balkansky 2002) have found evidence of annexation by Monte Albán during the Late Monte Albán I phase. Finsten (1996) also reported evidence of an intrusion into the northern part of her Peñoles study region by the Monte Albán state during

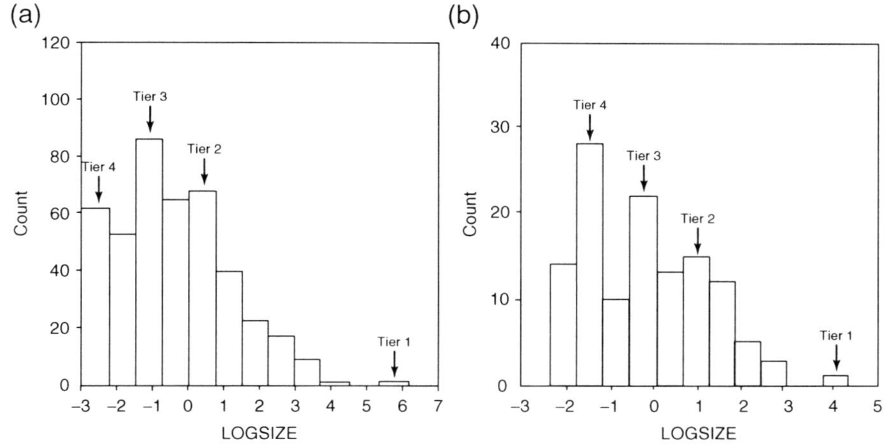

Fig. 5.7 Late Monte Albán I phase (300–100 B.C.) site size for the (**a**) Etla-Central subvalley and (**b**) Ocotlán-Zimatlán subvalley, both showing four modes of site size, indicating a four-tier settlement hierarchy. LOGSIZE is the natural logarithm of site size in hectares; original data from Kowalewski et al. (1989:Appendix I)

the Late Monte Albán I phase. In sum, I suggest that Monte Albán's political domain at this time (300–100 B.C.) comprised, at minimum, the Etla-Central subvalley, the Cañada, northern Peñoles, and Sola, a territory whose farthest boundaries reached as far as 100 km from the center (Fig. 5.8).

Although Monte Albán extended its political territory well beyond a day's round trip from the political center by 300–100 B.C., it is noteworthy that this expansion did not occur in an incremental, concentric fashion. Instead, we observe a nonuniform, strikingly asymmetric pattern of territorial growth. While such extra-valley regions as the Cañada, northern Peñoles, and Sola were annexed, the Ocotlán-Zimatlán and Tlacolula areas apparently resisted Monte Albán's aggressions at this time. Our recent investigations in the San Martín Tilcajete locality, seat of the capital of the Ocotlán-Zimatlán subvalley, have documented some of these strategies of resistance, including the shifting of the settlement's main plaza around 300 B.C. from the El Mogote site uphill to the more defensible El Palenque site, the construction of fortifications at El Palenque, a greater nucleation of population, and the development of secondary state political organization (Spencer and Redmond 2003). Among the manifestations of secondary state development is the four-tier settlement hierarchy that developed in the Ocotlán-Zimatlán subvalley during the Late Monte Albán I phase (Fig. 5.7b). There is also evidence that a minimally four-tier settlement hierarchy appeared in the Tlacolula subvalley at this time (Spencer and Redmond 2003:Fig. 15). At El Palenque, the presence of state institutions is supported by our excavation of a royal palace and a multiroom temple (Spencer and Redmond 2004b, 2005). Then, sometime in the first century B.C., in the early years of the Monte Albán II phase (100 B.C.–A.D. 200), the El Palenque site was burned and completely abandoned; a new site, Cerro Tilcajete, was founded on a nearby

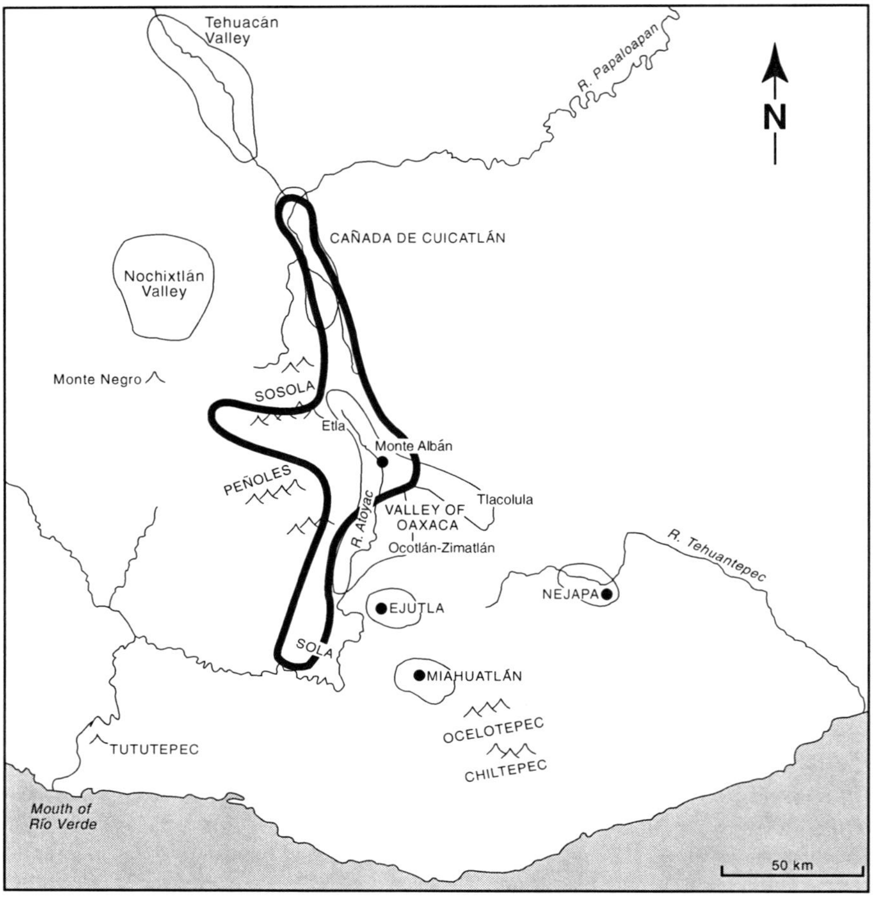

Fig. 5.8 Proposed maximum territory size of the Monte Albán polity during the Late Monte Albán I phase (300–100 B.C.)

ridge. Elson's (2003) research at Cerro Tilcajete revealed that it functioned as a secondary center of the Monte Albán state during the Monte Albán II phase, by which time Monte Albán had probably succeeded in annexing the Ocotlán-Zimatlán and Tlacolula subvalleys, as well as several other regions outside the Valley, bringing the territory dominated by the Monte Albán state to nearly 20,000 km^2 (Fig. 5.9) (Elson 2003; Marcus and Flannery 1996:206–207; Spencer and Redmond 2005).

As we review the entire sequence of development, the Early Monte Albán I phase is especially intriguing. At this time, a new center had been founded at Monte Albán; yet it does not show evidence of state organization. The founders of Monte Albán had clearly made a bold move, shifting the capital of the Etla-Central polity from San Jose Mogote squarely into what had previously been a political no-man's land

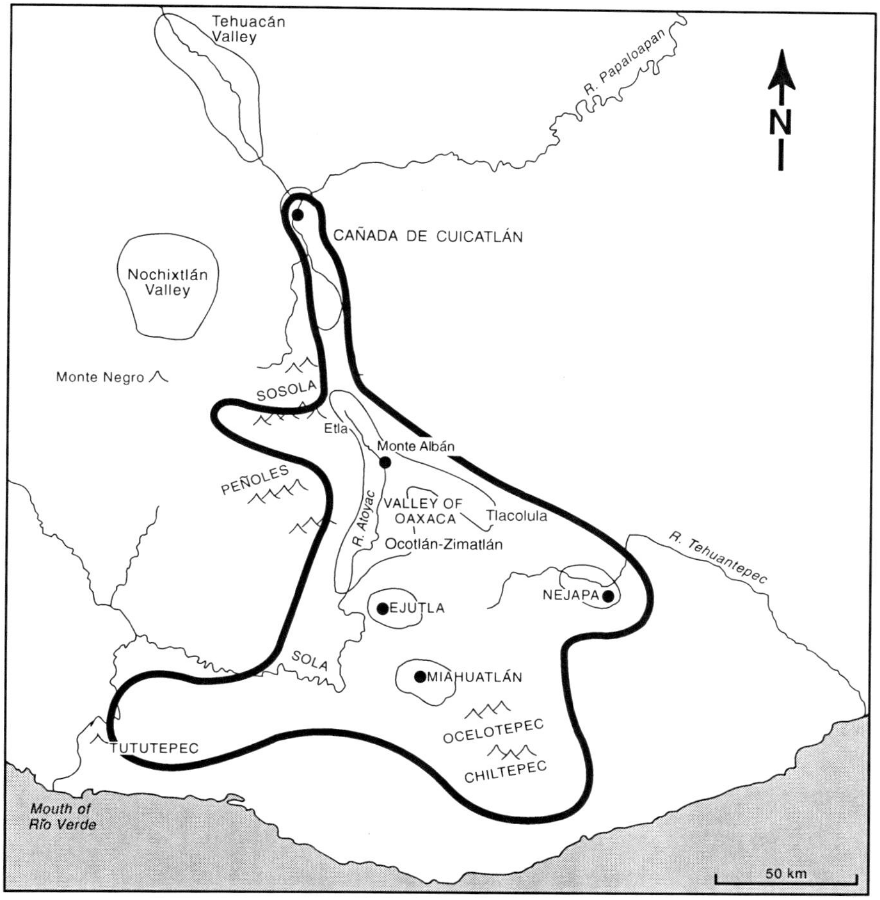

Fig. 5.9 Proposed maximum territory size of the Monte Albán polity during the Monte Albán II phase (100 B.C.–A.D. 200) (adapted from Marcus and Flannery 1996:Fig. 242)

between rival polities; moreover, this new center had an unprecedented concentration of population in a location not handy for good farmland. Although the new leaders of Monte Albán evidently failed in the Early Monte Albán I phase to annex the Tlacolula and Ocotlán-Zimatlán subvalleys, they achieved a modest expansion northward into the mountains.

Let us consider the Etla-Central polity's situation during Early Monte Albán I in terms of our adaptive landscape model (Fig. 5.2). While a settlement hierarchy of just two tiers can be identified for the Etla-Central polity, the "reach" of the polity was extended to some 40 km north of the capital of Monte Albán. This is somewhat greater than the 28-km radius that I have suggested (Spencer 1990:7) would be compatible with an administration lacking internal specialization and the capacity to delegate partial authority. Consequently, we might suspect that the Monte Albán polity was pursuing some less-than-optimal strategies during the Early Monte Albán I phase and was moving into the adaptive valley lying between the peaks, as shown in Fig. 5.2.

To test the adaptive valley expectation, we need some sort of relative measure of "adaptedness" that we can use to compare the Etla-Central polity to the other two polities, for the Early Monte Albán I phase as well as for earlier and later time periods. I suggest that one way to measure the overall "adaptedness" of an agrarian, preindustrial settlement system is to compute the correlation coefficient between the distribution of agricultural resources and the distribution of human population across the landscape. The relevant data are those compiled by Nicholas (1989) for her study of agricultural production and population, organized according to 4-km-square grid units by phase for all subvalleys. For each phase and for each grid unit, she presents estimates of potential population, based on a reasonable series of assumptions concerning agricultural production, and archaeological population, drawing on the field results of the Oaxaca Settlement Pattern Survey project (Kowalewski et al. 1989).

Figure 5.10 shows how the correlation coefficient between these two measures (archaeological population and potential population) varies over time—from the Rosario phase through the Late Monte Albán I phase—for the Etla-Central, Ocotlán-Zimatlán, and Tlacolula subvalleys. During the Early Monte Albán I phase, the Etla-Central area experienced a significant dip, which I suggest indicates a temporary decline in overall adaptedness for the political formation centered at Monte Albán.

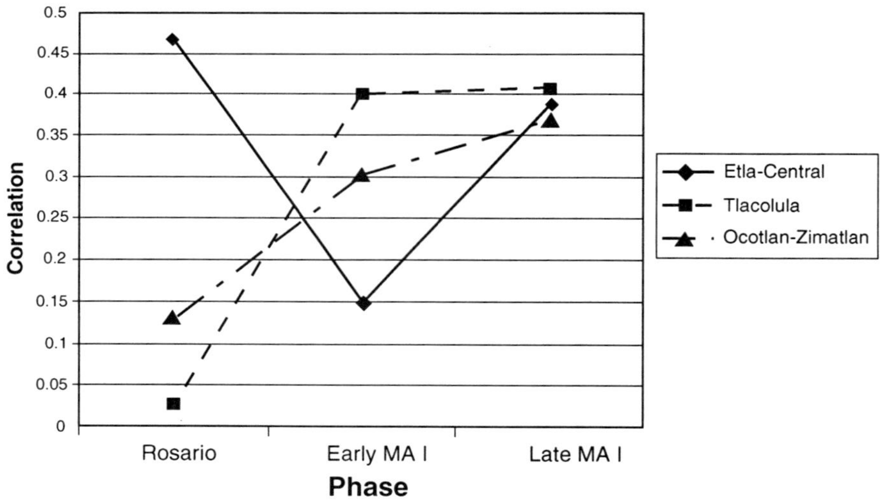

Fig. 5.10 Graph of correlation coefficient between potential population and archaeological population (data from Nicholas 1989) for the Rosario phase through the Late Monte Albán I phase for the Etla-Central, Tlacolula, and Ocotlán-Zimatlán subvalleys

We should bear in mind that the Etla-Central polity was pursuing a new, extreme, and risky initiative at that time. As Blanton (1978) has pointed out, the concentration of people at Monte Albán meant that many farmers lived far from their fields, which greatly increased the overall costs of provisioning. By contrast, the Tlacolula and Ocotlán-Zimatlán polities did not reconfigure their patterns of settlement in such a risky way, and they also seem to have maintained their three-tier settlement hierarchies during Early Monte Albán I. Yet, they *did* eventually respond to Monte Albán's political development and aggressive tactics by evolving into independent secondary states by the Late Monte Albán I phase. I would argue that, in the case of the Monte Albán polity in Etla-Central, state formation occurred through true innovation. In the Ocotlán-Zimatlán and Tlacolula polities, however, state development was the product of both resistance and emulation; the fact that their overall adaptedness (according to Fig. 5.10) did *not* take a dip in the previous Early Monte Albán I phase probably helped these rival polities resist Monte Albán. By contrast, the primary state in Etla-Central *did* move through a period of lowered adaptedness in the Early Monte Albán I phase, just prior to the appearance of primary state organization in the Late Monte Albán I phase.

Figure 5.11 is a graph of the same correlation coefficient between potential population and archaeological population in all three subregional polities, *but* after removing the first-order elite center in each case—that is, looking *only* at the local villages. The graph reveals that the villages of the Etla-Central polity showed a much higher correlation coefficient than the villages of the other two polities during

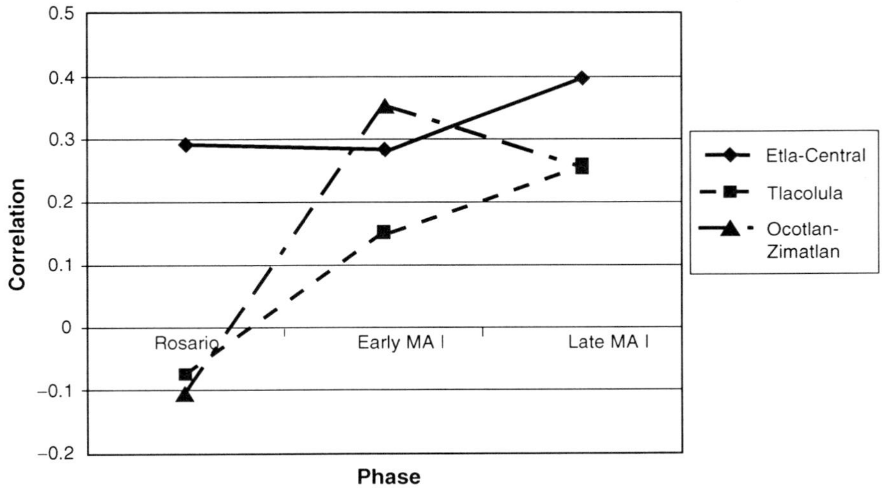

Fig. 5.11 Graph of correlation coefficient between potential population and archaeological population (data from Nicholas 1989) for the Rosario phase through the Late Monte Albán I phase for the Etla-Central, Tlacolula, and Ocotlán-Zimatlán subvalleys, but after removing the grid unit with the first-order center in each subvalley

the Rosario phase. The Etla-Central villages lost most of this relative advantage during the Early Monte Albán I phase, but then regained the advantage in the Late Monte Albán I phase. Thus, both at the local village level and at the level of the entire political formation, there is evidence that the Etla-Central polity experienced a decline in relative adaptedness during the Early Monte Albán I phase.

Recalling our analogy of Wright's adaptive landscape (Fig. 5.2), I suggest that the Etla-Central polity, led by the newly founded center of Monte Albán, was beginning a major and extremely risky strategy shift in Early Monte Albán I times, concentrating the polity's population with scant regard for the distribution of agricultural land. As a consequence, the Monte Albán polity entered an adaptive valley and even lingered there for a while—not an optimal place to be in. How did Monte Albán's early leaders make it through the Early Monte Albán I phase? What measures did they take to keep their polity together and their power intact?

The carved stones at Monte Albán might be giving us a clue here. It is notable that more than 80% of *all* the carved stones from *all* periods at Monte Albán date to the Early Monte Albán I phase, and most of them are depictions of sacrificed captives (the Danzantes carvings). Also, many of these carvings of captives seem to have been mounted as an impressive building façade. This display has been interpreted as a striking piece of visual propaganda. As Marcus has pointed out: "it was during the initial occupation of Monte Albán that the effort devoted to carving monumental figures was the greatest. This early effort probably coincides with the time when the rulers of Monte Albán would have felt the greatest need to legitimize their power" (Marcus 1974:90).

Of course, as the somewhat later conquest inscriptions on Building J indicate, another strategy for getting through the adaptive valley could have been to launch a campaign of interregional conquest. And we have seen that Monte Albán eventually did just that, beginning around 300 B.C. at the onset of the Late Monte Albán I phase. In order for these distant actions to succeed, the Monte Albán polity would have had to develop a full-time military along with internal administrative specialization and the concomitant capacity to delegate partial authority effectively—i.e., it would have had to become a state (Spencer 1998; Spencer and Redmond 2004a).

During the Late Monte Albán I phase in the core Etla-Central area, the distribution of population was brought much more into line with the distribution of agricultural resources (Fig. 5.10). Thus, by this measure, the Monte Albán polity had crossed the valley in the adaptive landscape and was climbing up the newly created peak—by now also armed with tributary extra-Valley territories and an increasingly powerful state administration, all of which would have improved its selective advantage at the polity level. By the end of the Late Monte Albán I phase, Monte Albán was ready to turn its sights on the Ocotlán-Zimatlán and Tlacolula polities. In Monte Albán II times, all three major branches of the Oaxaca Valley were integrated into the Monte Albán state and the regional settlement hierarchy for the entire Valley evinces four clear tiers, with Monte Albán in the top position (Marcus and Flannery 1996:173–174).

Conclusion

The transition from the Early Monte Albán I phase to the Late Monte Albán I phase in Oaxaca witnessed the first appearance of a four-tier regional settlement hierarchy, the first appearance of key state institutions such as the palace and multiroom temple, and the first great expansion of political territory beyond a day's round-trip from the capital. This coincident appearance of novel features is consistent with the morphogenesist model of primary state formation and, more generally, with macroevolutionary culture theory. Similarly consistent is the evidence that the Monte Albán polity went through a period of lowered adaptedness in the phase just prior to state formation. I have argued that the Monte Albán leadership dealt with this time of difficulty by experimenting with some innovative and transformative strategies that turned out to be successful, resulting in primary state formation. This was an emergent organizational form, a new peak in the adaptive landscape of political strategy. By contrast, two nearby polities that developed into secondary states (Tlacolula and Ocotlán-Zimatlán) did not go through a comparable period of relatively low adaptedness. Instead, they withstood annexation by Monte Albán and developed into secondary states as part of their resistance strategy, shifting quickly to the new adaptive peak of statehood that appeared around 300 B.C. Considerable experimentation, risk-taking, and some time were required for the Monte Albán polity to create this new peak, which is probably why it lingered for as long as it did in the adaptive valley during the Early Monte Albán I phase.

The peak shift was initiated, I suggest, by the decision to move the capital of the Etla polity from San José Mogote to Monte Albán around 500 B.C., a political action that is better seen as an expression of creative human agency (directed variation) than as simple, reactive adaptation to the physical and/or cultural environment of the times. After a period of uncertainty and challenge, the peak shift was accomplished by 300–100 B.C., when Monte Albán not only developed state organization but also succeeded in linking the Etla-Central subvalley to at least three other extra-Valley regions, creating a new and larger polity that competed successfully with the secondary states of Tlacolula and Ocotlán-Zimatlán, eventually bringing them under Monte Albán's control. Thus, both directed variation and multilevel selection were instrumental in the rise of the Monte Albán state.

As a final point, the difference in trajectory between the two secondary states and the Monte Albán polity leads me to hypothesize that a period of lowered adaptedness is a necessary precondition for primary state formation in general. When we examine other developmental sequences around the world, we should look for evidence of a similar time of challenge, of difficulty, of risky experimentation in the trajectory of a chief polity just before it evolves into a primary state.

References

Acosta, J.R. (1965). Preclassic and classic architecture of Oaxaca. In R. Wauchope and G.R. Willey (Eds.), *Handbook of Middle American Indians, Vol. 3: Archaeology of southern Mesoamerica, Part 2* (pp. 814–836). Austin: University of Texas Press.

Algaze, G. (1993). Expansionary dynamics of some early pristine states. *American Anthropologist* 95, 304–333.
Arnold, S.J., Pfrender, M.E., Jones, A.G. (2001). The adaptive landscape as a conceptual bridge between micro- and macroevolution. *Genetica*, 112–113:9–32.
Balkansky, A.K. (1998). Origin and collapse of complex societies in Oaxaca (Mexico): Evaluating the era from 1965 to the present. *Journal of World Prehistory* 12, 451–493.
Balkansky, A.K. (2002). *The Sola Valley and the Monte Albán state: A study of Zapotec Imperialism*. Museum of Anthropology, University of Michigan, Memoirs 36. Ann Arbor.
Blanton, R.E. (1978). *Monte Albán: Settlement Patterns at the Ancient Zapotec Capital*. New York: Academic Press.
Blanton, R.E., Feinman, G.M., Kowalewski, S.A., Nicholas, L.M. (1999). *Ancient Oaxaca: The Monte Albán State*. Cambridge: Cambridge University Press.
Blanton, R.E., Feinman, G.M., Kowalewski, S.A., Peregrine, P.N. (1996). A dual-processual theory for the evolution of Mesoamerican civilization. *Current Anthropology* 37, 1–14.
Boyd, R., and Richerson, P.J. (1985). *Culture and the Evolutionary Process*. Chicago: University of Chicago Press.
Carneiro, R.L. (1970). A theory of the origin of the state. *Science*169, 733–738.
Carneiro, R.L. (1981). The chiefdom: Precursor of the state In G. Jones and R. Kautz (Eds.), *The Transition to Statehood in the New World* (pp. 37–79). Cambridge: Cambridge University Press.
Caso, A. (1938). *Exploraciones en Oaxaca: Quinta y Sexta Temporadas 1936–1937*. Instituto Panamericano de Geografía e Historia, Publicación 34. México, D.F.
Caso, A. (1947). Calendario y escritura de las antiguas culturas de Monte Albán. In *Obras Completas de Miguel Othón de Mendizábal, Vol. 1*. México, D.F.: Talleres de la Nación.
Dobzhansky, T. (1937). *Genetics and the Origin of Species*. Reprinted edition, 1982. New York: Columbia University Press.
Drennan, R.D. (1989). The mountains north of the Valley. In Kowalewski, S.A., Feinman, G.M., Finsten, L., Blanton, R.E., Nicholas, L.M., *Monte Albán's hinterland, Part II: Prehispanic settlement patterns in Tlacolula, Etla, and Ocotlán, the Valley of Oaxaca, Mexico* (pp. 367–384). Museum of Anthropology, University of Michigan, Memoirs 23. Ann Arbor.
Dunnell, R.C. (1980). Evolutionary theory and archaeology. *Advances in Archaeological Method and Theory* 3, 35–99.
Earle, T.K. (1987). Chiefdoms in archaeological and ethnohistorical perspective. *Annual Review of Anthropology* 16, 279–308.
Eldredge, N. (1989). *Macroevolutionary Dynamics: Species, Niches, and Adaptive Peaks*. New York: McGraw-Hill.
Eldredge, N. (1995). *Reinventing Darwin: The Great Debate at the High Table of Evolutionary Theory*. New York: John Wiley and Sons.
Eldredge, N., and Gould, S.A. (1972). Punctuated equilibria: An alternative to phyletic gradualism. In Schopf, T.J.M. (Ed.), *Models in Paleobiology* (pp. 82–115). San Francisco: Freeman, Cooper.
Elson, C.M. (2003). *Elites at Cerro Tilcajete: A secondary center in the Valley of Oaxaca, Mexico*. Ph.D. dissertation, University of Michigan, Ann Arbor.
Feinman, G.M., and Neitzel, J. (1984). Too many types: An overview of sedentary prestate societies in the Americas. *Advances in Archaeological Method and Theory* 7, 39–102.
Finsten, L. (1996). Periphery and frontier in southern Mexico: The Mixtec Sierra in highland Oaxaca. In P. Peregrine and G.M. Feinman (Eds.), *Pre-columbian World Systems* (pp. 77–95). Monographs in World Archaeology 26. Madison, WI: Prehistory Press.
Flannery, K.V. (1972). The cultural evolution of civilizations. *Annual Review of Ecology and Systematics* 3, 399–426.
Flannery, K.V. (1995). *Prehistoric Social Evolution*. Research Frontiers in Anthropology. Needham Heights, MA: Simon and Schuster.
Flannery, K.V. (1998). The ground plans of archaic states. In G.M. Feinman and J. Marcus (Eds.), *Archaic States* (pp. 15–57). Santa Fe: School of American Research Press.

Flannery, K.V., and Marcus, J., Eds. (1983a). *The Cloud People: Divergent Evolution of the Zapotec and Mixtec Civilizations*. New York: Academic Press.

Flannery, K.V., and Marcus, J. (1983b). The origins of the state in Oaxaca: Editors' introduction. In K.V. Flannery and J. Marcus (Eds.), *The Cloud People: Divergent Evolution of the Zapotec and Mixtec Civilizations* (pp. 79–83). New York: Academic Press.

Flannery, K.V. and Marcus, J. (1983c). The earliest public buildings, tombs, and monuments at Monte Albán, with notes on the internal chronology of Period I. In K.V. Flannery and J. Marcus (Eds.), *The Cloud People: Divergent Evolution of the Zapotec and Mixtec Civilizations* (pp. 87–91). New York: Academic Press.

Flannery, K.V. and Marcus, J. (1990). *Borrón y cuenta nueva*: Setting Oaxaca's archaeological record straight. In J. Marcus (Ed.), *Debating Oaxaca Archaeology* (pp. 17–69). Museum of Anthropology, University of Michigan, Anthropological Papers 84. Ann Arbor.

Gould, S.J. (1997). Self-help for a hedgehog stuck on a molehill. *Evolution* 51, 1020–1023.

Kottak, C. (1977). The process of state formation in Madagascar. *American Ethnologist* 4, 136–155.

Kowalewski, S.A., Feinman, G.M., Finsten, L., Blanton, R.E., Nicholas, L.M. (1989). *Monte Albán's hinterland, Part II: Prehispanic settlement patterns in Tlacolula, Etla, and Ocotlán, the Valley of Oaxaca, Mexico*. Museum of Anthropology, University of Michigan, Memoirs 23. Ann Arbor.

Lande, R. (1976). Natural selection and random genetic drift in phenotype evolution. *Evolution* 30, 314–334.

Lande, R. (1986). The dynamics of peak shifts and the pattern of morphological evolution. *Paleobiology* 12, 343–354.

Lewis, H. (1968). Typology and process in political evolution. In J. Helm (Ed.), *Essays on the Problem of Tribe* (pp. 101–110). Proceedings of the 1967 Annual Spring Meeting of the American Ethnological Society. Seattle: University of Washington Press.

Marcus, J. (1974). The iconography of power among the Classic Maya. *World Archaeology* 6, 83–94.

Marcus, J. (1976). The iconography of militarism at Monte Albán and neighboring sites in the Valley of Oaxaca. In H.B. Nicholson (Ed.), *The Origins of Religious Art and Iconography in Preclassic Mesoamerica* (pp. 123–139). Los Angeles: Latin American Center, University of California at Los Angeles.

Marcus, J. (1992). Dynamic cycles of Mesoamerican states: Political fluctuations in Mesoamerica. *National Geographic Research and Exploration* 8, 392–411.

Marcus, J. and Flannery, K.V. (1996). *Zapotec Civilization: How Urban Society Evolved in Mexico's Oaxaca Valley*. London: Thames & Hudson.

McGuire, R.H. (1983). Breaking down cultural complexity: Inequality and heterogeneity. *Advances in Archaeological Method and Theory* 6, 91–142.

Newman, C.M., Cohen, J.E., Kipnis, C. (1985). Neo-Darwinian evolution implies punctuated equilibria. *Nature* 315, 400–401.

Nicholas, L.M. (1989). Land use in prehispanic Oaxaca. In Kowalewski, S.A., Feinman, G.M., Finsten, L., Blanton, R.E., Nicholas, L.M., *Monte Albán's Hinterland, Part II: Prehispanic Settlement Patterns in Tlacolula, Etla, and Ocotlán, the Valley of Oaxaca, Mexico* (pp. 449–505). Museum of Anthropology, University of Michigan, Memoirs 23. Ann Arbor.

O'Brien, M.J., and Lyman, R.L. (2000). *Applying Evolutionary Archaeology: A Systematic Approach*. New York: Kluwer Academic/Plenum.

Plog, F.T. (1974). *The Study of Prehistoric Change*. New York: Academic Press.

Price, B. (1978). Secondary state formation: An explanatory model. In R. Cohen and E.R. Service (Eds.), *Origins of the State: The Anthropology of Political Evolution* (pp. 161–186). Philadelphia: Institute for the Study of Human Issues, Inc.

Renfrew, C. (1979). Transformations. In C. Renfrew and K.L. Cooke (Eds.), *Transformations: Mathematical Approaches to Culture Change* (pp. 3–44). New York: Academic Press.

Rosen, R. (1982). On a theory of transformations for cultural systems. In C. Renfrew, M.J. Rowlands, B.A. Seagraves (Eds.), *Theory and Explanation in Archaeology: The Southampton Conference* (pp. 301–313). New York: Academic Press.

Sanders, W.T. (1974). Chiefdom to state: Political evolution at Kaminaljuyú, Guatemala. In C.B. Moore (Ed.), *Reconstructing Complex Societies: An Archaeological Colloquium* (pp. 97–116). Supplement to the Bulletin of the American Schools of Oriental Research 20. Cambridge: Department of Anthropology, Harvard University.

Service, E.R. (1975). *Origins of the State and Civilization*. New York: Norton.

Simpson, G.G. (1944). *Tempo and Mode in Evolution*. Reprinted edition, 1984. New York: Columbia University Press.

Spencer, C.S. (1982). *The Cuicatlán Cañada and Monte Albán: A Study of Primary State Formation*. New York: Academic Press.

Spencer, C.S. (1990). On the tempo and mode of state formation: Neoevolutionism reconsidered. *Journal of Anthropological Archaeology* 9, 1–30.

Spencer, C.S. (1997). Evolutionary approaches in archaeology. *Journal of Archaeological Research* 5, 209–264.

Spencer, C.S. (1998). A mathematical model of primary state formation. *Cultural Dynamics* 10, 5–20.

Spencer, C.S., and Redmond, E.M. (1997). *Archaeology of the Cañada de Cuicatlán, Oaxaca*. Anthropological Papers 80. New York: American Museum of Natural History.

Spencer, C.S., and Redmond, E.M. (2001a). Multilevel selection and political evolution in the Valley of Oaxaca, 500–100 B.C. *Journal of Anthropological Archaeology* 20, 195–229.

Spencer, C.S., and Redmond, E.M. (2001b). The chronology of conquest: Implications of new radiocarbon analyses from the Cañada de Cuicatlán, Oaxaca. *Latin American Antiquity* 12, 182–202.

Spencer, C.S., and Redmond, E.M. (2003). Militarism, resistance, and early state development in Oaxaca, Mexico. *Social Evolution & History* 2, 25–70.

Spencer, C.S., and Redmond, E.M. (2004a). Primary state formation in Mesoamerica. *Annual Review of Anthropology* 33, 173–199.

Spencer, C.S., and Redmond, E.M. (2004b). A Late Monte Albán I Phase (300–100 B.C.) palace in the Valley of Oaxaca. *Latin American Antiquity* 15, 441–455.

Spencer, C.S., and Redmond, E.M. (2005). Institutional development in Late Formative Oaxaca: The view from San Martín Tilcajete. In T.G. Powis (Ed.), *New Perspectives on Formative Mesoamerican Cultures* (pp. 171–182). British Archaeological Reports, International Series 1377. Oxford: Archaeopress.

Teltser, P.A., Ed. (1995). *Evolutionary Archaeology: Methodological Issues*. Tucson: University of Arizona Press.

Upham, S. (1987). A theoretical consideration of middle range societies. In R.D. Drennan and C.A. Uribe (Eds.), *Chiefdoms in the Americas* (pp. 345–368). Lanham, MD: University Press of America.

Webster, D. (1975). Warfare and the evolution of the state: A reconsideration. *American Antiquity* 40, 464–470.

Wilson, D.S., and Sober, E. (1994). Reintroducing group selection to the human behavioral sciences. *Behavioral and Brain Sciences* 17, 585–654.

Winter, M. (1994). El Proyecto Especial Monte Albán 1992–1994: Antecedentes, intervenciones y perspectivas. In M. Winter (Ed.), *Monte Albán: Estudios recientes: Contribución No. 2 del Proyecto Especial Monte Albán 1992–1994* (pp. 1–24). Oaxaca: Centro INAH Oaxaca.

Wright, H.T. (1977). Recent research on the origin of the state. *Annual Review of Anthropology* 6, 379–397.

Wright, H.T. (1986). The evolution of civilizations. In D. Meltzer, D. Fowler, J. Sabloff (Eds.), *American Archaeology: Past and Future* (pp. 323–365). Washington, D.C.: Smithsonian Institution Press.

Wright, H.T., and Johnson, G.A. (1975). Population, exchange, and early state formation in southwestern Iran. *American Anthropologist* 77, 267–289.

Wright, S. (1931). Evolution in Mendelian populations. *Genetics* 16, 97–159.

Yoffee, N. (1979). The decline and rise of Mesopotamian civilization: An ethnoarchaeological perspective on the evolution of social complexity. *American Antiquity* 44, 5–35.
Yoffee, N. (1993). Too many chiefs? (or, safe texts for the 90s). In N. Yoffee and A. Sherratt (Eds.), *Archaeological Theory: Who Sets the Agenda?* (pp. 60–78). Cambridge: Cambridge University Press.

Chapter 6
Evolutionary Biology and the Emergence of Agriculture: The Value of Co-opted Models of Evolution in the Study of Culture Change

Melinda A. Zeder

Introduction

The application of macroevolutionary theory to the study of cultural evolution is only one of a number of cross-disciplinary incursions by archaeologists seeking epistemological inspiration for higher order explanation of culture change. Archaeologists have been rummaging around in the closets of evolutionary biologists since Darwin, trying on a wide variety of different approaches to the study of biological evolution in the hope of finding overarching theories that can be applied directly to the evolution of human culture. The origin of agriculture, along with the sweeping reordering of economy, social organization, and belief systems that generally accompanied agricultural emergence, is frequently used as a case study opportunity for the application of co-opted evolutionary models. One of a handful of major transitions in human history, the Neolithic Revolution has been promoted as an ideal proving ground for establishing the efficacy of these different explanatory approaches.

In this chapter, I explore the application of three different co-opted evolutionary models to this watershed in cultural evolution—neo-Darwinian or selectionist evolutionary archaeology, macroevolutionary archaeology, and human behavioral ecology. I examine the basic tenets of each evolutionary model as it has been applied to the study of human cultural evolution, reviewing in particular how researchers have used the model in the study of agricultural origins. I then turn to the archaeological record of agricultural emergence in the Near East, arguably the "Mother of all Neolithic Revolutions," where we have the deepest and most comprehensive documentation of the panoply of culture change that accompanies agricultural origins. This consideration allows an assessment of how well, or how poorly, each of these evolutionary approaches matches this empirically rich example of cultural evolution.

M.A. Zeder (✉)
Archaeobiology Program, National Museum of Natural History, Smithsonian Institution, Washington, DC 20560, USA
e-mail: zederm@si.edu

A.M. Prentiss et al. (eds.), *Macroevolution in Human Prehistory*,
DOI 10.1007/978-1-4419-0682-3_7,

Neo-Darwinian "Selectionist" Archaeology

Archaeologists enrolled in the neo-Darwinian or selectionist school of evolutionary archaeology tend to be the most evangelical in their application of biological evolutionary theory to the study of culture change (Dunnell 1978, 1980; Leonard and Jones 1987; O'Brien and Lyman 2000; see discussion in Zeder 2009). These archaeologists draw inspiration from the evolutionary models generated by the "modern synthesis" school of biological evolution which brought together Darwinian concepts of natural selection with principles of population genetics (e.g., Dawkins 1986; Huxley 1942; Mayr 1942, 1963; Simpson 1944). Neo-Darwinian models of biological evolution portray evolution in terms of changes in allele frequencies within individual organisms that are shaped by gene flow, genetic drift, and natural selection operating on phenotypic expression of random genetic variation. Selectionist archaeologists acknowledge that human cultures differ from biological systems in that the units of variability on which selection acts in cultural evolution are human behaviors, not genes, and that variability is transmitted through cross-generational behavioral emulation and instruction, or social learning, and not by genetic transmission in the course of sexual reproduction (following Boyd and Richerson 1985; Lyman and O'Brien 1998: 619; Rindos 1984:54). Nevertheless, they embrace the neo-Darwinian "trait level" point of view, arguing that the study of cultural evolution requires tracing the operation of directly analogous processes of selection, drift, and transmission operating on the discrete attributes of material culture, argued to represent "the hard parts of the behavioral segment of [past] phenotypes" (Dunnell 1989:44; Leonard and Jones 1987; O'Brien et al. 1998: 490). And while the possibility that selection may also operate at the level of groups of individuals is acknowledged, especially in the case of complex societies (Dunnell 1989; Rindos 1984:49), emphasis here is placed on the individual culture bearer and the array of behaviors (and their artifact trait proxies) he or she engages in. Understanding the course of cultural evolution to these archaeologists involves tracing the genealogies of traits, specifically the variation of traits through time (Lipo et al. 2006; O'Brien et al. 1998:487). The primary task in this endeavor is to chart the trajectories of trait variability and to determine whether the phylogeny of genealogically related traits are shaped by selection operating on "functional" traits, whose "replicative fitness" is judged by their differential persistence through time, or the result of random factors of drift and transmission errors operating on "stylistic" traits without adaptive significance (Dunnell 1978; Leonard and Jones 1987: 214).

While selectionist archaeologists acknowledge the role of deliberate human intent in the invention of new behaviors and in decisions of whether to adopt one behavior over another, they adamantly deny the role of human intentionality in cultural evolution. They argue that the intentions behind such choices are unknowable, and that, at least in the archaeological record, the behaviors resulting from such deliberations are indistinguishable from behavioral variability arising from unintentional transcription errors and recombinations of behaviors in the transmission from one generation from the next (Lyman and O'Brien 1998:619; O'Brien and Holland 1992:45; Rindos 1989a, 1989b). In any event, it is not the source of behavioral variability that is significant in the selectionist scheme of things. Instead, selectionists focus on the ways in which selection, drift, or transmission ("meme flow"), as the ultimate drivers of cultural evolution, shape the trajectory of behaviors (represented in artifact traits) through time (Lyman and O'Brien 1998:644; Rindos 1985).

Selectionist archaeologists also recognize that the different modes of transmission in biological and cultural systems affect the tempo of change, acknowledging, in particular, that the transmission of culture traits through social learning can quicken the rate of culture change to the extent that even "punctuated" events of rapid evolutionary change are possible (Lyman and O'Brien 2001:408), Nevertheless selectionist archaeologists are more comfortable with the neo-Darwinian model of phyletic gradualism as the dominant mode of culture change, which is characterized as continuous and proceeding at a gradual and incremental pace (Rindos 1985). Apparent instances of punctuational culture change, they argue, are more likely artifacts of an incomplete archaeological record or the result of faulty analytical methods directed at an inappropriate scale of change (Lyman and O'Brien 1998: 627; O'Brien et al. 1998:487).

This issue of scale of change also leads this school of evolutionary archaeology to reject the classical typological groupings of cultures followed by most archaeologists of the earlier "processualist" school of archaeology (i.e. the band, tribe, chiefdom, and state classificatory system of Service [1962] and Fried [1967]. Such higher order groupings, they maintain, serve mostly to blur variability or mask the selective forces shaping variability, arbitrarily "(c)arving a continuum of variation into chunks" that are then classified in terms of their "central tendencies" (Lyman and O'Brien 1998:627). They contend that these false cultural taxonomies are especially pernicious, since they can be confused with evolutionary stages and the arbitrary divisions between these false stages mistaken for instances of rapid transition from one to another—transitions that essentially require explanation through appeal to nomothetic, universal, prime-mover forces (Dunnell 1978, 1980, 1988; Leonard and Jones 1987). To selectionists, the burden in the explanation of culture change lies not in accounting for general patterns of macroevolutionary change that cut across many unrelated cultural entities but instead on tracing "historical patterns of differential trait representation" within individual genealogically linked cultural lineages and then identifying the microevolutionary processes responsible for the patterns of variation observed (Jones et al. 1995; Lyman and O'Brien 1998:615, 2000). Efforts at isolating drivers of evolution across unrelated cultural lineages (outside of selection, drift, and transmission) are considered to be a waste of time, since each lineage is a unique product of a particular set of historical contingencies—of the product of selection, drift, and transmission operating on randomly generated behavioral variability.

The first and best known application of "cultural selectionism" to the study of agricultural origins can be found in David Rindos' book "Origins of Agriculture: An Evolutionary Perspective" (Rindos 1984) which uses agriculture as the ideal case study application of neo-Darwinian theories to the study of culture change. Rindos rejects earlier models of agricultural origins that cast this development in progressivist terms as the product of conscious human inventiveness, arguing instead that agriculture is the product of a co-evolutionary process between humans and plants (and presumably animals) akin to co-evolutionary symbiotic relationships found in nature—in effect no different from co-evolutionary mutualism between social insects and plants. He also rejects earlier models that portray agricultural origins as revolutionary events caused by prime-mover catalytic forces. Instead Rindos advances a gradualist model that sees agriculture as an outgrowth of a relatively slow incremental process involving at least three different "modes" of relationship characterized as "incidental domestication," "specialized domestication," and "agricultural domestication." And while these modes are sometimes mistaken for progressivist stages or phases in the process of agricultural emergence (i.e. O'Brien 1987), Rindos maintains that they were instead meant to portray types of human/plant relationships that may or may not, depending on highly localized conditions, pave the way for the next modal relationship (Rindos 1987:192).

An application of this model to a specific case study example, O'Brien's consideration of sedentism, population growth, and resource utilization in the Midwestern United States, traces the complicated and reinforcing co-evolutionary events that led to the intensification of plant use and the origin and/or adoption of domestic plants into the resource base of Middle and Late Woodland peoples (O'Brien 1987). He argues that no one factor (i.e., population pressure, sedentism, or climate change) can be singled out as causal in this sequence of events and that agricultural origins in this region evolved in a gradual, incremental co-evolutionary way. This early attempt at applying the tenets of selectionist archaeology to agricultural origins drew fire from Dunnell (1987), the patriarch of the selectionist school (and a more gentle rebuke from Rindos [1987]) for failing to demonstrate how selection operates on variation to produce culture change in this case, a failure that O'Brien, in response, attributed to the lack of data of sufficient resolution to allow such a demonstration (O'Brien 1987:193).

Macroevolutionary Archaeology

The selectionist perspective on culture change drew fire from early advocates of macroevolutionary archaeology who, like proponents of macroevolutionary theory in biology (Eldredge 1989a; Eldredge and Gould 1972; Gould 1989 2002; Vrba and Eldredge 1984), often framed their ideas about cultural evolution in terms of a critique of neo-Darwinian evolutionary approaches (Chatters and Prentiss 2005; Rosenberg 1994a, 1998; Spencer 1997; see discussion in Zeder 2009). Macroevolutionary archaeologists reject the selectionist view of cultures as discrete, independently varying traits whose persistence over time is determined solely by processes of selection, drift, and transmission. They embrace, instead, the alternative macroevolutionary view that sees organisms (and by extension cultures) as integrated wholes or constellations of interacting traits whose form is shaped as much by historical contingencies and constraints to change as by the specific adaptive attributes of individual traits (Gould and Lewontin 1979; Seilacher 1972).

Many working within this paradigm have co-opted the term "*Bauplan*" from evolutionary biology to stand for the basic designs, or blueprints, that organize linked constellations of culture traits into coherent and enduring forms (Chatters and Prentiss 2005:47; Rosenberg 1994a:308; Spencer 1997: 234). Used with more uniformity in biology as a series of homologous characters nested in a series of ever more general structural plans at higher taxonomic levels (Hall 1996:223–227), macroevolutionary archaeologists are less consistent in how they define cultural *Baupläne*. Chatters and Prentiss, for example, define *Baupläne* in terms of socioeconomic variables, as sets of resource management strategies (RMS) that consist of "constellation[s] of shared ideas directed at the acquisition, distribution, and consumption of energy and resources". These constellations of increasingly more generalized "*Bauplänal*" blueprints are arranged in a hierarchy of levels which at the highest scale are the constellations of traits that separate very broad sets of resource strategies like hunter-gatherers and food producers (Chatters and Prentiss 2005:48–50). Spencer, on the other hand characterizes the structural plans that undergird cultural forms in term of sociopolitical factors, arguing that the admittedly flawed cultural typologies of Service and Fried still have utility in defining the basic structural features that define cultural units with different leadership formats (Spencer 1990, 1993, 1997). Rosenberg (1994a) takes a different tack, restricting the term *Bauplan* to mean the ideational structure or *ethos* of a culture that provides the highest-order and most conservative organizational framework for a culture. These ideational organizational rubrics sit at a "superstructural" level subsuming lower-order "structural" levels consisting of a society's economic and political systems, which themselves sit above a society's "infrastructural" level that consists of its "productive capacity".

Macroevolutionary archaeologists also find the selectionist trait-based focus of culture change insufficient to account for major cultural change (Spencer 1997:225–226). They acknowledge that selective pressures operating at a localized inter-individual or inter-group level may help cultures respond to changes in the physical or social environment, but, they contend, such microevolutionary forces are more likely to preserve culture norms and cannot on their own account for major shifts scale or complexity (Chatters and Prentiss 2005:48; Rosenberg 1994a:333; Spencer 1997:226). Instead, and once again following the lead of macroevolutionary biologists, these archaeologists adopt a more hierarchical approach which acknowledges that evolution works at multiple levels—from individuals, to families, to local communities, to regional polities (Rosenberg 1994a:320–321; Spencer and Redmond 2001:201). Just as these higher order forces are held to be responsible for major cladistic change in biological systems (Gould 2002:726; Gould and Eldredge 1993:224; Vrba and Eldredge 1984), macroevolutionary archaeologists also argue that major cultural changes in which one cultural *Bauplan* is replaced by another are driven by the combined forces of evolutionary processes operating across hierarchical levels of social organization (especially those that affect higher-order levels above that of competing individuals).

The conditions under which new cultural *Baupläne* arise to replace old ones differ depending on how a cultural *Bauplan* is defined. Chatters and Prentiss, for example, hold that diversification of RMS and the creation of new socioeconomic *Baupläne* are most likely under conditions of economic opportunity rather than under periods of stress and heightened intergroup competition (Chatters and Prentiss 2005:51). They characterize cultural diversification in terms of allopatric speciation (following Mayer 1963) which occurs when groups that are either geographically or effectively isolated from other groups experience conditions that reduce the risk of experimenting with strategies that deviate from the dominant cultural behaviors. Periods of cultural diversification are usually followed by a period of "culture sorting" (again a concept co-opted from evolutionary biology, [Elredge and Gould 1972]) in which new *Baupläne* come into increasing competition with each other and with parent groups, resulting in the decimation (or abandonment) of less fit strategies and the proliferation or adoption of *Baupläne* more suited to the natural and social environments (Prentiss and Chatters 2003). Rosenberg, on the other hand, sees cultures as more conservative and resistant to change with stress being the major precipitating factor behind culture change (Rosenberg 1994a, 1998). Stress-driven innovations in behaviors, he argues, are more likely to be accepted if they conform with cultural norms, or at least if they can be seen as relieving stresses at the level of the productive capacity of the system, especially if they can be accommodated by the logic of its political and economic system and do not threaten the underlying societal ideological norms that, to Rosenberg, form the society's cultural *Bauplan*. Such responses are seen as helping the society maintain itself, keeping the group in a state of equilibrium rather than inducing new cultural forms. Old cultural *Baupläne* are only abandoned and replaced by new ones, according to Rosenberg, under conditions of severe stress that cannot be alleviated by such lower level adjustments, but that require responses that seriously deviate from the societies' conceptual underpinning at this highest organizational level. Spencer, in contrast, allows that both opportunity and stress are capable of propelling scalar shifts in organization and control within societies which drive the emergence of new sociopolitical *Baupläne*. Such shifts, he argues, happen under circumstances in which enhanced organization capacity allows the group to either take advantage of opportunities or defend itself against external or internal threats. They come about through a process he terms "extrapolation" in which one unit of organization control begins to assert authority over another resulting in a more inclusive political entity (Spencer 1990, 1993, 1997:238–239; Spencer and Redmond 2001:199).

Regardless of how one characterizes cultural *Baupläne* and the ways in which new ones arise and replace old ones, all macroevolutionary archaeologists subscribe to the basic principle that evolutionary change from one *Bauplan* to another proceeds as a punctuational process in which periods of rapid transition (that according to Rosenberg [1994a:314] typically occur in a century or less) are followed by longer periods of relative stasis (Chatter and Prentiss 2005:50; Rosenberg 1994a:318; Spencer 1990, 1997:237). Such periods of abrupt, revolutionary change in the archaeological record are not, as selectionists hold, artifacts of gaps in the record or the product of artificial typologies (Lyman and O'Brien 1998: 627) but represent actual times in which there is a major restructuring of cultural norms (whether in resource management, administration, or ideology) driven by major macroevolutionary forces of change. Periods of apparent stasis that separate broadly different cultural forms are also real and represent periods of relative stability in which cultures actively preserve basic structural continuity by making small microevolutionary adjustments in response to various external and internal pressures (Rosenberg 1994a:333).

While selectionists acknowledge that the forces of selection operating on adaptive traits and the transmission of behaviors through social learning may lend an appearance of directionality to cultural change, they steadfastly reject the importance of directed variation in cultural evolution claiming that it explains nothing about how that change came about (Dunnell 1980; Lyman and O'Brien 1998:621). Macroevolutionary archaeologists disagree stressing the importance of directionality in both cultural and biological evolution. They argue that the hierarchical nature of culture process operating simultaneously within and across different levels within a society lends an important element of directionality to culture change (Spencer and Redmond 2001:201), as does the element of historical contingency (or path dependency) when non-conflicting, neutral, or at least not maladaptive cultural attributes are carried over from the ancestral form into the descendent form (Rosenberg 1994a:329). The retooling of behaviors that evolved to serve one function to serve a new adaptive function (analogous to the macroevolutionary concept of "exaptation" [Gould 1988 2002:49, 726–730]) is also an important factor that lends directionality to culture change.

These archaeologists also argue that the capacity for directionality in culture change is significantly enhanced over anything found in biological systems by the human ability to evaluate outcomes of behavior and to abandon, adjust, and perpetuate behaviors based on this evaluation. This capacity for conscious human decision making in the invention and transmission of the behaviors (following Boyd and Richerson 1985; Eerkens et al. 2006; Richerson and Boyd 2005:69; Spencer 1993:46–47, 1997:239) introduces a Lamarkian dimension to cultural evolution that greatly enhances its potential for rapid and radical directional change (Rosenberg 1990:399–400, 1994a:313; Spencer 1997:230). And while, as we have seen, selectionists acknowledge that the mode of transmission is a major feature that separates cultural from biological evolution and that it may have an impact on the tempo and pace of culture change, they nevertheless deny human intent any significant role in cultural evolution (Rindos 1984:6 1985:72). Rather than as an incidental and noncritical aspect of cultural change, macroevolutionary archaeologists see this unique element of human agency as perhaps the most important driver of cultural evolution—one that imparts entirely new and unique characteristics and capacities to cultural change, which, in turn, clearly and definitively distinguishes cultural from biological evolution (Boyd and Richerson 1985; Spencer 1993:46).

The focus of explanation of culture change following this paradigm involves isolating the processes operating across multiple levels in a cultural system that are responsible for the abandonment of old cultural behaviors and the adoption of new ones. Macroevolutionary archaeologists are, however, much less likely to embrace the prime-mover nomothetic, one-size-fits-all explanations of culture change of earlier evolutionary anthropologists, focusing instead on teasing out the variables that shape the trajectory of individual historically linked ancestral and descendent cultural forms (Chatters and Prentiss 2005:52). At the same time, they are likely to employ a comparative approach that looks across convergent examples of cultural forms both for processes held in common and for those that are unique to individual expressions of these forms (Spencer 1990:233–234).

Insights into how a macroevolutionary perspective might be applied to the origins of agriculture are found in Prentiss and Chatters' initial applications of macroevolutionary theory to archaeology, in which they characterize the shift from mobile foraging to storage dependent collectors in the Pacific Northwest as prime case study example of punctuated evolutionary change (Chatters and Prentiss 2005; Prentiss and Chatters 2003). Following their macroevolutionary model, they argue that the relaxation of intergroup competition in the Pacific Northwest during a period of more optimal climatic conditions from about 5600 to 4200 cal. B.P. allowed the diversification of previously ubiquitous foraging strategies, resulting in a proliferation of different combinations of mobile foraging and residential mobility strategies across the region. An ensuing climatic downturn beginning about 4200 cal. B.P. resulted in a period of increased intergroup competition that effectively weeded out less well adapted strategies which were subsequently replaced by relatively uniform storage-dependent collector strategies throughout the Pacific Northwest. Prentiss and Chatters revisit these issues in their respective chapters in this volume. Here, they expand the scope of this earlier work to include macroevolutionary scale concepts of "emergent character" and "emergent fitness" (Arnold et al. 2001; Vrba and Eldredge 1984) applied to Wright's (1931, 1932) fitness landscape model as a way of mapping, in Prentiss's chapter, the punctuational shifts in resource strategies in the Pacific Northwest and, in Chatter's chapter, the emergence of agrarian Mississippian chiefdoms in the southeastern US. And while these studies do not specifically address issues of plant and animal domestication or agricultural origins within this macroevolutionary perspective, the changes in resource management strategies that result in reduced logistical mobility, intensification of exploitation strategies, and storage of surplus in the Pacific Northwest example, and the crystallization of stratified social groups, based on intensive maize agriculture in the Mississippian case study, do bracket these developments and provide a template of how macroevolutionary models might be applied to domestication and agricultural origins.

Kuijt and Prentiss's contribution to this volume applies a similar model in their consideration of changing resource management strategies during the Late Epipaleolithic in the Near East. Here they develop the concept of cultural niches to describe the ways in which humans manage habitats under various socioeconomic *Baupläne*. They trace an alternating tempo of stability and rapid change, in which, over the course of the climatically dynamic Late Epipaleolithic, periods of sudden niche diversification and equally sudden constriction are each followed by longer periods of general stability—a process that leads up to, but that does not include, the domestication of plants (and in other places animals) and the emergence of agricultural economies in the subsequent Pre-Pottery Neolithic.

Also operating within a macroevolutionary framework, Rosenberg (1990, 1994a, 1998) casts these same developments in a very different light. While Chatters,

Prentiss, and Kuijt see novel resource strategies arising in times of opportunity, Rosenberg believes that changes in mobility and intensification of resource extraction can only arise in periods of stress, in particular demographic stress, which in turn leads to territorial compression, and increasingly restrictive and proprietary rights to resources. And yet while the mechanics and context of change are different in Rosenberg's model, he too envisions the shift from a non-surplus-producing mobile hunter-gatherer system to a surplus-producing (if not food producing) one of punctuated revolutionary change.

Human Behavioral Ecology

The most recent co-opted evolutionary theory to enter the arena of archaeological explanation is behavioral ecology. Targeted at discovering the principles that shape adaptive behaviors in animals, especially with respect to resource acquisition or foraging (Charnov 1976; Krebs and Davies 1978; MacArthur and Pianka 1966; Stephens and Krebs 1986), concepts of behavioral ecology were quickly and eagerly adopted by both archaeologists and ethnographers interested in human foraging behavior, particularly within hunter-gatherer societies (Hawkes and O'Connell 1992; Hawkes et al. 1997; Kelly 1995, 2000;Winterhalder and Smith 1981). This fast growing school of human behavioral ecology (HBE) occupies at least some common ground with the earlier described neo-Darwinian or selectionist school of evolutionary archaeology. The overlap between these two approaches is especially evident in the HBE emphasis on the "microscale" of evolutionary change, which focuses on the role of individual actors in the process of microevolutionary change (Bettinger this volume, Winterhalder and Kennett in press). HBE practitioners and selectionist evolutionary archaeologists also share a deep suspicion of macroscale, universalist, prime-mover explanations of culture change (Layton et al. 1991; Winterhalder and Goland 1997). Moreover, both selectionist and HBE approaches to the study of culture change feature human behavior as the primary unit of cultural evolution and stress the role of social learning in the transmission of behaviors. As we have seen, however, selectionists profess no interest in the ways in which novel behaviors are generated and transmitted. Adherents of HBE, on the other hand, concentrate almost entirely on the decisions that shape these behaviors. Thus while selectionist archaeologists focus on tracing the genealogy of related behavior proxies (traits) and identifying the hand of selection, drift, and transmission in shaping these genealogies, archaeologists working with an HBE framework focus instead on isolating the underlying principles responsible for these behaviors in the first place.

An extreme application of behavioral ecology to human systems might argue that the behaviors of both birds and humans are guided by the same set of optimizing principles and that the human evaluation of energy expenditures and returns in making foraging decisions involves no more conscious intent than do the foraging choices of nonhuman animals. Most HBE practitioners, however, are quite comfortable with incorporating intentionality and conscious human decision making into their models (Piperno 2006:164) and, in fact, tend to feature rational human decision making as a primary driver of culture change (Bettinger 2006). This emphasis on the role of conscious deliberation in cultural evolution is something that human behavioral ecologists hold in common with macroevolutionary archaeologists and is a major point of departure from the selectionist view of culture change.

The underlying principles held to guide behavioral choices in behavioral ecology are grounded as much in microeconomics as they are in biology (Bettinger 1991:83–84, 2006:306; Winterhalder and Kennett in press) and are laid out with exemplary clarity and precision in Winterhalder and Kennett (2006:11–17). The central assumption behind both behavioral ecology and human behavioral ecology is that of optimization, usually measured in terms of net energy returns of a particular foraging behavior over acquisition and processing costs. Behaviors in both human and nonhuman animals are, however, never fully optimal but are constrained by certain physiological, morphological, cognitive, and, in the case of humans, technological parameters that limit the range of behavioral choices. Optimality may also be impinged upon by delays in the development and spread of adaptive behaviors or in the sorting out of different sets of competing goals. Other basic concepts that shape adaptive behaviors include considerations of "marginal value" and "opportunity costs" that help determine the decision when to shift from one behavior to another by weighing the benefits of engaging in the new behavior against the diminishing benefits of the original behavior. "Discounting" (weighing of future returns against immediate returns) and "risk assessment" (assessing the probability of realizing expected returns) are additional factors that guide optimizing behavior.

These general concepts are operationalized in the form of various HBE models, which apply these fundamental behavioral features to a cost/benefit assessment of different courses of action under different circumstances. Most familiar to archaeologists are diet breadth models, which are based on the assumption that optimal resources (those with the greatest net energy gain over acquisition and processing costs) are always selected over less optimal resources. The inclusion of a lower ranked resource in the diet, following a strict application of the diet breadth model, has nothing to do with their ubiquity but is instead determined by the abundance and availability of higher ranked resources (Bettinger 1991:84–86; Hawkes and O'Connell 1992). Diet breadth add-ons discussed in Winterhalder and Kennett (2006:14–17) include considerations of the distribution of resources over the landscape, the ranking of resource patches, and the computation of the variables that will determine the amount of time dedicated to harvesting resources in one patch before moving to another. Other HBE models with a spatial dimension include the "ideal free distribution" model which examines habitat selection choices and the rate and direction of new habitat colonization, as well as central place foraging models that model habitation choice by adding travel time from foraging site to home site into the computation of resource net energy returns.

Ethnologists working within a behavioral ecology framework use these models to isolate the factors that guide the choices that individuals make in day-to-day foraging and other productive (and reproductive) activities. HBE archaeologists use them to tease out the factors responsible for shaping the long-term trajectory of such choices over time. There are substantial methodological challenges to marshalling empirical data with the kind of kilocalorie precision required in these cost/benefit analyses. This is especially the case for archaeologists who must infer behavior from material culture and time average the decisions of many individuals over temporal spans of hundreds of years or more (Smith 2006). Despite these limitations, HBE practitioners see great value in the dialectic their models promote between HBE principles and real-life behaviors which they maintain helps hone the effectiveness of their models in capturing the driving forces behind human behavior and, by extension, cultural evolution.

The emergence of agricultural economies presents special challenges to behavioral ecologists. Agriculture involves an emphasis on the utilization of plants, whose generally smaller energetic returns and higher processing costs put them lower down on forager's ranked resource list (Bettinger 1991:98–100; Gremillion and Piperno in press; Hawkes et al. 1982; Kennett et al. 2006a:127; Piperno 2006:142, Tables 7.1 and 7.2). In addition, agriculture inherently incurs heavy discounting costs with lots of up-front energy expenditure for a deferred and often quite unpredictable return (Alvard and Kunzner 2001; Gremillion and Piperno in press; Hawkes and O'Connell 1992: 64; Tucker 2006; Winterhalder and Kennett in press). Explaining the adoption of agricultural strategies under the central optimization principles of behavioral ecology, then, requires figuring out why people chose to focus on lower ranked resources with long deferred, risky returns.

The first serious attempts at applying principles of foraging theory to agricultural origins can be found in an extended conversation among various HBE practitioners that unfolded in *Current Anthropology* in the early 1990s (Hawkes and O'Connell 1992; Layton et al. 1991, Layton and Foley 1992; Winterhalder and Goland 1993). The discussion began with Layton et al.'s proposal that behavioral ecology (specifically diet breadth models) offers a superior approach to the study of agricultural origins than can be found in mainline directional and progressivist, cultural evolutionary approaches. Hawkes and O'Connell (1992) agreed and expanded upon this position, while also critiquing the Layton et al's. application of diet breadth models to agricultural emergence. Following the classic tenets of optimal foraging diet breadth models (Charnov and Orians 1973; Emlen 1966; MacArthur and Pianka 1966), Hawkes and O'Connell argued that regardless of the abundance of lower ranked plant resources, foragers will always ignore them when there is a sufficiently good chance of encountering and being able to procure higher ranked resources. The broadening of a diet to include lower ranked resources, then, can only occur when there is a decline in the abundance and, subsequently, in the encounter rates of higher ranked resources. They go on to explore conditions under which the costs of searching for declining high-ranked prey begin to outweigh the net returns of these energy-rich resources, making it worthwhile to invest in lower ranked plant resources despite their relatively higher handling costs. Under conditions of continued resource depression, this trade-off between search and handling costs results in a broad-based diet where the primary foraging costs are derived from handling, or processing, lower ranked resources rather than searching for higher ranked resources. The domestication of plant resources within such a subsistence economy is seen as a product of increases in handling efficiency (either from improvements in technology or from changes in the morphology of the plants themselves) that further boosts net returns. Domestication of animals is characterized not as a means of compensating for declining game as Foley earlier proposed (Foley 1982) but as a way of further reducing the handling costs of lower ranked plant resources through their conversions into meat and milk. Winterhalder and Goland (1993, 1997) added to this discussion with a consideration of the conditions under which changes in the density, reproductive capacity, and net yield of lower ranked plant resources might raise their rank in the menu of available resources, setting foraging groups on the path to farming by initiating co-evolutionary relationships between humans and plants that boost net return rate along the lines explored by Rindos (1984).

Initial applications of diet breadth models to archaeological case studies can be found in Piperno and Pearsall's study of agricultural origins in the Neotropics (Piperno and Pearsall 1998) that holds the changing environmental conditions in the Early Holocene responsible for altering the menu of resources available to foraging populations in lowland Central and South America. Expansion of forests into previously open areas during this time increased the search costs of previously more abundant, higher ranked open-habitat resources (including large and medium-sized grazers like mammoths, glyptodonts, flat-headed peccaries, and horses, as well as high-quality, low-cost plant resources like dry-land cacti and legumes). The decreasing availability of these high-ranked resources caused by the post-Pleistocene changes in Neotropical environments resulted in an increasing emphasis on a widening array of previously little exploited lower ranked tropical forest plant and animal resources (i.e. smaller, often arboreal, game animals with higher search and capture costs and plants like squashes, manioc, yams, and other wild roots and tubers which at low densities provided too little return and were too costly to process to make them a worthwhile dietary component). Increasing human focus on certain plant species (those with a combination of particular life history, nutritional, and genetic attributes) eventually led to their domestication through a co-evolutionary cycle of enhanced returns and increasing human dependence (Piperno and Pearsall 1998, see also Piperno 2006).

Another early advocate of the application of behavioral ecology to the transition from foraging to farming, Gremillion (1996, 1998) drew on a range of HBE concepts (i.e. optimization, risk management, and opportunity costs) to frame various hypotheses for the shift in resource utilization and the adoption of small seeded plant domesticates in eastern North America at around 1000 BC. In the first archaeological application of behavioral ecology to animal domestication, Alvard and Kunzner (2001) considered the circumstances under which the immediate benefits of slaughtering an animal on encounter might be deferred long enough for a herd of managed animals to reach a sustainable size. They argue that the relatively rapid reproductive rate of medium-sized mammals addressed this discounting dilemma most effectively which is the reason why animals like sheep and goats were domesticated before larger meat package animals like cattle.

Kennett and Winterhalder's (2006) recent edited volume greatly expands upon the range and scope of HBE applications to the study of agricultural origins and diffusion. Rather than a punctuated switch for one dominant subsistence mode to another, chapters in this influential volume portray the transition from foraging to farming as an attenuated process in which managed resources (either locally domesticated or introduced from elsewhere) gradually infiltrate subsistence economies based on wild plant and animal resources. In each case, authors marshal a variety of different HBE principles and models to help explain the timing and the scope of the incorporation of domestic resources into these economies. The delayed embrace of maize agriculture in both the Pacific coast of southern Mexico (Kennett et al. 2006a) and in southeastern Arizona (Diehl and Waters 2006), for example, are both viewed from the perspective of diet breadth, cost-benefit analysis in which initially more profitable foraging strategies in these regions are only replaced when maize agriculture becomes more productive. In the Mexican case, a more robust embrace of maize agriculture is deferred until evolved varieties of higher yield maize become available. In the Arizonan example, the introduction of ceramic storage technology that mitigates storage losses is the added feature that makes an investment in maize more worthwhile. Denham and Barton (2006) similarly use diet breadth concepts to show how increasing human encouragement of starch-rich tropical plants elevated their rank within the subsistence round of foragers in highland New Guinea, resulting in decreased mobility and increased dependence on plant cultivation. Concepts of marginal valuation are used to sort out why certain foragers in the Fremont region of the eastern Great Basin and northern Colorado Plateau decided to farm, while others did not (Barlow 2006).

Central place foraging models are brought to bear in Gremillion's expanded consideration of agricultural emergence in eastern North America which examines how considerations of travel time between collection/cultivation localities and habitation sites may have affected decisions about which plant resources to exploit and the evaluation of the respective costs of the cultivation of domesticates and harvesting wild plants (Gremillion 2006). Ideal free distribution models are applied to changes in herd management, agricultural strategies, and land use by Neolithic farming populations in eastern Spain (McClure et al. 2006) and to predict the pace and direction of island colonization in Oceania (Kennett et al. 2006). Aldenderfer introduces the concept of costly signaling, a male reproductive strategy that has been linked to hunting behavior (Hawkes and Bliege-Bird 2002; O'Connell et al. 2002), as a means of explaining how hunters in the Andean highlands overcame discounting impediments to become camelid herders (Aldenderfer 2006). In many cases, the authors of these thought-provoking essays acknowledge that they lack the kind of empirical data needed to rigorously test the application of behavioral ecology to archaeological case studies of agricultural origins (i.e. Kennett et al. 2006a, and see Smith 2006), concentrating instead on developing the test implications of these models as applied to these particular case study examples and describing the kinds of data that would be needed to evaluate their models.

Agricultural Origins in the Near East

Each of these different applications of evolutionary biology to cultural evolution has looked to the emergence of agricultural economies as a way of demonstrating the efficacy of these co-opted evolutionary models in the study of culture change. Selectionist archaeologists cast this major transition in human history in terms of a gradual unfolding of events in which universal Darwinian forces, operating independently of sentient human agency, shaped the trajectory of discrete attributes within unique cultural lineages. Behavioral ecologists see agricultural emergence as the cumulative product of day-to-day human decisions guided by a set of microeconomic optimizing principles. Macroevolutionary archaeologists, on the other hand, characterize the origins of agriculture as a revolutionary right-angled departure from previous hunting and gathering adaptations guided by deliberate and directed human responses to stresses and opportunities whose impact reverberated across multiple societal levels.

As we have seen, case study examples used to advance these divergent positions have been drawn from around the world, from New Guinea to eastern North America. Curiously very few of these case studies have been based in the Near East, arguably the oldest and best known example of agricultural origins and the one that produced many of the major crops and livestock animals found in agricultural economies today. Even the Rosenberg (1990, 1994a, 1998) and the Kuijt and Prentiss (this volume) case studies discussed above stop just short of the actual domestication of plant and animal species and do not consider the subsequent coalescence of agricultural economies based on plant and animal domesticates.

This is particularly surprising given the increasing resolution of the empirical record of agricultural origins in this region. After more than 50 years of concentrated research there is now a detailed and comprehensive archaeological record of agricultural origins from all parts of the broad Fertile Crescent arc that spans the Near East. The regional scope of this work includes the far eastern Fertile Crescent of the Zagros Mountains and its piedmont where, from the 1950s through the 1970s, the first interdisciplinary expeditions examining Near Eastern agricultural origins were conducted (Braidwood and Howe 1960, Braidwood et al. 1983; Hole and Flannery 1967; Hole et al. 1969; Morentsen 1974, 1975; Smith 1972; Solecki 1981; Solecki 1965, (Figs. 6.1–6.2, Table 6.1). The southern and northern Levant stretching from the Sinai to the Upper Euphrates Valley has been the focus of more than three decades of intensive investigation into agricultural origins (Aurenche et al. 1988, 1989; Bar-Yosef 1982, 1990; Bar-Yosef et al. 1991; Cauvin 1978; Goring-Morris 1987; Moore 1991; Perrot 1983). Most recently, there has been a great deal of high profile work in the foothills and highlands of the eastern Taurus Mountains of southeastern Anatolia (Hauptmann 1993, 1999; Hours and Copeland 1983; Kozlowski 1999; Rosenberg 1994b; Rosenberg and Davis 1992), in central Anatolia (Esin and Harmankya 1999; Özbasaran 2000; Watkins 1996), and even on Cyprus 60 km off the Levantine coast (Guilaine and Briois 2001; Peltenburg 2003), which seems to be the beachhead of the first diffusion of Neolithic economy out of the Near East and into the Mediterranean Basin (Vigne et al. 2003; Zeder 2008a). Moreover,

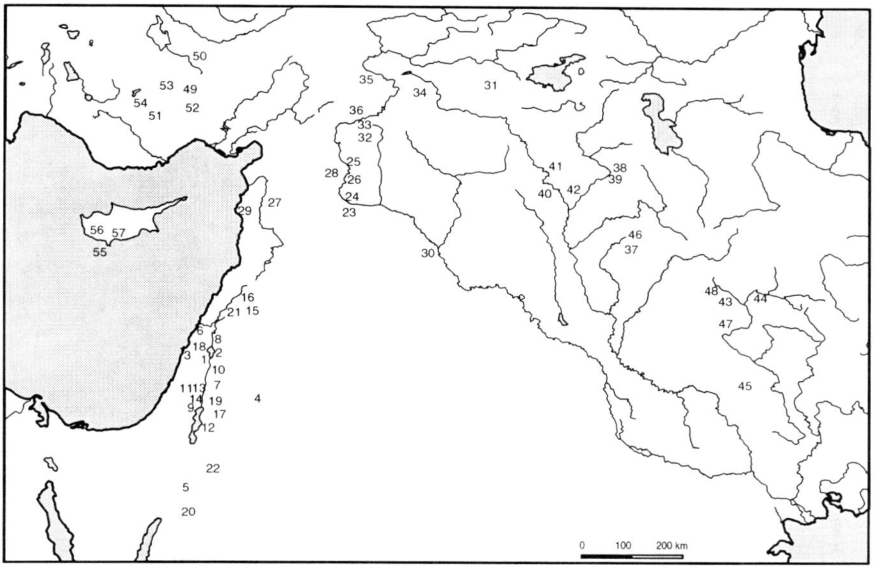

Fig. 6.1 Map of the Near East showing location of major sites. See Table 6.1 for names of sites

the past decade has witnessed remarkable advances in the documentation of domestication (see Zeder et al. 2006a, b) with increasingly well-developed archaeological and genetic analytical methods providing new insights into the process of domestication of many of the major Near Eastern crops and livestock. An increasingly refined temporal framework for these developments has also been built using precise and accurate small sample radiometric methods and calibration equations (Byrd 2005; Zeder 2009).

The Near Eastern record of the transition from foraging to farming has something to offer evolutionary anthropologists of all persuasions and would seem the ideal case study example for the evaluation of the efficacy of different evolutionary approaches to the study of culture change. The remarkable decades-long study of Kowzlowski and Aurenche (2005), for example, which uses a wide array of artifact types and architectural styles to construct deep-time subregional cultural lineages provides the ideal data set for the application of a selectionist cladistic analysis (i.e. Lipo et al. 2006). The depth and scope of this region-wide record, and its increasing temporal resolution is certainly complete enough to confirm that evidence of punctuated change is real and not the artifact of a spotty empirical record or the product of a faulty cultural typology based on an inadequate understanding of cultural variability. And although even with this exemplary empirical record we are still some way from being able to monitor daily decisions of individual actors or to tabulate accurate kilocalorie estimates of the energy costs in procuring Near Eastern resources and their returns, we do have a fairly robust understanding of the range

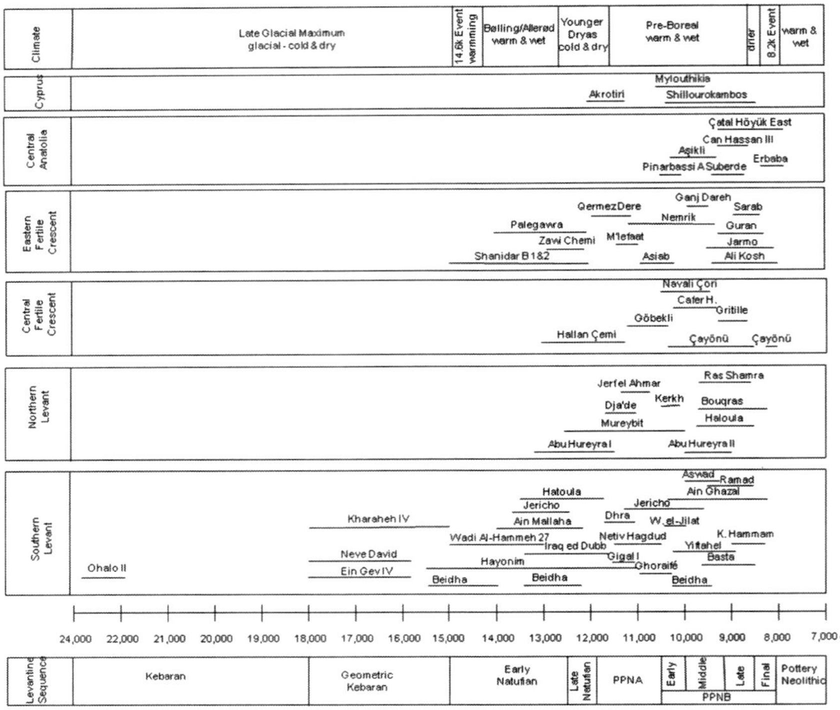

Fig. 6.2 Time line of Near Eastern sites, Levantine chronology, and climatic conditions. Compiled using information from Bar-Yosef and Meadow, 1995; Aurenche et al. 2001; Kuijt and Goring-Morris, 2002; Nesbitt, 2002, and Byrd, 2005

of resources utilized and the ways in which human interaction with these resources evolved over time, information well suited to a behavioral ecological approach to the study of culture change.

Although the empirical details of his model have not stood the test of time, V. Gordon Childe's conception of the basic nature of Neolithic emergence and its key distinguishing features outlined in his classic essay on the Neolithic Revolution (Childe 1951) provides an ideal rubric for the evaluation of the efficacy of these co-opted evolutionary models in characterizing the unfolding events that surrounded the emergence of agricultural economies in the Near East. The ten components Childe identified as key constituent elements of the Neolithic Revolution include the economic, social, and ideological aspects of this transition, as well as key material culture attributes needed to evaluate each of the different evolutionary frameworks considered here—whether these components are viewed as discrete attributes shaped by Darwinian forces, as parts of a constellation of interacting traits that comprise the Neolithic *Bauplan*, or as the set of external and internal constraints that structured resource selection over the course of the transition from foraging to farming. Childe's ten components are the following: (1) an agricultural economy based on domesticated plants and animals, (2) exponential population growth,

Table 6.1 Major Near Eastern Sites (from Bar-Yosef and Meadow, 1995; Aurneche et al. 2001; Nesbitt, 2002, Byrd, 2005)

Ref. No	Site	Region	Modern country	Dates kya cal. BP
1	Oahalo II	Southern Levant	Israel	24.0–22.0
2	Ein Gev IV	Southern Levant	Israel	18.0–16.0
3	Neve David	Southern Levant	Israel	18.0–16.0
4	Kharaheh IV	Southern Levant	Jordan	18.0–15.0
5	Beidha Early Natufian	Southern Levant	Israel	15.5–14.0
6	Hayonim Cave & Terrace	Southern Levant	Israel	15.5–11.0
7	Wadi al-Hammeh 27	Southern Levant	Jordan	15.0–13.0
8	Ain Mallaha	Southern Levant	Israel	14.0–12.0
5	Beidha Late Natufian	Southern Levant	Israel	13.4–12.3
9	Jericho Natufian	Southern Levant	Israel	13.7–12.3
10	Iraq ed Dubb	Southern Levant	Jordan	13.2–11.5
11	Hatoula	Southern Levant	Israel	13.7–12.9
12	Dhra	Southern Levant	Jordan	11.6–11.2
13	Netiv Hagdud	Southern Levant	Israel	11.5–10.8
14	Gigal I	Southern Levant	Israel	11.4–11.2
9	Jericho PPNA&B	Southern Levant	Palestine	11.2–9.5
15	Aswad	Southern Levant	Syria	10.5–9.3
16	Ghoraife	Southern Levant	Syria	10.8–10.3
17	Wadi el-Jilat 7	Southern Levant	Jordan	10.5–10.0
18	Yiftahel	Southern Levant	Israel	10.8-8.8
19	‘Ain Ghazal	Southern Levant	Jordan	10.4–8.2
5	Beidha PPNB	Southern Levant	Jordan	10.2–9.5
20	Basta	Southern Levant	Jordan	9.5–8.7
21	Ramad II	Southern Levant	Syria	9.4–8.6
22	Khirbet Hammam	Southern Levant	Jordan	9.0–8.5
23	Abu Hureyra I	Northern Levant	Syria	13.3–11.5
24	Mureybit Ia-IV	Northern Levant	Syria	12.6–10.0
25	Dja’de	Northern Levant	Syria	11.6–11.0
26	Jerf el Ahmar	Northern Levant	Syria	11.3–10.9
27	Tel Kerkh	Northern Levant	Syria	10.5–10.2
23	Abu Hureyra II	Northern Levant	Syria	10.0–9.0
28	Haloula	Northern Levant	Syria	9.7–8.5
29	Ras Shamra	Northern Levant	Syria	9.5–8.6
30	Bouqras	Northern Levant	Syria	9.5–8.2
31	Hallan Çemi	Central Fertile Crescent	Turkey	13.0–11.3
32	Göbekli Tepe	Central Fertile Crescent	Turkey	11.2–10.5
33	Navali Çori	Central Fertile Crescent	Turkey	10.7–9.7
34	Cayönü Aceramic	Central Fertile Crescent	Turkey	10.5–8.5
35	Cafer Höyük	Central Fertile Crescent	Turkey	10.2–9.5
36	Grittle	Central Fertile Crescent	Turkey	9.5–8.8

Table 6.1 (Continued)

Ref. No	Site	Region	Modern country	Dates kya cal. BP
34	Cayönü Ceramic	Central Fertile Crescent	Turkey	8.4–8.3
37	Palegawra	Eastern Fertile Crescent	Iraq	16.0–12.0
38	Shanidar B1&2	Eastern Fertile Crescent	Iraq	15.0–12.3
39	Zawi Chemi Shanidar	Eastern Fertile Crescent	Iraq	13.0–12.3
40	Qermez Dere	Eastern Fertile Crescent	Iraq	12.0–11.2
41	Nemrik	Eastern Fertile Crescent	Iraq	11.5–9.2
42	M'lefaat	Eastern Fertile Crescent	Iraq	11.5–11.0
43	Asiab	Eastern Fertile Crescent	Iran	11.0–10.5
44	Ganj Dareh	Eastern Fertile Crescent	Iran	10.0–9.7
45	Ali Kosh	Eastern Fertile Crescent	Iran	9.5–8.0
46	Jarmo	Eastern Fertile Crescent	Iraq	9.5–8.0
47	Guran	Eastern Fertile Crescent	Iran	9.3–8.5
48	Sarab	Eastern Fertile Crescent	Iran	9.0–8.5
49	Pinarbassi A	Central Anatolia	Turkey	10.5–10.2
50	Aşikli Höyük	Central Anatolia	Turkey	10.4–9.4
51	Suberde	Central Anatolia	Turkey	9.6–8.8
52	Can Hasan III	Central Anatolia	Turkey	9.4–8.7
53	Çatal Höyük East	Central Anatolia	Turkey	9.4–8.2
54	Erbaba	Central Anatolia	Turkey	8.6–8.4
55	Aetokremnos	Cyprus	Cyprus	12.0–11.7
56	Mylouthikia	Cyprus	Cyprus	10.3–9.5
57	Shillourokambos	Cyprus	Cyprus	10.2–8.5

(3) storage of surplus and a system of delayed returns of productive resources, (4) sedentism, (5) trade networks focusing on nonessential items, (6) decentralized social mechanisms for the coordination of collective activities, (7) associated and enabling magico-religious traditions that focus on the promotion of fertility, (8) ground stone implements, (9) pottery, and (10) weaving implements (e.g., spindle whorls). The following discussion presents a relatively abbreviated summary of the Near Eastern empirical record for each of these key features, considering both the timing of their appearance in the archaeological record and the regional variability in how they are manifested. A more comprehensive review of this very large data set can be found in Zeder 2009.

Material Culture Attributes (Components 8–10)

The appearance of material culture attributes in Childe's model provide convenient temporal brackets for Neolithic emergence in the Near East. Weaving implements in the form of clay spindle whorls are the last of the ten components to appear. First reported about 9,500 cal. B.P., they become quite ubiquitous, especially in the central and eastern Fertile Crescent, at about 9,000 cal. B.P. (Kowzlowski and Aurenche 2005: 31, 258)—a development which indicates that the genetically driven changes that transformed sparse secondary coat hairs of caprines into dominant wooly fleece had taken place by this time (Bökönyi 1977; Ryder 1983).

Pottery is also a relatively late constituent element of Neolithic emergence in the Near East. Clay-crafted ceramics are first documented at the site of Ganj Dareh in highland Iran at about 10,000 cal. B.P. (Smith 1976). The spread of undecorated ceramics is quickly followed by painted wares found in the Zagros, southeastern Anatolia, the northern Levant, and central Anatolia by about 9,000 B.P. with each region displaying distinctive stylistic variants (Kowzlowski and Aurenche 2005:32–34, 270–273). Pottery was not used in the southern Levant until about 8,400 cal. B.P. (Aurenche et al. 2001).

Ground stone tools, on the other hand, are the earliest of Childe's ten key components to appear in the Near East. Both bedrock and portable ground stone mortars clearly associated with cereal processing are found in the southern Levant as early as 24,000 years ago (Piperno et al. 2004; Weiss et al. 2004). They are ubiquitous throughout the southern Levant, along with pestels, querns, stone bowls, and hand stones by about 14,000 cal. B.P. (Bar-Yosef 1990, Bar-Yosef and Meadow 1998:57), with the full complement of ground stone tools widespread across the Fertile Crescent by about 13,000 cal. B.P. (Kowzlowski and Aurenche 2005:23, 145–150).

The appearance of these different material culture attributes, from 24,000 to 9,500 cal. B.P., brackets a very long temporal span of more than 14 millennia. Even if one starts the Neolithic countdown clock with the widespread appearance of ground stone tools at 14,000 cal. B.P. and Neolithic emergence with the appearance of pottery at about 10,000 cal. B.P., this is still a quite extended 4,000-year period during which the other eight constituent components make an appearance at different times and places in the Near East.

Sedentism and Storage (Components 3 and 4)

The ground stone tools whose appearance brackets the beginning of the temporal span considered here were used by mobile Early Epipaleolithic (Kebaran) period foraging populations in the Southern Levant (24,000–18,000 cal. B.P.). Even during the height of the Late Glacial Maximum (c. 23,000 cal. B.P.), small groups of foragers seem to have congregated, possibly for extended periods of time, in favored locations like Ohalo II on the shores of the Sea of Galilee. While residing in these sheltered based camps they utilized a broad spectrum of plant resources including a wide variety of small and large seeded wild cereals, legumes, nuts, and fruits, and a diverse array of game animals (Nadel 2004; Piperno et al. 2004; Weiss et al. 2004). Mobile foraging adaptations proliferated in the ensuing Geometric Kebaran period (18,000–14,600 cal. B.P.) when, under conditions of climatic amelioration, populations expanded into more desertic areas (Bar-Yosef and Meadow 195:54). These people seem to have followed a similar seasonal round, dispersing into upland areas in spring and summer and coalescing into larger social units at sites like Neve David and in Gev IV, in the autumn and winter. Although there seems to have been a heavy emphasis on plant processing in these base camp settlements, there is no evidence for storage in Middle Epipaleolithic sites, although this may be attributable to the lack of large-scale horizontal excavation of these sites or the use of perishable baskets for plant storage (Bar-Yosef 1990; Byrd 2005:254; Henry 1989: 169–179).

Sedentary communities are clearly evident in the southern Levant in the following Early Natufian (ca. 14,000–12,0500 cal. B.P.) in the form of large (c. 1200 m^2) base camps situated in strategic locations with access to several different resource zones that housed as many as 60 people in small oval structures (Byrd 2005:255). Year round occupation of at least some of these base camps is indicated by the seasonality of plant and animal resources utilized and by the presence of commensal animals that exploit the niche created by long-term human habitation (Tchernov 1991). Limited evidence for storage in these base camps can be found in the plastered pits that occur outside of house structures at the site of Ain Mallaha in the Upper Jordan Valley and the paved bins at Hayonim Terrace (Bar-Yosef 1982, Kuijt in press). The smaller, ephemeral sites that radiate from larger base camps suggest continued seasonally mobile foraging activities during this first phase of the Late Epipaleolithic (Henry 1989:219).

The resumption of more mobile settlement patterns in the subsequent Younger Dryas climatic down turn of the Late and Final Natufian southern Levantine sequence (12,900–11,600 cal. B.P.) is followed by a return and proliferation of sedentary settlement patterns in the following Pre-Pottery Neolithic (PPNA) that overlaps with the end of the Younger Dryas and the ensuing climate amelioration in the Early Holocene (c. 12,000–10,500 cal. B.P.). PPNA settlements like Nativ Hagdud, Gigal, and Dhra' were often located near springs or marshy areas. They range in size from 2000 m^2 hamlets to 2.5 ha "villages" comprising numerous circular or oval semi-subterranean houses (Bar-Yosef 1992:34, Bar-Yosef and Meadow 1995:62). In contrast to Early Natufian settlements, there is ample evidence for storage both inside and outside of houses in the form of pits, bins, and even what have been interpreted as "silos" (Bar-Yosef and Meadow 1995:62; Kuijt 2009). There is, however, also evidence for special temporary campsites likely used for the extraction of special resources during the PPNA (Bar-Yosef and Meadow 1995:62).

The following Early PPNB period (10,500–10,000 cal. B.P.) marks the beginning of the shift from round or oval to small-scale subrectangular architecture (Kuijt and Goring-Morris 2002:385). During the ensuing Middle PPNB (10,000–9,200 cal B.P.), fully sedentary villages of up to 3 ha, situated in better watered parts of the region, comprised of free-standing, multiroom rectangular structures (Byrd

1994:659; Kuijt 2000:85). Smaller more ephemeral settlements with small rounded buildings are still found in more arid zones, however (Kuijt and Goring-Morris 2002:388–389). In the Late PPNB (9,200-8,700 cal. B.P.) settlement shifts eastward with a number of 10–14 ha "mega-villages" like Ramad, Basta, and Khirbet Hammam established in the Eastern Jordan Valley, characterized by closely packed, increasingly compartmentalized, and often two-story rectangular houses. While storage facilities can be found outside of houses in the PPNA, over the course of the PPNB storage functions shift to inside houses, with lower storey areas often dedicated exclusively to storage (Kuijt 2000:81; Kuijt 2009). A period of climatic deterioration culminating in the abrupt global pulse of cold, dry conditions at 8,200 cal. B.P. coincides with the Final PPNB/PPNC which saw the collapse of the large Late PPNB villages and the dispersal of population into more diffuse, less-nucleated settlements (Bar-Yosef and Meadow 1995:45; Kuijt and Goring-Morris 2002:413). A new village system appears in the southern Levant in the following Pottery Neolithic Period (ca. 8000–7800 cal B.P.) (Gopher and Gophna 1993).

The sparser record from regions outside the southern Levant sees the first evidence of large, semi or fully sedentary settlement during the Late Epipaleolithic climatic downturn at sites like Mureybit and Abu Hureya I (Cauvin 1978; Moore et al. 2000) in the Middle Euphrates Valley, at Hallan Çemi in the Taurus region of southeastern Anatolia (Rosenberg and Redding 2000), and at Zawi Chemi Shanidar in the Zagros (Solecki 1981). All these sites are characterized by the presence of relatively densely packed round and oval semi-subterranean structures, numerous heavy ground stone tools, and some evidence of storage. Round houses in these regions also transitioned into rectangular houses, with the earliest evidence of this transition seen in the northern Levant at about 10,900 cal. B.P. (roughly 500 years earlier than in the southern Levant) and the latest in the Zagros evidenced at the site of Ganj Dareh at about 10,000 cal. B.P. Throughout the region the transition from round to rectangular houses is accompanied by an increasing elaboration of storage facilities within houses best exemplified by the sequence of storage configurations seen at Çayönü in southeastern Anatolia (Özdoğan and Özdoğan 1989). Colonizing populations that arrived in Cyprus at about 10,500 cal. B.P., in contrast, never seem to have made the transition in architectural styles, retaining round houses throughout the Neolithic (Peltenburg 2004).

Population Growth (Component 2)

Estimating the rate and amplitude of population growth from archaeological data is never easy, which is why people often resort to using trends in resource intensification and mobility as indirect proxy measures of population size and degree of packing (Binford 2001; Rosenberg 1998)—a problematic practice when one is testing models that hold demographic pressures responsible for changes in a broadening of the resource base and sedentism (Zeder 2006a, 2009 Zeder and Smith in press). Though admittedly incomplete in many regards, survey data recording the number, size, and spacing of settlements and the intensity and duration of their occupation are a preferable alternative to such proxy data. The evidence for population growth in the southern Levant comes from comprehensive studies by Henry (1989, 2002) for the Epipaleolithic up to the Early Natufian and by Kuijt and Goring-Morris (2002:424, Kuijt 2000)for the Late Natufian through the PNNB. Despite the marked increase in settlement size and thickness of deposits in Early Natufian settlements, population growth over the course of the Epipaleolithic and into the PPPNB seems to have been a rather slow and incremental process without clear evidence of packing or pressure on resources (Henry 1989). The exponential spike in population levels which is one of Childe's ten components of Neolithic emergence is not seen in the southern Levant until the Late PPNB, with a subsequent crash in population in the ensuing Pottery Neolithic A (PNA) periods at about 8,300 cal. B.P (Kuijt and Goring-Morris 2002).

Population measures in the rest of the Fertile Crescent suggest even lower population densities in the Epipaleolithic through the beginning of the Pre-Pottery Neolithic, with possible relatively empty areas separating settlement zones in between the southern and northern Levant, southeastern Anatolia, central Anatolia, and the Zagros (Kowzlowski and Aurenche 2005:85). Evidence for demographic trends across this broad area echo those in the southern Levant suggesting a slow and gradual process of population growth that spikes toward the latter part of the sequence at about 9,500 to 9,000 years ago (Hole 1990a; 1990b; Hole and Flannery 1967; Morentsen 1974, 1975; Özbaşaran and Buitenhuis 2002; Özdoğan 2002).

Mechanisms for Social Cohesion (Component 6)

The transition from small round or oval houses to rectangular multiroom structures witnessed throughout the region provides special insight into changes in social mechanisms operating at the household level. While originally read as indicative of a shift from a system of partilineal, polygamous households to that of monogamous, nuclear families (Flannery 1972), Byrd has subsequently argued that both the smaller round houses and the later rectangular ones were occupied by nuclear families (Byrd 1994, 2005). Flannery (1993, 2002), Byrd (2000) and Kuijt (2000,2009) all agree, however, that the move from storage structures located outside of houses to the increasing formalization of storage and other activity areas within households witnessed during this architectural transition signals the increasing importance of notions of restricted ownership over resources and resultant increases in individual household autonomy. This shift, though different in timing and in the precise details of household arrangement, is seen in all regions of the Fertile Crescent.

At the same time household-level organization was looking progressively inward, and there seems to have been corollary developments in community organization directed at leveling differences among households and promoting community cohesion. Elements of community-level organization can be traced as far back as the Early Natufian (c 13,000 cal. B.P.) in a large structure with plaster and painted benches at Ain Mallaha argued to have served a nondomestic function

(Byrd 1994:259). Late Natufian sites of Hallan Çemi and Zawi Chemi Shanidar (c. 12,500 cal. B.P.) provide several indications of structures, public areas, and possible feasting activities all believed to help promote community cohesion (Rosenberg and Redding 2000, Solecki 1981:53–54; Solecki and McGovern 1980).

The best known example of concerted communal activity in the major stone constructions built by inhabitants of PPNA Jericho (ca. 11,000 cal. BP) (Kenyon 1981). These massive structures included a wall, that may have helped divert seasonal flash flooding (Bar-Yosef 1986), and a tower, which at one point served as a repository for collective burials providing a ritual focus for the community that expended so much energy building this major stone construction (Bar Yosef 1986; Bar-Yosef and Meadow 1995:63; Kuijt and Goring-Morris 2002:373–376). Additional ritual focal points are found in Southeastern Anatolia in the large, symmetrically arranged, T-shaped stone monoliths decorated with bas-relief images of animals at the site of Göbekli Tepe and Karahan used over the course of the PPNA and into the early PPNB (11,000–9,500 cal. B.P) (Çelik 2005; Schmidt 2005). Special purpose constructions of varying form and elaboration are found throughout the Fertile Crescent in the clearly nondomestic buildings found in every phase of occupation at Çayönü in southweastern Anatolia (Özdoğan and Özdoğan 1989), the "shrines" at 'Ain Ghazal in the southern Levant (Rollefson 2000), and the "cult buildings" at Aşikli Höyük in central Anatolia (Esin 1998). Despite variations in the form of these special function constructions, the general underlying picture that emerges from across the region is one of essentially self-contained independent communities in which highly autonomous households were bound together by community-level social events and ritual (Hole 2000)—a picture of social life that corresponds well to Childe's image of the decentralized social mechanisms used to coordinated of collective activities in emergent Neolithic communities.

Further insight into social mechanisms that held communities together can be found in burial treatments. The first real concentration of human burials is found in the Early Natufian sedentary communities of the southern Levant. The grave goods, body placement, and burial architecture of often collective Early Natufian burials has been interpreted by Byrd and Monahan (1995:283; Byrd 2005:257) as indicative of horizontal rather than vertical differentiation between individuals, with burial practices reinforcing affiliations between different age, grade, and sex distinctions that cut across kinship distinctions and promote community cohesion. More mobile Late Natufian people in the southern Levant returned to abandoned Early Natufian camps to bury their dead, often as secondary burials with few or no grave goods, perhaps as a way of connecting with the community structure that emerged during easier times. Despite the lack of any form of social distinction in Late Natufian burials, Byrd and Monahan (1995:283) see the practice of selective skull removal of certain individuals from these graves as a signal of emergent notions of individual leadership. The primary burial of unornamented bodies and selective secondary removal of skulls, which may be decorated and reburied in extra-household more public places, continues into the PPNA and intensifies in the ensuing PPNB in the southern Levant. This practice is interpreted by Kuijt (2000; Kuijt and Goring-Morris 2002) as evidence of ritual practice directed at minimizing differences between households, while emphasizing a collective community ethos built around lineage lines and a sense of shared leadership among households.

While once again different in many details, burial practices elsewhere in the Fertile Crescent seem to echo themes seen in the southern Levant. The necropolis at Çayönü containing the remains of some 300–400 individuals, including up to 90 disarticulated skulls, and some evidence of animal and perhaps even human sacrifice (or less dramatically, but perhaps more likely, post-mortem dismemberment) provides a particularly good example of the dead forming a focus point for the cohesion of the community of the living (Hole 2000; Özdoğan 1997). Skull removal and decoration also finds its way to central Anatolia at the well-known Late PPNB site of Çatal Höyük. Instead of being placed in a central community focal point as done in the southern Levant, the decorated skull found at Çatal Höyük is placed within a house—a practice interpreted as reinforcing household-level ritual at a site that is argued to lack evidence of corporate ritual or civic leadership (Asouti 2006).

Magico-religious Traditions Emphasizing Fertility (Component 7)

Insight into ideology and views of the cosmos can be sought in the small portable clay or stone figurines often thought to be vehicles of magic or charms used in ritual acts (Ucko 1968; Voigt 2000). In a radical departure from Natufian figurative art dominated by representations of animals or genderless humans, the PPNA (c. 12,000–10,500 cal. B.P.) sees a proliferation of female figurines and other clay objects emphasizing fecundity and reproduction (Bar-Yosef and Meadow 1995:64; Kuijt and Goring-Morris 2002:377) that seem to correspond quite well with Childe's seventh component of Neolithic emergence. While much has been made of the ascendancy of "mother goddess" symbols during the PPNA (as well as the later rise of bull symbols in the ensuing PPNB) (Cauvin 2000a, b), it is important to note that there is ample representation of male and genderless imagery in portable art, as well as continued representation of a wide variety of animal symbols. Moreover, while female figurines can be found across the Fertile Crescent during the PPNA and later, there is a great deal of regional variation in posture, ornamentation, and style in female representation (Kowzlowski and Aurenche 2005:28-29). The degree of regional variation in the use of these symbols runs counter to the Cauvin's evangelizing scenario of the spread of northern Levantine symbols and other aspects of Neolithic life across and, eventually, out of the Fertile Crescent. Although certain symbolic elements recur in both portable and nonportable art across the region (female figurines, bull representations and animal crania set into architecture, snakes and birds of prey in glyptic art, and wall murals), there is enough regional variability in this art to suggest that peoples of the central Fertile Crescent, with their distinctive megaliths and figurative art, shared a common symbolic system (and possibly a common language) that was distinct from people in the southern Levant, Zagros, and central Anatolia (Kowzlowski and Aurenche 2005; Stordeur 2004).

Trade Networks (Component 5)

Although Childe maintained that emergent Neolithic communities were essentially self-sufficient, he also predicted that trade among food-producing communities, especially in nonessential or luxury items, established essential communication channels that were responsible for the spread of Neolithic lifestyles. Trade in exotic items (obsidian, marine shells, rare stones) is seen in the Near East as far back as the Epipaleolithic (Byrd 2005:254), overlaying a process of increasing regionalization in lithic industries. Over the course of the PPNA, the process of localization in many elements of material culture continues, with a corollary increase in the range and amount of long distance exchange of traceable trade items (Bar-Yosef and Meadow 1995, Bar-Yosef Mayer 2000; Kuijt and Goring Morris 2002).

An exponential expansion and elaboration of interregional trade, however, is not seen until the PPNB, during which time a vast "interaction sphere" seems to have existed linking all areas of the Fertile Crescent. These far flung trade networks are sometimes portrayed as the primary vehicles for the diffusion of Neolithic lifeways across the region, whether emanating from the southern Levant (Bar-Yosef 2001; Bar-Yosef and Belfer-Cohen 1989), from the Euphrates Valley (Cauvin 2000a b), or from the region between the upper Euphrates and the Tigris (Kowzlowski and Aurenche 2005). However, there is also considerable evidence for a more polycentric view of the PPNB world as made up of more fragmented distinct local cultures that both evolved and remained relatively independent of one another (Gebel 2002, 2004; Rollefson 2004; Rollesfson and Gebel 2004). Despite the persistence of local traditions in different parts of the Fertile Crescent, there can be no denying that the PPNB witnessed an explosion of trade items across the region that included exotic goods, lithics, and other elements of domestic technology (including plant and animal domesticates), as well as various elements of social and religious behavior that can be found from the southern Levant to the Zagros and into central Anatolia. The spread of so many aspects of Neolithic life, when combined with the continued regional identity of material culture across the region, suggests the existence of a pan-regional social, economic, and ideological interaction sphere into which local communities selectively subscribed, adopting and tailoring certain foreign elements to meet localized needs (Asouti 2006).

Agricultural Economy Based on Plant and Animal Domesticates (Component 1)

The final and central component of Childe's Neolithic model, the domestication of plants and animals and the development of agricultural economies, is coming into increasingly high-resolution focus in the Near East. Documenting the process of domestication and the emergence of agriculture has been a primary objective of research in the region for more than a half a century (Braidwood and Howe 1960; Braidwood et al. 1983). The last decade, in particular, has witnessed remarkable advances in our understanding of how this process unfolded in the Near East, thanks in large measure to the development of breakthrough archaeological, genetic, and chronometric methods for detecting and dating the process of plant and animal domestication (see discussion in Zeder et al. 2006a, 2006b). Perhaps the greatest breakthroughs have been in the development of powerful new markers of domestication that are not only transforming our understanding of how this process unfolded but are also causing a reconsideration of the very concept of domestication and the relationships between domestication and agricultural emergence (see discussion in Zeder 2006a, 2006b).

The earliest morphologically altered plant domesticates in the region have been argued to be the handful of plump grains of rye recovered among the large quantity of wild cereals and other wild plants remains from Late Natufian (c. 13,000–12,000 cal B.P.) levels at Abu Hureyra in the northern Levant (Hillman 2000). The domestic status of these rye remains is controversial (see Nesbitt 2002), but even if domestic rye were present at Abu Hureyra, this potential domesticate appears to have had an ephemeral existence, since domestic rye is not seen again in Near Eastern assemblages for another 2,000 years when it appears at the site of Can Hasan III in southeastern Anatolia (Hillman 1978). Modern domestic rye appears to have had a European origin (Zohary and Hopf 2000).

Barley is another possible candidate for the earliest Near Eastern plant domesticate based on the non-shattering, tough rachises found in small numbers among wild, brittle rachis barley at PPNA Nativ Hagdud (Bar-Yosef and Kislev 1989). The small percentage of tough rachis grains in this assemblage (about 4%), however, is within the range of occurrence of this domestic morphotype in modern wild strands of barley, raising questions whether these grains represent true domesticates or the precursors of varieties eventually selected in the domestication process (Kislev 1989 1997; Zohary 1992). Another proposed PPNA domesticate (c. 11,400 and 11,200 cal. B.P.), the parthenocarpic figs (non-seed-bearing) found at Gigal I in the lower Jordan Valley (Kislev et al. 2006a), have also been argued to have been the mutant forms known to occur in wild figs (Denham 2007; Lev-Yadin et al. 2006a; but see Kislev et al. 2006b.)

The earliest securely identified and dated evidence of the arrival of tough rachis domestic einkorn wheat in the Near East was not found until the Early PPNB (c.10,500 cal. B.P.) at sites in southeastern Anatolia: Navali Çori, Cafer Höyük, and, possibly, Çayönü (Nesbitt 2002; Tanno and Wilcox 2006b). Firm evidence of tough rachis, domestic two-row barley is not found until the Middle PPNB (c. 10,000 cal B.P.) at which time it was recovered from sites throughout the Fertile Crescent and central Anatolia (Nesbitt 2002). The Middle and Late PPNB also saw the development of various forms of free threshing wheat and barley (Nesbitt 2002). The addition of domestic varieties seems to have been very gradual, with tough rachis varieties comprising only 10% of the einkorn recovered from Early PPNB levels at Navali Çori and domestic morphotype barley only 35% of the barley recovered from Middle PPNB levels at Aswad , and 50% of the Late PPNB barley at Ramad (Tanno and Wilcox 2006).

Evidence for initial domestication of non-cereal crop plants like pulses is less clear since morphological markers of pulse domestication are either absent or not yet identified. However, based on the habits of wild pulses (their high rate of seed dormancy, low yield, and the tendency of their seed pods to shatter when ripe) it has been argued that large quantities of lentils recovered from sites like Jerf el Ahmar and Netiv Hagdud make it likely that lentils were under cultivation and well on their way to full domestication in both the northern and southern Levant by the PPNA (Weiss et al. 2006). Early PPNB (c. 10,200 cal. B.P.) chickpeas and fava beans recovered in northwestern Syria at the Tel Kerkh are also thought to represent an early stage of domestication of these staples of the modern Near Eastern diet (Tanno and Wilcox 2006b).

The possibility that cereal and pulse crop plants were subjected to a prolonged period of intensive cultivation prior to displaying characteristic morphological markers of domestication is becoming increasingly more likely. A variety of methods are now being used to monitor this process, including the presence of weed complexes characteristic of cultivation (Colledge 1998, 2002), the likely movement of plants outside their preferred habitats, the progressive decrease of other gathered plants in archaeobotanical assemblages, and increases in the breadth (but not the length) of seeds (Wilcox et al. 2008), all which are taken to be signs of human management of morphologically wild cereals and pulses. These markers have been used to argue that intensive cultivation of wild cereals and pulses stretches back perhaps as far the Late Natufian at sites like Mureybit and Abu Hureyra (Colledge 2002) and was certainly practiced by the PPNA in both the southern and northern Levant (Tanno and Wilcox 2006a; Wilcox 2005; Wilcox et al. 2008). The delay between intensive cultivation of future crop plants and the display of morphological change may, in fact, be as much as several millennia (Weiss et al. 2006; Wilcox et al. 2008). The timing of the leading edge of this process is still uncertain, but both macro-and micro-botanical evidence from Early Epipaleolithic Ohalo II definitively demonstrates the utilization of wild cereals in the southern Levant reaches back at least 24,000 years ago (Piperno et al. 2004; Weiss et al. 2004).

Dogs with characteristic shortened jaws and crowded teeth found in burial contexts at Ain Mallaha (c. 13,000 cal B.P.) in the southern Levant (Davis and Valla 1978; Tchernov and Valla 1997) and a dog jaw found in roughly contemporary levels at Palegawra cave in the Zagros (Turnbull and Reed 1974) provide the earliest evidence of animal domestication in the Near East. Believed to have been a nonfood domesticate in the Near East, dogs probably entered into association with humans through the same commensal relationship as mice, rats, and sparrows drawn to human habitations to feed off refuse (Zeder in press).

Evidence for the appearance of morphologically altered domestic livestock species is much less clear. For more than 30 years, the industry standard for documenting initial livestock domestication has been a marked and sudden reduction in overall body size said to occur in all livestock species including sheep and goats, pigs, and cattle (Grigson 1969; Helmer 1992. Hongo and Meadow 2000; Legge 1996; Meadow 1989; Peters et al. 1999 2005; Uerpmann 1978, 1979). Based on this marker, goat domestication was set at about 10,000 B.P, with sheep domestication occurring a bit later between about 10,000 to 9,750 cal. B.P (Bar-Yosef and Meadow 1995: 89). This size reduction marker put both cattle and pig domestication somewhat later still at about 9,500-9,000 cal. B.P. (Bar-Yosef and Meadow 1995:90). Thus the domestication of animals was thought to postdate the appearance of plant domesticates in the region (at that point dated to between about 11,000–10,500 cal. B.P.) by as much as millennia or more.

Recent work, however, has called the utility of this marker into question. A study assessing the impact of various factors on body size in modern wild and domestic caprines from the Zagros (sex, regional variation, age, and domestic status) has demonstrated that the signature reduction in the size of goat bones detected in the archaeofaunal assemblage from Ganj Dareh around 10,000 cal. B.P. was not a reflection of body size reduction in these animals as originally argued (Bar-Yosef and Meadow 1995:87; Uerpmann 1979) but was instead attributable to a shift in the demographic composition of the adult portion of managed herds (Zeder 2001, 2005, 2006c; Zeder and Hesse 2000). This demographic change has been linked to a change in harvest patterns from a hunter's strategy that maximizes immediate off-take (thus focusing on large adult males) to a herder's interest in the growth of herds which dictates the early harvest of young males and the delayed slaughter of females until they have passed peak reproductive males—a harvest pattern clearly evident in sex-specific harvest profiles constructed for goat assemblages from the Zagros region (Zeder 2001, 2005, 2006c, 2008b). A similar demographic shift is also responsible for the apparent reduction in the size of sheep skeletal elements in the Zagros at about 9,000 years ago (Zeder 2008b). Morphological change that can unequivocally be tied to domestication is not seen in goats until at least 500 years later, once managed herds were moved out of the natural habitat of wild goats. As seen in the goat remains from Ali Kosh in lowland Iran (Hole et al. 1969), the horns of these animals demonstrate changes in size and shape indicative of the relaxation of pressure for large horns in mate competition which seems to have come into play once these animals were genetically isolated from wild herds and humans assumed complete control over their breeding (Zeder 2001, 2005 2006c, 2008b).

Although archaeozoologists have been slow to adopt methods for the construction of high-resolution sex-specific harvest profiles, lower resolution demographic data from southeastern Anatolia suggests that the management of morphologically wild sheep and goats dates back at least as far as the end the Early PPNB (c. 10,500 cal. B.P.) (Peters et al. 1999, 2005). Similar patterns are seen among sheep specimens from Aşikli Höyük in central Anatolia at about 10,400–9,400 cal. B.P (Buitenhuis 1997; Vigne et al. 1999) and among Middle PPNB (10,000–9,200 cal. B.P.) goats in the northern Levant (Legge 1996) and in the southern Levant (Horwitz 1993, 2003). These data suggest that sheep and goat were brought under human control, possibly independently of one another, in the region stretching from southeastern Anatolia into northwestern Iran by at least 10,500 cal. B.P, if not earlier (Zeder 2008a). Clear-cut morphological markers of caprine domestication are not seen then for at least a millennia after the initial management of caprine herds. Demographic data from Late Epipaleolithic (ca. 12,5000 cal. B.P.) contexts at Hallan Çemi and Zawi Chemi Shanidar may capture the leading edge of this process, initiated by the development of hunting strategies geared at enhancing the availability of local stock of wild sheep under increasing hunting pressure from increasingly sedentary communities (Redding 2005; Zeder 2008b).

A similar extended process is also becoming increasing likely for pigs, the leading edge of which may also be seen in archaeofaunal data from Hallan Çemi during the Late Epipaleolithic (Rosenberg et al. 1998). More definitive evidence of the process of pig domestication can be found in the modifications in molar lengths and the change in demographic profiles first detected in Early PPNB levels (c. 10,300 cal. B.P.) at Çayönü (Ervvnck et al. 2001). Demographic shifts detected among cattle remains from the Upper Euphrates Valley (Helmer et al. 2005) suggest that cattle were coming under management between 11,000–10,000 cal. B.P.

Thus for both plant and animal domesticates we are beginning to be able to detect a prolonged process of increasingly intensive human management that precedes the appearance of traditional morphological markers of domestication by a 1,000 years or more. This long delay between intensive human management and subsequent morphological change in both plants and animals raises real questions about when the threshold from wild to domestic was crossed—or even if it is worthwhile to try to determine a specific turning point in what is in effect a gradual, continuous process (Zeder 2006a). It is interesting to note, however, that early pioneers who colonized Cyprus at about 10,500 cal. B.P arrived with morphologically domesticated wheat and barley, as well as morphologically wild but likely managed sheep, goats, cattle, and pigs (along with other game animals like fallow deer and foxes that were never domesticated) (Colledge 2004; Murray 2003; Vigne et al. 1999, 2000, 2003). The wholesale exportation of this domestic subsistence base, as well as mainland wild resources, strongly suggests that the process of domestication on the mainland was well along by this date. A more concerted effort at applying the new suite of non-morphological markers of plant and animal domestication in the Near East will likely detect that the leading edge of this process was much earlier than currently thought (Zeder 2008a).

Genetic data from modern wild and domestic crops plants and animals is providing further important insight into this process. Recent genetic studies suggest that wild einkorn was brought under domestication (perhaps multiple times) in the north-central Fertile Crescent (Kilian et al. 2007, though see Heun et al. 1997, 2008), the region where, as we have seen, archaeological evidence has produced the earliest domestic einkorn. Genetic evidence also points to barley domestication occurring both in the western (Badr et al. 2000) and the eastern arms (Morrell and Clegg 2007) of the Fertile Crescent, again a finding that corresponds to archaeological evidence from these regions (Wilcox 2002:137). Emmer wheat domestication seems to have been concentrated in the northern Levant according to genetic data (Ozkan et al. 2002; Salamini et al. 2002), although archaeological evidence points to there being another center of emmer domestication in the southern Levant which may have no living descendents (Wilcox 2002). The wild variety of lentil most closely related to modern domestic lentils is found in southeastern Turkey and northern Syria (Ladizinsky 1989), pointing perhaps to the early spread of cultivated lentils from the northern to the southern Levant in the PPNA (Weiss et al. 2006). The variety of wild chickpea most closely related to modern domestic chickpeas (Sudupak et al. 2004) is the most westerly variety sampled and the closest to the site of Tel Kerkh where Wilcox believes he has evidence of chickpea cultivation at 10,200 cal. B.P (Tanno and Wilcox 2006b).

Genetic analyses of livestock species domesticated in the Near East provide a similar picture of multiple lineages of domesticates having been brought under domestication in different places across the region. A remarkable recent study of genetic variability among modern wild bezoar goats (Naderi et al. 2007, 2008) indicates that all six currently known lineages of domestic goat were brought under domestication in the Zagros/Taurus region. Genetic signatures of the wild goats belonging to at least two of these domestic lineages (the A and the C lineages) suggest a process of rapid population growth and geographic translocation which may represent the imprint of human mitigated reproductive isolation and movement of these animals within the natural habitat of wild goats during the early phases of goat domestication—a genetic signature perhaps of the long period of human management detected in the harvest profiles of archaeological assemblages of morphologically wild goats in this region. These same two lineages appear to have moved out of this natural habitat domestication incubator together somewhere in southeastern Anatolia (Naderi et al. 2008), traveling together as far as the coast of southern France where animals belonging to these lineages have been detected among archaeological goat bone assemblages dating back to about 7,000 cal. B.P. (Fernández et al. 2006). While still lacking the geographical precision of the goat genetic data, at least two of the three domestic lineages of sheep are likely to have been domesticated in this region (Bruford and Townsend 2006; Guo et al. 2005). Three and perhaps four of the five lineages of domestic taurine cattle were also probably domesticated here (Bradley 2006), with one lineage (the T3 lineage) the major variety that spread throughout Europe and another (the T1 lineage) the variety that spread across North Africa (Achilli et al. 2008). Similarly, at least four of the many domestic lineages of pigs seem to have originated in the Near East (Larson et al. 2005, 2007).

Thus earlier scenarios that suggested a single center of plant domestication and a 1,000- year delay between the timing of plant and animal domestication can no longer be supported. The new picture that is emerging from these combined archaeological and genetic data point to a process in which people throughout the Fertile Crescent were actively experimenting with various plant and animal resources (including some, like gazelle, which never made it any further than the first tentative steps down the pathway to domestication [Zeder in press]). Over a period of several millennia beginning by at least 11,000 cal. B.P., these experiments resulted in a fully formed domestic Partnership between morphologically and genetically altered crop and livestock species and humans increasingly invested in their propagation. These different managed crops and livestock species, however, did not coalesce into fully developed agricultural economies in which domesticates are the dominant source of calories (after Smith 2001) until the Middle PPNB (Helmer et al. 1998; Nesbitt 2002), where they first emerge in the central Fertile Crescent at about 10,000 cal. B.P., taking another 1,500–2,000 years to reach the farthest extremities of its eastern and western arms. Even long after the emergence of agricultural economies based on herding and farming, we still see a great deal of variability in the mix of wild and domestic resources at different sites in different social and environmental contexts in the region (Zeder 1994, 2006d:140).

The Utility of Co-opted Evolutionary Models in Explaining Agricultural Emergence in the Near East

The Near Eastern archaeological record has enough detail and chronological control to evaluate how well each of the earlier considered co-opted evolutionary models (neo-Darwinian selectionism, macroevolution, and human behavioral ecology) account for this keystone case study example of agricultural emergence. It is especially useful to consider this record in terms of four fundamental areas where these models differ most from one another - (1) on the scale of change (macro vs. micro-evolutionary); (2) on the tempo of change (punctuated vs. gradual); (3) on the degree of directedness of change (directed vs. undirected); and (4) on the role of human intent in culture change (major vs. none).

Locus of Change (Macro vs. Micro)

Both selectionist and human behavioral ecological approaches to culture change stress the micro-level forces that operate on the level of individual actors, whether shaped by Darwinian forces of selection, drift, and transmission in the selectionist approaches or by the rules that govern human decision making in the human behavioral ecology approach. In contrast, macroevolutionary approaches stress the importance of forces that work at larger scales, above the level of individuals, operating simultaneously in a hierarchical fashion within and across different levels of the society. The Near Eastern case study has elements that lend support to both a macro and a micro view of evolutionary change.

The increasing resolution of our understanding of the archaeological record of agricultural origins in the Near East has made it increasingly difficult to accommodate one-size-fits-all explanatory models for agricultural emergence that spotlight single macroscale forces as the causal agents of agricultural emergence (Zeder 2006a, 2009; Zeder and Smith in press)—whether it be climate change (Richerson et al. 2001), demographic pressure (Rosenberg 1990 1998), social aggrandizement (Hayden 1992 1995 2003, in press), or religious conversion (Cauvin 2000a, b). However, the record does support a model which characterizes agricultural emergence in terms of a series of mutually reinforcing macro-level economic, environmental, social, and ideological factors operating in concert across multiple levels of Near Eastern society—a model which we have seen is very much in line with a macroevolutionary approach to the study of culture change.

Following this model, the climatic amelioration after the Late Glacial Maximum that resulted in a spread of plant and animal resources out of protected refugia can be seen as creating conditions that made it possible for larger groups of people to congregate together for longer periods of the year. An elaboration of social mechanisms and ritual practices that helps to preserve bonds of community in these more sedentary Early Natufian settlements was accompanied by a broadening resource base that allowed people to continue living in these more permanent base camps despite any localized pressures on resources incurred by reduced mobility. The powerful bonds of community forged in these first sedentary settlements may account for their use as cemeteries by Late Natufian people in the southern Levant who continued to exploit the same range of resources, though in a more mobile way, during the climatic downturn of the Younger Dryas (Munro 2003, 2004). In the northern Levant and in the central and Eastern Fertile Crescent, which may not have been as profoundly affected by these climatic conditions (Wilcox 2005), sedentary communities were able to weather the impact of climate change and localized resource depression through the intensification of resource strategies, which may have involved cultivation of plants and game management strategies to promote the stocks of prey animals. With stabilization of climate in the Early Holocene, the pull of these social bonds is again seen in the proliferation of sedentary communities throughout the western and central Fertile Crescent whose continued viability required the increasingly active manipulation of resources that, in the case of certain species, moved them farther along the pathway to full domestication. At the same time, this process resulted in the creation of resources amenable to ownership, surplus production, and restricted access, concepts that threatened the egalitarian ethos that drew these communities together in the first place. Forces that worked to divide households in these communities were countered with an increasing number of measures directed at leveling economic and social differences and promoting community cohesion. New views of the relationship between humans and the cosmos helped people rationalize their place in this new social and economic order. Increased channels of interaction among communities forged expanded economic and social bonds that provided additional buffers against resource shortfalls. Under the combined pressures of human-induced degradation of local landscapes and the climatic downturn of the mid-ninth millennium cal. B.P., the Late PPNB mega villages in the southern Levant were no longer able to maintain the delicate balance between these competing centrifugal and centripedal social forces and fragmented into an array of sedentary and more mobile groups following mixed agro-pastoral and hunter-gatherer strategies (Kuijt and Goring-Morris 2002). The nucleated villages that followed the 8.2 kya climate pulse in the Pottery Neolithic were reconstituted along very different rules of social ranking and leadership (Gopher and Gophna 1993).

There is no way that a selectionist approach which restricts one's focus to the level of individual actors and the forces of selection, drift, and transmission operating on randomly generated discrete human behaviors (monitored through their artifact trait proxies) could account for these highly interrelated developments that clearly operate at the level of groups of individuals—from households to communities to whole regional interaction spheres. Even if one had all the data (and the several generations of archaeologists) needed to piece together cladogenetic trees tracking material culture lineages over the more than 10 millennia encompassed by this transition, this myopic focus on the smallest level of culture change could never begin to capture the complex hierarchy of interacting forces responsible for agricultural emergence in the Near East. Moreover, even with the remarkably robust Near Eastern archaeological record of resources and resource strategies, the human behavioral ecology playbook of rules for decision making does not seem up to the task of accounting for the range and complexity of collective decisions made during this long transitional period over this large region, especially if these rules are structured solely around energy optimization.

At the broadest most macro-level, the process of Neolithic emergence in the Near East seems to have been driven by very general economic goals oriented toward promoting a predictable and secure resource base, as well as social goals oriented toward binding groups of people together, both operating in concert within a framework of environmental variability and climate change—forces that cannot be easily accommodated under either a selectionist or a human behavioral ecology approach.

And yet these macro-level forces in and of themselves do not provide a full account of the course of agricultural emergence in the Near East. The increasing resolution of the archaeological record for agricultural emergence in the Near East makes it clear that although there are general similarities that cut across the region, there is still a tremendous amount of variability in the ways in which people responded to these larger forces in the southern Levant, northern Levant, southeastern Anatolia, central Anatolia, the Zagros, and on Cyprus. A wide variety of more microscale, localized forces operating at the level of individuals and small groups of individuals clearly played pivotal roles in shaping the way in which the process played out in each of these subregions. The solutions that people across the region found to meet overarching economic and social goals within the context of global climate change were shaped by local parameters and constraints. The variable responses to climate change of regions at different latitudes with different topographies and weather patterns, the differential density and diversity of different plant and animal resources in different parts of the region, the range of raw materials present, the demographic history of colonization and population growth, the localized human pressures on landscapes, the degree communities engaged in a broader sphere of interaction, and the ways in which they incorporated borrowed elements into their highly localized way of life; all of these localized factors helped lend a highly regional flavor to the emergence of agricultural economies in different parts of the Fertile Crescent.

Consideration of both the general macroscale forces that set the basic parameters of change and the microscale highly localized factors that determined how these higher-level trends played out at the local level are needed, if one is to understand the process of agricultural emergence in the Near East. Looking beyond the Near East, these same higher level macroscale forces were also likely at play in other world regions where agriculture emerged, as were a wide range of microscale factors particular to each case. Identifying the locus of culture change, and by so doing identifying the causes of change, cannot, then, stop at defining the single or even the multiple macrolevel forces that set the general parameters of change. Nor can it focus entirely on the particular histories of instances of change within single cultural lineages. Instead, a comparative approach is required which seeks to identify both the unique factors that shape the trajectories of different instances of agricultural emergence, while also identifying the most macroscale forces that pertain to all cases as well as the microscale forces that explain how these forces played out in each instance (Spencer 1990:6, 1997:233–234).

Tempo of Change (Punctuated vs. Gradual)

The admittedly oversimplified diagram of the chronological occurrence of Childe's ten key components of the Neolithic Revolution in the Near East shown in Figure 6.3 does not seem, at first glance, to be consistent with a model of punctuated process in which change happens in a rapid burst, occurring on a centenary or even decadal scale (following Rosenberg 1994a:314). Instead, the first stirrings of Childe's revolution can be traced back to the evidence of ground stone tools and plant processing in the Late Glacial Maximum, while the final fading away of foraging ways of life and their replacement by fully formed agricultural economies does not happen for another ten plus millennia. This extended time frame would seem more consistent

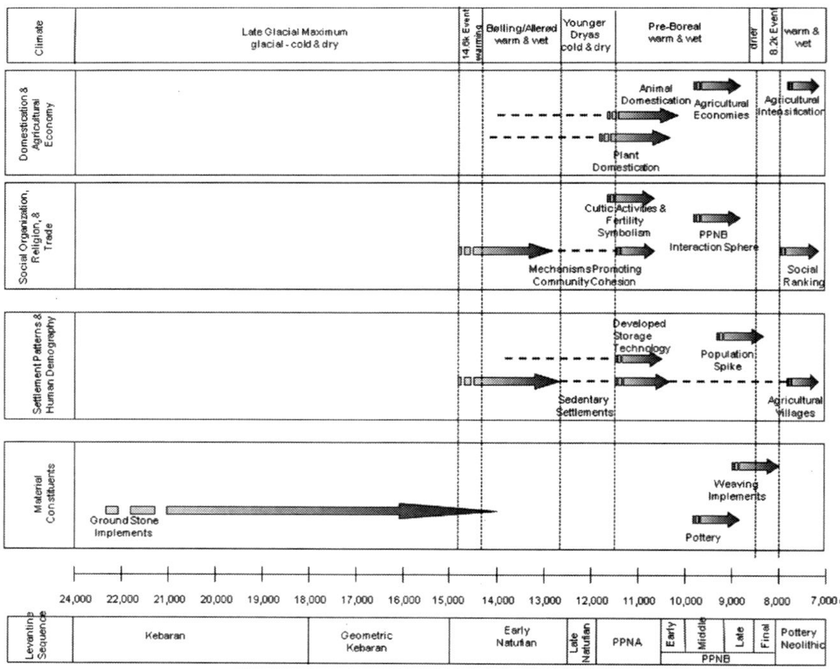

Fig. 6.3 Time line of the appearance of major components of the 10 key components of Childe's Neolithic society in the Near East

with the gradual model of incremental change favored by selectionist archaeologists and human behavioral ecologists.

And yet a closer examination of this chart (as well as those found in Kuijt and Prentiss' contribution to this volume and in Byrd 2005: Figure 6) suggests a different story. The timing of the appearance of various components of Childe's model relative to the history of climatic mood swings that marks the Late Pleistocene/Early Holocene in the region shows a developmental staircase consisting of short periods of rapid change often coinciding with these hot and cold flashes separated by periods of more gradual incremental change, if not stasis. Over time, the tempo of these changes quickens, and the number of Childean components that crystallize at any one time increases. The long period of expansion of the niches exploited by foraging Kebaran and Geometric Kebaran peoples in the waning days of the Late Glacial Maximum is interrupted by the rapid adoption of sedentary lifestyles in the Early Natufian coinciding with the 14.6 warming event (Byrd 2005). Over the next 2,000–3,000 years these communities were supported by an increasingly broad utilization of wild resources and bound together by emergent mechanisms of social cohesion. The climatic downturn of the Younger Dryas is followed by another 1000 years of, in the southern Levant, more mobile foraging strategies focused on the same range of resources as in the Early Natufian and, in the northern Levant, the establishment of sedentary settlements supported by increasingly intense manipulation of wild plant and animal resources. The advent of stable Holocene climates in the following PPNA sees a large number of rapid revolutionary changes in the incorporation of emergent domesticates into subsistence bases, in household and community organization, and in cosmology. The continued more gradual elaboration of all these emergent cultural components over the next 2,000–3,000 years culminates in a spike in population growth, expanding regional trade networks, and the coalescence of the different component elements into fully formed agricultural communities. In the southern Levant, Late PPNB mega communities seem to collapse under their own weight during another climatic downturn, only to re-form again some 500 or so years later in a socially much changed configuration of agricultural village life.

Throughout this sequence, each period of change builds on previous evolutionary events, with the overall process of agricultural emergence unfolding in stepwise fashion of innovation and consolidation as different components of the Neolithic package are established. It would seem then that the initial apparent lack of fit between Neolithic emergence in the Near East and a macro-evolutionary model of punctuated equilibrium results from a conflation of Childe's ten components into a single Neolithic blueprint or *Bauplan*. Instead of a single massive revolutionary upheaval, what is coming into focus is a transition comprising a sequence of smaller episodes of restructuring whose cumulative impact over the millennia leads to the establishment of fully formed agricultural economies in the region (an incremental process of culture change also seen in Chatters' and Eldredge's contributions to this volume).

Moreover, if one steps back from this close-grained view and examines these developments in terms of a broader perspective of human cultural evolution, the stepwise revolutionary events that took place over this time span mark a real departure from the preceding 30,000 years during which modern humans in this region had been hunting and collecting wild resources whose availability they could influence but never really control. It becomes, from this vantage point, an interval of profound change in essentially every aspect of social, economic, ideological, and political life in this region that gave birth to the increasingly sophisticated agro-pastoral societies based on an expanding array of genetically controlled domestic breeds and governed by increasingly stratified social and political arrangements that people in this region have lived under ever since.

Directedness of Change (Directed vs. Undirected)

As we have seen both human behavioral ecologists and, especially, selectionists go to great pains to disavow the role of directed change in cultural evolution. To selectionists, any appearance of direction in the course of change is likely a chimera, and even if real says nothing about how that change came about, which, according to their model, is solely attributable to the mechanism of either selection or transmission operating on randomly generated behavioral variability (Leonard and Jones 1987; Lyman and O'Brien 1998:621). And while human behavioral ecology advocates hold that basic optimizing covering laws provide direction to human decisions that govern the innovation and emulation of behaviors, they are steadfast in their rejection of models of culture change that bear any taint of progressive evolutionary viewpoints (Layton et al. 1991). Instead, they prefer to focus on the day-to-day localized and immediate decisions and their consequences (Winterhalder and Goland 1997; Winterhalder and Kennett 2006:8) and are made uncomfortable by the prospects of there being any degree of long-term directionality to culture change. Macroevolutionary archaeologists, like their biological brethren, also reject progressivist, unilineal frameworks of evolutionary development (Gould 1988; Rosenberg 1994a; Spencer 1997). But they hold that factors like historical contingency, exaptation, and the hierarchical nature of macroevolutionary change all provide a large measure of directionality of change over time (Rosenberg 1994a:329; Spencer 1997:231–232; Spencer and Redmond 2001:201).

As we have seen, the emergence of agriculture in the Near East comprised of a number of highly localized stories in which people living within particular environmental settings were working to meet overall social and economic goals with the resources (natural and cultural) at hand. However, the record from the region clearly does not support a view which sees the individual paths taken as the product of random variation filtered by selection, drift, and transmission. Nor can this record be interpreted solely as the cumulative effect of decisions made by hundreds of individuals coping with proximate cost-benefit considerations that just happen to follow similar trajectories across this broad region. Instead, what emerges from a review of the Near Eastern record is a more directional process, shaped by the mutually reinforcing macro-level economic, environmental, social, and ideological factors working across multiple levels within and between regions, by the legacy of past ways of coping that are carried over into new strategies for dealing with changing conditions, and by the continual retooling of technology, subsistence strategies, and ritual practice to meet new challenges. This is not to say that the pathway to agriculture was a one-way expressway that once embarked upon could not be exited (Smith 2001). There are plenty of examples of U-turns (the return to more mobile strategies in both the Late Natufian and the Terminal PPNB, for example), dead-ends (the early experiments with rye in the Late Natufian northern Levant), detours (the retention of round houses and, presumably, associated household-level social mechanisms, on Cyprus), and decisions to take more scenic alternate routes (exemplified by the different, sometimes quite circuitous, trajectories of agricultural emergence taken in different parts of the Fertile Crescent arc). And yet there also seem to have been a certain set of shared constraints and a limited number of general solutions, plus the important element of connectivity and sharing of solutions among populations across the Fertile Crescent, that resulted in striking similarities in the ways in which localized groups responded to economic goals through intensification of resource management and to social goals through community coping measures that, in the end, imparted a very real direction to the course of culture change in the Near East over this major transition.

Human Intent in Culture Change (Lots vs. None)

Disagreement over the role of human intent, of deliberate conscious decision making, in cultural evolution is the major schism between these different co-opted models of cultural evolution. To selectionists human intent is an anathema not to be allowed in any conception of culture change, with any acknowledgment of the role of human sentience in culture change a heresy of significant magnitude to warrant expulsion from this particular sect of evolutionary archaeology (Leonard and Jones 1987; Lyman and O'Brien 1998). And while they subscribe to many tenets of the neo-Darwinian approach to the study of culture change, human behavioral ecologists differ with selectionists in their central focus on the rules that govern decision making and their general acknowledgement of the role of conscious human intent in making these decisions (Bettinger 2006; Piperno 2006). Here, they are in agreement with macroevolutionary archaeologists for whom human agency is a central attribute of culture change. Indeed, to macroevolutionary archaeologists, human intent is the major agent responsible for directionality in cultural evolution and the primary way in which cultural evolution differs from biological evolution (Spencer 1997).

Any objective reading of the record of agricultural emergence in the Near East cannot escape the central role of intentional human action in the transition. Nowhere is this more evident than in the core process of plant and animal domestication. While it is true that the co-evolutionary interaction between humans and target plant and animal populations which lies at the heart of the domestication process shares certain commonalities with biological mutualisms between farmer ants and their fungal crops (Zeder 2006a), even the entomologists who study these relationships readily admit that the fundamental difference between these two processes lies in the element of human intent (Schultz et al. 2005). These biologists know what apparently selectionist archaeologists do not—that mutualisms in nature are driven by natural selection operating on mutation-induced behavioral and morphological variation in both partners, while the relationships between humans and target domesticates are largely driven by humans who are able to spontaneously and with deliberate intent modify their behaviors toward plant and animal partners through observation, imitation, and innovation based on a conscious and deliberative evaluation of the effects of previous actions. This element of deliberation, decision, and transmission of behaviors that arise through conscious human innovation ramps up the emerging mutualism between humans and target domesticates, imbuing the process with greater speed and magnitude than any mutualism found in nature. And while selectionists admit that the transmission of behaviors through social learning is a major difference between biological and cultural evolution, their unwillingness to let go of the anti-intent doctrine so central to the selectionist credo, to admit that the innovation and transmission of behaviors involves sentient human actions, makes it impossible for them to effectively explain domestication, or, I submit, any other aspect of human cultural evolution (though see a potential softening of this hard line position in Mesoudi and O'Brien 2008; O'Brien 2008).

Human intent is a central driving feature throughout the process of domestication. Inveterate tinkerers, humans had likely been deliberately manipulating environments to encourage the availability of plant and animal resources of interest (niche construction or ecosystem engineering *sensu* Odling-Smee et al. 2003; Smith 2007a, b) for thousands of years before the first morphologically altered domesticates appeared on the scene. This process involved a series of conscious and deliberate decisions to try out different things that might enhance productivity of these resources—to move wild einkorn plants to soils where they were more likely to flourish, to water and weed wild stands of cereals and pulses, to transplant branches from trees that produced tastier figs, and to selectively cull wild herds of animals in ways that provided sustainable long-term returns. The responses of certain plants and animals to these ministrations, both plastic and genetically driven, further encouraged human focus and elaboration of the behaviors and tools that helped these responsive partners prosper. And while it is highly unlikely that humans did these things with the goal of starting their own agro-businesses several millennia down the line, they were guided in these deliberate actions by proximate and quite consciously recognized goals of securing a predicable resource base to support their family and their community of families. To deny humans this quintessentially human attribute is to strip cultural evolution of its primary distinguishing feature and so doom the viability of any co-opted model of evolution that discounts human intent in the process of culture change.

And yet while macroevolutionary archaeologists readily embrace the element of human intent in their models of culture change (Spencer 1997), the macroscale of their models makes it difficult for them to monitor the impact of human agency on the course of cultural evolution. Macroevolutionary approaches simply do not allow for the detection or explanation of the motivations behind the actions of individual actors. Here, the human behavioral ecology focus on human decision making would seem to offer a particularly powerful tool, providing real possibility of linkage between the micro- and the macro-level forces that so clearly work together to drive evolutionary developments. And yet, the effectiveness of HBE models in providing the rules that guide human intentionality is open to debate.

Two basic elements of HBE optimizing models seem particularly hard to reconcile with the Near Eastern record: (1) the fundamental principle that lower ranked resources (based on net energetic returns) will only be utilized if high-ranked alternatives are not available and (2) that immediate returns are preferred over deferred returns. These essentially axiomatic HBE assumptions define a context of agricultural origins in which substantial conceptual obstacles lie in the way of a shift to an emphasis on plant resources or a willingness to accept the discounting costs of deferred harvest of plants and animals. Some combination of depression of higher ranked resources and enhancement of the density or desirability of lower ranked resources has to have occurred in order for humans to decide to focus on plants like cereals and pulses with their higher handling costs and lower returns and to suppress their blood-lust instinct to kill and consume any sizable meat package that comes along. And yet the improving archaeobotanical record from the Near East shows that plants were a prominent part of the resource menu from the very beginning of the sequence, well before any likely hunting depression on game animals could have manifested itself (Piperno et al. 2004; Weiss et al. 2004). In fact, there is increasing evidence for the importance of plant resources in Near Eastern diets stretching well into the Middle Paleolithic (Albert et al. 2003; Lev et al. 2005). Moreover, the notion that farming and herding requires a mind-set willing to risk waiting for deferred returns, while foraging and hunting confers immediate returns, cannot be supported. The increasing insights afforded by new markers of human manipulation of wild plant and animal resources show that people were making long-term investments in landscapes and associated plant communities in order to enhance down-the-line returns on their investments well before plants and animals were co-opted into domestic partnerships. People may also have been altering hunting strategies in such a way as to assure long-term availability of wild game. Moreover, the notion that herding involves deferring slaughter of animals until a certain sustainable herd size is reached (Alvard and Kuzner 2001) fails to acknowledge the fact that animals are continually culled from managed herds. The difference between hunting and herding is that the hunter kills to maximize immediate return while the herder kills in way that promotes herd growth. The herder is, however, still killing and eating animals, which are now perhaps a little bit more handy and more tractable than their unmanaged forbearers.

Sedentism and the associated broadening of the resource base in the Near East does not seem to be born out of necessity, arising either from human or climate induced environmental change. Nor can it be viewed as resulting from a microeconomic cost/benefit analysis weighing energetic returns against search and handling costs. Instead, as argued here, there seems to have been compelling social pulls that brought people together for longer periods of time and in greater numbers whenever conditions permitted, even if the net energetic returns of the resource strategies that permitted coalescence were less than optimal from an energetics point of view.

These same social ties can also be seen as a major motivating factor driving the progressive intensification of subsistence strategies aimed at sustaining aggregations of people for as long as prevalent ecological constraints, technological capacities, and mechanisms for social cohesion would allow.

The problem, then, may not lie in the HBE focus on human decision making, but on the basic optimizing premise and the cost/benefit assumptions that HBE archaeologists maintain guide these decisions. Recognition that human decision making may be guided by "something other than economic self-interest" is not new to human behavioral ecology (Bettinger 2006; Boyd and Richerson 1988; Winterhalder 1986;) and represents, perhaps, a willingness to embrace a more catholic approach that looks beyond optimizing as a central governing attribute of human behavior—a development that augers well for the utility of a human behavioral approach to understanding the role of human agency in the course of culture change.

Conclusions

This review of the Near Eastern record of agricultural origins suggests that all three co-opted evolutionary models considered here has something to offer in understanding the long sequence of events that unfolded over the transition from foraging to farming. Microevolutionary forces played an important role in shaping the trajectories taken in various parts of the region and in the incremental adjustments people made to accommodate changing environmental and social conditions. And yet there were also clearly macroscale linkages between individuals, communities, and regions that, often in rapid revolutionary bursts, lent this process a clear directionality. At the same time, the role of daily human decisions in how to best organize their lives and their activities to meet proximate goals of economic and social sustainability also played a part in this process, especially in the domestication of plants and animals and the development of agricultural economies based on them, which is the central element of this transition.

Where these models fall short, especially the microevolutionary and human behavioral ecology models, is when their proponents give too much deference to their biological underpinnings and attempt the wholesale application of models developed to explain biological evolution to the very different process of cultural evolution. This is especially a trap for selectionist archaeologists in their rejection of human intent as a factor in cultural evolution. Moreover, while selectionists may give lip service to group-level evolutionary forces, punctuated change, and hierarchy as possible components of cultural evolution, their dogged pursuit of trait-level cladogenesis of cultural lineages is an explanatory dead end. Deferring dealing with process until patterns are thoroughly mapped has not worked too well in biology, and it is even more doubtful that such an alpha-taxonomy approach to culture change will serve the study of cultural evolution any better. Moreover, the tendency of HBE advocates to treat optimizing principles drawn from behavioral ecology as immutable covering laws rather than testable hypotheses (Smith 2006) affords these principles more power than even biologists are willing to grant them.

Instead of being directly applicable to cultural evolution, these different approaches to biological evolution are best used as heuristic devices providing general models of evolutionary change that can be loosely applied to the study of culture change. The key role of human intentionality and the transmission of behaviors through social learning and the resultant capacity for Lamarkian change unique to human culture requires that archaeologists broaden their search for epistemological inspiration beyond biology to include theoretical frameworks grounded in the social and cognitive sciences (Boyd and Richerson 1985; Richerson and Boyd 2005; Spencer 1997:247).

None of these efforts, however, makes much sense in the absence of empirical data that can be used to both test and strengthen evolutionary models designed to explain the course of culture change. Too often, it seems, archaeologists working within these evolutionary frameworks focus more on the elegance of their models and their ability to come up with some cogent explanation or model addenda that accounts for the failure of the empirical record to conform to their predictions, while still showing that their general premise was right all along. As the record for agricultural emergence clearly shows (and as seen in many of the contributions to this volume), the growing precision with which we are able to monitor culture process through archaeological analysis makes model spinning in the absence of empirical reality check less and less justifiable. In fact, our increasing control of the empirical record and our ability to frame nuanced, multivariable explanatory scenarios to account for this record may be bringing us, at long last, to a tipping point where we may be able to actually contribute to the development of higher order theories of culture change rather than continuing to look to other disciplines for epistemological validation.

References

Achilli, A., Olivieri, A., Pellecchia, M., Uboldi, C., Colli, L., Al-Zahery, N., Accetturo, M., Pala, M., Kashani, B., Perego, U.A, Battaglia, V., et al. (2008). Mitochrondrial genomes of extinct aurochs survive in domestic cattle. *Current Biology* 18, 157–158.

Albert, R., Bar-Yosef, O., Meignen, L. (2003). Quantitative phytolith study of hearths from the Natufian and Middle Paleolithic levels of Hayonim Cave (Galilee, Israel). *Journal of Archaeological Sciences* 30, 461–480.

Aldenderfer, M. (2006). Costly signaling, the sexual division of labor and animal domestication in the Andean highlands. In D.J. Kennett and B. Winterhalder (Eds.), *Behavioral ecology and the transition to agriculture* (pp. 167–196). Berkeley: University of California Press.

Alvard, M.S., Kunzner, I. (2001). Deferred harvests: The transition from hunting to animal husbandry. *American Anthropologist* 103, 295–311.

Arnold, S.J., Pfrender, M.E., Jones, A.G. (2001). The adaptive landscape as a conceptual bridge between micro- and macroevolution. *Genetica* 112–113, 9–32.

Aurenche, O., Cauvin, J., Sanlaville, P. (Eds.). (1988). Préhistoire du Levant II (1re Partie). *Paléorient* 14(2), 5–345.

Aurenche, O., Cauvin, J., Sanlaville, P. (Eds.). (1989). Préhistoire du Levant II (2e Partie) *Paléorient* 15(1), 1–179.

Aurneche, O., Galet, P., Régagnon-Caroline, E., Évin, J. (2001). Proto-neolithic and neolithic cultures in the Middle East. The birth of agriculture, livestock raising and ceramics 12,500–5,500 cal. B.C. *Radiocarbon* 43, 1191–1202.

Asouti, E. (2006). Group identity and the politics of dwelling at Neolithic Çatalhöyük. In I. Hodder (Ed.), *Çatalhöyük Perspectives: Themes from the 1995-99 Seasons* (pp. 75–79). Çatalhöyük Research Project 6, Cambridge: British Institute of Archaeology at Ankara & McDonald Institute for Archaeological Research.

Badr, A., Müller, K., Schäfer-Pregl, R., El Rabey, H., Effgen, S., Ibrahim, H.H., Pozzi, C., Rohde, W., Salamini, F. (2000). On the origin and domestication history of barley (*Hordeum vulgare*). *Molecular Biology and Evolution* 17, 499–510.

Barlow, K.R. (2006). A formal model for predicting agriculture among the Fremont. In D.J. Kennett and B. Winterhalder (Eds.), *Behavioral ecology and the transition to agriculture* (pp. 87–102). Berkeley: University of California Press.

Bar-Yosef, O. (1982). The Natufian of the southern Levant. In T.C. Young, P.E.L. Smith, and P. Mortensen (Eds.), *The Hilly Flanks. Essays on the Prehistory of Southwest Asia* (pp. 11–42). Studies in Ancient Oriental Civilization, No. 36, Oriental Institute, Chicago: University of Chicago.

Bar-Yosef, O. (1986). The walls of Jericho: An alternative interpretation. *Current Anthropology* 27, 157–162.

Bar-Yosef, O. (1990). The last glacial maximum in the mediterranean levant. In C. Gamble and O. Soffer (Eds.), *The World at 18,000 BP, vol. 2: Low Latitudes* (pp. 58–72). London: Unwin Hyman.

Bar-Yosef, O. (1992). From foraging to farming in the Mediterranean Levant. In T.D. Price and A.-B. Gebauer (Eds.), *Transitions to Agriculture in Prehistory* (pp. 21–48). Madison: Prehistory Press.

Bar-Yosef, O. (2001b). PPNB interaction sphere. *Cambridge Archaeological Journal* 11, 114–120.

Bar-Yosef, O., and Belfer-Cohen, A. (1989). The Levantine "PPNB" interaction sphere. In I. Hershkovitz (Ed.), *People and Culture in Change: Proceedings of the Second Symposium on Upper Paleolithic, Mesolithic and Neolithic Populations of Europe and the Mediterranean Basin* (pp. 59–72), British Archaeological Reports International Series, No. 508. Oxford: Archaeopress.

Bar-Yosef, O., and Kislev, M. (1989). Early farming communities in the Jordan valley. In D. Harris and G. Hillman (Eds.), *Foraging and Farming: The Evolution of Plant Exploitation* (pp. 632–642), London:Unwin Hyman.

Bar-Yosef, O., and Meadow, R.H. (1995). The origins of agriculture in the Near East. In T.D. Price and A.-B. Gebauer (Eds.), *Last Hunters, First Farmers: New Perspectives on the Transition to Agriculture* (pp. 39–94). Santa Fe: School of American Research Press.

Bar-Yosef, O., Gopher, A., Tchernov, E., Kislev, M. (1991). Netiv Hagdud: An early Neolithic village site in the Jordan valley. *Journal of Field Archaeology* 18, 405–424.

Bar-Yosef Mayer, D.E. (2000). The economic importance of mollusks in the Levant. In M. Mashkour, A.-M. Choyke, H. Buitenhuis, and F. Poplin (Eds.), *Archaeozoology of the Near East 4A* (pp. 218–227). Arc-Publicatie 32, The Netherlands: Centre for Archaeological Research and Consultancy, Groningen.

Bettinger, R. (1991). *Hunter-gatherers: Archaeological and Evolutionary Theory*. New York: Plenum Press.

Bettinger, R. (2006). Agriculture, archaeology, and human behavioral ecology. In D.J. Kennett and B. Winterhalder (Eds.), *Behavioral ecology and the transition to agriculture* (pp. 304–322). Berkeley: University of California Press.

Binford, L.R. (2001).*Constructing frames of reference: an analytical method for archaeological theory building using ethnographic and environmental data sets*. Berkeley: University of California Press.

Bökönyi, S. (1977). *Animal remains from the Kermanshah Valley, Iran*. British Archaeological Reports Supplementary Series, No. 34, Oxford: Archaeopress.

Boyd, R., Richerson, P.J. (1985). *Culture and the Evolutionary Process*. Chicago: University of Chicago Press.

Bradley, D.G. (2006). Genetics and the origins of domestic cattle. In M.A. Zeder, E. Emshwiller, B.D. Smith, and D.G. Bradley (Eds.), *Documenting Domestication: New Genetic and Archaeological Paradigms* (pp. 317–328). Berkeley: University of California Press.

Braidwood, R.J., and Howe, B. (1960). *Prehistoric Investigations in Iraqi Kurdistan*, The Oriental Institute of the University of Chicago Studies in Ancient Oriental Civilization, No. 31. Chicago: University of Chicago Press.

Braidwood, L.R., Braidwood, B., Howe, B., Reed, C.A., and Watson, P.J. (1983). *Prehistoric archaeology along the Zagros flanks*, The Oriental Institute of the University of Chicago Studies in Ancient Oriental Civilization, No. 105, Chicago: University of Chicago Press.

Bruford, M., and Townsend, S.J. (2006). Mitochondrial DNA diversity in modern sheep: Implications for domestication. In M.A. Zeder, E. Emshwiller, B.D. Smith, and D.G. Bradley (Eds.), *Documenting Domestication: New Genetic and Archaeological Paradigms* (pp. 306–316). Berkeley: University of California Press.

Buitenhuis, H. (1997). Aşikli Höyük: A "protodomestication" site. *Anthropozoologica* 25–26, 655–662.

Byrd, B.F. (1994). Public and private, domestic and corporate: The emergence of the Southwest Asian village. *American Antiquity* 56, 639–666.

Byrd, B.F. (2000). Households in transition: Neolithic social organization within Southwest Asia. In Kuijt, I. (Ed.), *Life in Neolithic Farming Communities: Social Organization, Identity, and Differentiation* (pp. 63–98). New York: Academic/Plenum Press.

Byrd, B. (2005). Reassessing the emergence of village life in the Near East, *Journal of Archaeological Research* 13, 231–290.

Byrd, B., Monahan, C. (1995). Death, mortuary ritual, and natufian social structure. *Journal of Anthropological Archaeology* 14, 251–287.

Cauvin, J. (1978). *Les premiers villages de Syrie-Palestine du IXeme au VIeme millenaires avant J.C.* Paris: CNRS.

Cauvin, J. (2000a). The symbolic foundations of the Neolithic Revolution in the Near East. In Kuijt, I. (ed.), *Life in Neolithic farming communities, social organization, identity and differentiation* (pp. 235–254). London: Kluwer Academic/Plenum.

Cauvin, J. (2000b). *The birth of the Gods and the origins of agriculture*. Cambridge: Cambridge University Press.

Çelik, B. (2005). A new Early Neolithic settlement: Karahan Tepe. Neo-lithivs 2–3/00, 6–8.

Charnov, E.L. (1976). Optimal foraging, the marginal value theorem. *Theoretical Population Biology* 9(2), 129–136.

Charnov, E.L., and Orians, G. (1973). *Optimal foraging: Some theoretical explorations*. Salt Lake City: Department of Biology, University of Utah.

Chatters, J.C., Prentiss, W.C. (2005). A Darwinian macro-evolutionary perspective on the development of hunter-gatherer systems in northwestern North America. *World Archaeology* 37, 46–65.

Childe, V.G. (1951). *Man makes himself*. New York: The New American Library of World Literature, LTD.

Colledge, S. (1998). Identifying pre-domestication cultivation using multivariate analysis. In A. Damania, J. Valkoun, G. Willcox, and C. Quallset (Eds.), *The origins of agriculture and plant domestication* (pp. 121–131). Aleppo: ICARDA, Aleppo.

Colledge, S. (2002). Identifying pre-domestication cultivation using multivariate analysis: Presenting the case for quantification. In R.T.J. Cappers and S. Bottema (Eds.), *The Dawn of Farming in the Near East* (pp. 141–152). Studies in Near Eastern Production, Subsistence, and Environment, No. 6, Berlin: Ex Oriente.

Colledge, S. (2004). Reappraisal of the archaeobotanical evidence for the emergence and dispersal of the "founder crops". In E. Peltenburg and A. Wasse (Eds.), *Neolithic Revolution. New perspectives on Southwest Asia in light of recent discoveries in Cyprus* (pp. 49–60). Levant Supplementary Series 1, Oxford: Oxbow.

Davis, S.J.M., Valla, F. (1978). Evidence for domestication of the dog 12,000 years ago in the Natufian of Israel. *Nature* 276, 608–610.
Dawkins, R. (1986). *The blind watchmaker*. New York: W. W. Norton & Company, Inc.
Dawkins, R. (1986). *The Blind Watchmaker*. London: Harlow, Longman.
Denham, T. (2007). Debate. Early fig-domestication, or gathering of wild parthenocarpic figs? *Antiquity* 81, 457–461.
Denham, T., and Barton, H. (2006). The emergence of agriculture in New Guniea: A model of continuity from pre-existing foraging practices. In D.J. Kennett and B. Winterhalder (Eds.), *Behavioral ecology and the transition to agriculture* (pp. 237–264). Berkeley: University of California Press.
Diehl, M.W., and Waters, J.A. (2006). Aspects of optimization and risk during the early agricultural period in southeastern Arizona. In D.J. Kennett and B. Winterhalder (Eds.), *Behavioral ecology and the transition to agriculture* (pp. 63–86). Berkeley: University of California Press.
Dunnell, R. (1978). Style and function: A fundamental dichotomy. *American Antiquity* 43, 192–202.
Dunnell, R. (1980). Evolutionary theory and archaeology. *Advances in Method and Theory* **3**: 35–99.
Dunnell, R. (1987). Comment on "Sedentism, population growth, and resource selection in the Woodland Midwest: A review of coevolutionary developments. *Current Anthropology* 28, 191–192.
Dunnell, R. (1988). The concept of progress in cultural evolution. In M.H. Nitecki (Ed.), *Evolutionary Progress* (pp. 169–194). Chicago: University of Chicago Press.
Dunnell, R. (1989). Aspects of the application of evolutionary theory in archaeology. In C.C. Lamberg-Karlovsky (Ed.), *Archaeological thought in America*, (pp. 35–49). Cambridge: Cambridge University Press.
Eerkens, J.W., Bettinger, R.L., and McElreath, R. (2006). Cultural transmission, phylogentics, and the archaeological record. In C.P. Lipo, M.J. O'Brien, M. Collard, and S.J. Shennan (Eds.), *Mapping Our Ancestors: Phylogentic Approaches in Anthropology and Prehistory* (pp. 169–183). New Brunswick: Aldine Transaction.
Elredge, N. (1989a). *Macro-evolutionary dynamics: species, niches, and adaptive picks*, New York: McGraw Hill.
Elredge, N., and Gould, S.J. (1972). Punctuated equilibria: An alternative to phyletic gradualism. In T.J.M. Schopf (Ed.), *Models in Paleobiology* (pp. 82–115). San Francisco: Cooper.
Emlen, J. (1966). The role of time and energy in food preference. *American Naturalist* 100, 611–617.
Ervynck, A., Dobney, K, Hongo, H., Meadow, R.H. (2001). Born free!: New evidence for the status of pigs from Çayönü Tepesi, Eastern Anatolia. *Paléorient* 27(2), 47–73.
Esin, U. (1998). The aceramic site of Aşikli Höyük and its ecological conditions based on its floral and faunal remains. *Tüba-Ar* 1, 95–104.
Esin, U., and Harmankya, S. (1999). Aşikli. In M. Özdoğan and N. Başgelen (Eds.), *Neolithic Turkey, the Cradle of Civilization New Discoveries* (pp. 115–132). Istanbul: Arkeoloji ve Sanat Yauinlari.
Fernández, H., Hughes, S., Vigne, J.-D., Helmer, D., Hodgins, G., Miquel, C., Hanni, C., Luikart, G., Taberlet, P. (2006). Divergent mtDNA of goats in an Early Neolithic site, far from the initial domestication areas. *Proceedings of the National Academy of Sciences, U.S.A.* 103, 15375–15379.
Flannery, K.V. (1972). The cultural evolution of civilizations. *Annual Review of Ecology and Systematics* 3, 399–426.
Flannery, K.V. (1993). Will the real model please stand up: Comments on Saidel's round House or square? *Journal of Mediterranean Archaeology* 6(1), 109–117.
Flannery, K.V. (2002). The origins of the village revisited: From nuclear to extended households. *American Antiquity* 67, 417–433.

Foley, R. (1982). A reconsideration of the role of large mammal predation in tropical hunter-gatherer adaptation. *Man* 17, 393–403.

Fried, M. (1967). *The Evolution of Political Society*. New York: Random House.

Gebel, H.G.K. (2002). The Neolithic of the Near East. An essay on a "polycentric evolution" and other current research problems. In A. Hausleiter, S. Kerner, and B. Müller-Neuhof (Eds.), *Material Culture and Mental Spheres. Rezeption archäologischer Denkrichtungen in der Vorderasiatischen Altertumskunde. Internationales symposium für Hans J. Nissen* (pp. 214–324). *Berlin 23–24 Juni 2000*, Münster: Alter Orient und Altes Testament 293, Ugarit-Verlag.

Gebel, H.G.K. (2004). There was no centre: the polycentric evolution of the Near Eastern Neolithic. *Neo-Lithics* 1, 28–32

Gopher, A., Gophna, R. (1993). Culture of the eighth and seventh millennia BP in the southern Levant: A review for the 1990s. *Journal of World Prehistory* 7, 297–353.

Goring-Morris, A.N. (1987). *At the Edge: Terminal Pleistocene Hunter-Gatherers in the Negev and Sinai*. British Archaeological Reports International Series 361, Oxford: Archaeopress.

Gould, S.J. (1988). On replacing the idea of progress with an operational notion of directionality. In H. Nitecki, (Ed.), *Evolutionary Progress* (pp. 319–338). Chicago: University of Chicago Press.

Gould, S.J. (1989). Punctuated equilibrium in fact and theory. *Journal of Biological and Social Structures* 12, 117–136.

Gould, S.J. (2002). *The Structure of Evolutionary Theory*. Cambridge: Belknap.

Gould, S.J., Eldredge, N. (1993). Punctuated equilibrium comes of age. *Nature* 366, 223–227.

Gould, S.J., Lewontin, R.C. (1979). The spandrels of San Marco and the Panglossian paradigm: A critique of the adaptationalist programme. *Proceedings of the Royal Society London Series B* 205, 581–598.

Gremillion, K.J. (1996). Diffusion and adoption of crops in evolutionary perspective. *Journal of Anthropological Archaeology* 15, 183–204.

Gremillion, K.J. (1998). Changing roles of wild and cultivated plant resources among early farmers of eastern Kentucky. *Southeastern Archaeology* 17, 140–157.

Gremillion, K.J. (2006). Central place foraging and food production on the Cumberland Pleateau, Eastern Kentucky. In D.J. Kennett and B. Winterhalder (Eds.), *Behavioral ecology and the transition to agriculture* (pp. 41–62). Berkeley: University of California Press.

Gremillion, K.J., and Piperno, D.R. (in press). Human behavioral ecology, phenotypic plasticity, and agricultural origins: insights from the emerging evolutionary synthesis. *Current Anthropology*.

Grigson, C. (1969). The uses and limitations of differences in absolute size in the distinction between the bones of aurochs (*Bos primigenius*).and domestic cattle (*Bos Taurus*). In P.J. Ucko, and G.W. Dimbleby (Eds.), *The Domestication and Exploitation of Plants and Animals I* (pp. 277–294). Chicago: Aldine-Atherton, Inc.

Guilaine, J., and Briois, F. (2001). Parekkilisha Shillourokambos. An early Neolithic site in Cyprus. In E. Swiny (Ed.). *The Earliest Prehistory of Cyprus: From Colonization to Exploitation* (pp. 37–53). Cyprus American Archaeological Research Institute Monograph Series 12, Boston: American Schools of Oriental Research.

Guo, J., Du, L.-X., Ma, Y.-H., Guan, W.J., Li, H.-B., Zhao, Q.-J., Li, X., Rao, S.-Q. (2005) A novel maternal lineage revealed in sheep (*Ovis aries*). *Animal Genetics* 36, 331–336.

Hall, B.K. (1996). *Baupläne*, phylotypic stages, and constraint: Why there are so few types of animals. *Evolutionary Biology* 29, 215–261.

Hauptmann, H. (1993). Ein Kultegebäube in Nevali Çori. In M. Frangipane, H. Hauptmann, M. Liverani, P. Matthiae, and M. Mellink (Eds.), *Between the Rivers and over the Mountains* (pp. 37–69). Archaeologica Anatolica et Mesopotamica Alba Palmieri Dedicata, Rome: Università di Roma "La Sapienza".

Hauptmann, H. (1999). The Urfa region. In M. Özdağan and N. Basgelen (Eds.), *Neolithic in Turkey*, vol. 1 (pp. 65–86). Istanbul: Arkeoloji ve Sanat Yay.

Hawkes, K., Bliege-Bird, R. (2002). Showing off, handicap signaling, and the evolution of Men's work. *Evolutionary Anthropology* 11, 58–67.

Hawkes, K., O'Connell, J.F. (1992). On optimal foraging models and subsistence transitions. *Current Antropology* 33, 63–66.

Hawkes, K., Hill, K., O'Connell, J.F. (1982). Why hunters gather: Optimal foraging and the Aché of eastern Paraguay. *American Ethnologist* 9, 379–398.

Hawkes, K., O'Connell, J.F., Rogers, L. (1997). The behavioral ecology of modern hunter-gatherers and human evolution. *Trends in Ecology and Evolution* 12, 29–32.

Hayden, B. (1992). Models of domestication. In A.-B. Gebauer and T.D. Price (Eds.), *Transition to Agriculture in Prehistory* (pp. 11–19). Madison: Prehistory Press.

Hayden, B. (1995). A new overview of domestication. In T.D. Price and A.-B. Gebauer (Eds.), *Last Hunters, First Farmers: New Perspectives on the Transition to Agriculture* (pp. 273–300), Santa Fe: School of American Research Press.

Hayden, B. (2003).Were luxury foods the first domesticates? Ethnoarchaeological perspectives from Southeast Asia. *Journal of World Prehistory* 34, 458–469.

Hayden, B. (in press). Is the proof in the pudding?: Feasting and the origins of domestication. *Current Anthropology.*

Helmer, D. (1992). *La Domestication des Animaux par Les Hommes Préhistoriques*. Paris: Masson.

Helmer, D., Roitel, V., Sana, M., and Wilcox, G. (1998). Interprétations environnementales des données archaéozoologiques et archéobotanique en Syrie du Nord de 16,000 BP à 7000 BP, et les débuts de la domestication de plantes et des animaux. In M. Fortin and O. Aurenche (Eds.), *Natural Space. Inhabited Space in Northern Syria (10th–2nd mill. BC)* (pp. 9–33). Lyon and Toronto: Maison de l'Orient and Canadian Society for Mesopotamian Studies.

Helmer, D., Gourichon, L., Monchot, H., Peters, J., and Saña Segui, M. (2005). Identifying Early Domestic Cattle from Pre-Pottery Neolithic Sites on the middle Euphrates using Sexual Dimorphism. In J.-D. Vigne, J. Peters and D. Helmer (Eds.), *The First Steps of Animal Domestication* (pp. 86–95). Oxford: Oxbow Books.

Hemmer, H. (1988). *Domestication: The Decline of Environmental Appreciation*. Cambridge: Cambridge University Press.

Henry, D.O. (1989). *From Foraging to Agriculture: The Levant at the End of the Ice Age*, Philadelphia: University of Pennsylvania Press.

Henry, D.O. (2002). Models of agricultural origins and proxy measures of prehistoric demographics. In R.T.J. Cappers, and S Bottema (Eds.), *The Dawn of Farming in the Near East* (pp. 15–26). Studies in Early Near Eastern Productions, Subsistence, and Environment 6, Berlin: ex orient2.

Heun, M., Schafer-Pregl, R., Klawan, D., Castagna, R., Accerbi, M., Borghi, B., Salamini, F. (1997). Site of einkorn wheat domestication identified by DNA fingerprinting. *Science* 278, 1312–1314.

Heun, M., Haldorsen, S., Vollan, K. (2008). Reassessing domestication events in the Near East: Einkkorn and *Triticum urartu*. *Genome* 51, 444–451.

Hillman, G.C. (1978). On the origins of domestic rye – *Secale cereale*: The finds from Aceramic Can Hasan III in Turkey. *Anatolian Studies* 28, 157–174.

Hillman, G.C. (2000). Abu Hureyra I: The Epipaleolithic. In A.T.M. Moore, G.C. Hillman, and A.J. Legge (Eds.), *Village on the Euphrates: From Foraging to Farming at Abu Hureyra* (pp. 327–398). Oxford: Oxford University Press.

Hole, F. (1990a). Archaeology of the village period. In F. Hole (Ed.), *The Archaeology of Western Iran: Settlement and Society from Prehistory to the Islamic Conquest* (pp. 29–78). Washington, D.C.: Smithsonian Institution Press.

Hole, F. (1990b). Settlement and society in the village period. In F. Hole (Ed.), *The Archaeology of Western Iran: Settlement and Society from Prehistory to the Islamic Conquest* (pp. 78–106). Washington, D.C.: Smithsonian Institution Press.

Hole, F. (2000). Is size important? Function and hierarchy in Neolithic settlements. In I. Kuijt (Ed.), *Life in Neolithic Farming Communities. Social Organization, Identity and Differentiation* (pp. 192–209). London: Kluwer Academic/Plenum.

Hole, F., and Flannery, K.V. (1967). The prehistory of southwestern Iran: A preliminary report. *Proceedings of the Prehistoric Society* 33, 147–206.

Hole, F., Flannery, K.V., and Neely, J.A. (1969). *Prehistory and Human Ecology on the Deh Luran Plain*. Memoirs of the Museum of Anthropology, No. 1. Ann Arbor: The University of Michigan Press.

Hongo, H., and Meadow, R.H. (2000). Faunal remains from pre-pottery levels at Çayönü, southeastern Turkey: a preliminary report focusing on pigs (*Sus* sp.). In M. Mashkour, A.M. Choyke, H. Buitenhuis, and F. Poplin (Eds.), *Archaeozoology of the Near East 4A* (pp. 121–139). Arc-Publicatie 32, Groningen, The Netherlands: Centre for Archaeological Research and Consultancy.

Horwitz, L.K. (1993). The development of ovicaprine domestication furing the PPNB of the southern Levant. In H. Buitenhuis and A.T. Clason (Eds.), *Archaeozoology of the Near East I* (pp. 27–36), Leiden: Universal Book Service.

Horwitz, L.K. (2003). Temporal and spatial variation in Neolithic caprine exploitation strategies: a case study of fauna from the site of Yiftah'el (Israel). *Paléorient* 29(1), 19–58.

Hours, F., and Copeland, L. (1983). Les rapports entre l'Anatoli et la Syrie du Nord à l'eépoque des premières communautés villageoise de bergers et de paysans, 7600-5000 B.C. In T.C. Young Jr., P. Smith, and P. Mortensen (Eds.), *The Hilly Flanks. Essays on the Prehistory of Southwest Asia* (pp. 75–90), Studies in Ancient Oriental Civilization, No. 36, Oriental Institute, Chicago: University of Chicago Press.

Huxley, J.S. (1942). *Evolution: The Modern Synthesis*. London: Allen and Unwin.

Jones, G.T., Leonard, R.D., and Abbott, A.L. (1995). The structure of selectionist explanations in archaeology. In P.A. Tesler (Ed.), *Evolutionary archaeology: Methodological issues* (pp. 13–32). Tuscon: University of Arizona Press.

Kelly, R.L. (1995). *Foraging spectrum: Diversity in hunter-gatherer lifeways*. Washington D.C.: Smithsonian Institution Press.

Kelly, R.L. (2000). Elements of a behavioral ecological paradigm for the study of prehistoric hunter-gatherers. In M.B. Schiffer (Ed.), *Social theory in archaeology* (pp. 63–78). Salt Lake City: University of Utah Press.

Kennett, D.J., and Winterhalder, B. (Eds.) (2006). *Behavioral ecology and the transition to agriculture*. Berkeley: University of California Press.

Kennett, D.J., Voorhies, B., and Martorana, D. (2006a). An ecological model for the origins of maize-based food production on the Pacific coast of southern Mexico. In D.J. Kennett and B. Winterhalder (Eds.), *Behavioral ecology and the transition to agriculture* (pp. 103–136). Berkeley: University of California Press.

Kennett, D.J. Andersen, A., and Winterhalder, B. (2006b). The ideal free distrivution, food production, and the colonization of Oceania. In D.J. Kennett and B. Winterhalder (Eds.), *Behavioral ecology and the transition to agriculture* (pp. 265–288). Berkeley: University of California Press.

Kenyon, K. (1981). *Excavations at Jericho, Vol 3: The Architecture and Stratigraphy of the Tell* . London: British School of Archaeology in Jerusalem.

Kilian, B., Özkan, H., Walther, A., Kohl, J., Dagan, T., Salamini, F., Martin, W. (2007). Molecular diversity at 18 loci in 321 wild and 92 domesticate lines reveal no reduction of nucleotide diversity during *Triticum monococcum* (einkorn) domestication: Implications for the Origin of Agriculture. *Molecular Biology and Evolution* 24, 2657–2688.

Kislev, M.E. (1989). Pre-domesticated cereals in the Pre-Pottery Neolithic A period. In I. Hershkovitz (Ed.), *People and Culture in Change* (pp.147–152). British Archaeological Reports International Series 508, Oxford: Archaeopress.

Kislev, M.E. (1997). Early agriculture and paleoecology of Netiv Hagdud. In O. Bar-Yosef and A. Gopher (Eds.), *An early Neolithic village in the Jordan Valley* (pp. 209–236). Peabody Museum of Archaeology and Ethnology, Cambridge: Harvard University.

Kislev, M.E., Hartmann, A. Bar-Yosef, O. (2006a). Early domesticated fig in the Jordan Valley, *Science* 312, 1374–1375.

Kislev, M.E., Hartmann, A., Bar-Yosef, O. (2006b). Response to Comment on "Early Domesticated Fig in the Jordan Valley". *Science* 314, 1683b.

Kozlowski, S.K. (1999). *The Eastern Wing of the Fertile Crescent*, British Archaeological Reports, International Series 760, Oxford: Archaeopress.
Kowzlowski, S.K., and Aurenche, O. (2005). *Territories, Boundaries and Cultures in the Neolithic Near East*, Lyon:Maison de l'Orient et de la Méditerranée Jean Pouilloux.
Krebs, J.R. and Davies, N.B. (1978). Decision-making. In J.R. Krebs and N.B. Davies (Eds.), *Behavioral Ecology: An Evolutionary Approach* (3rd edition) (pp. 105–136). Oxford: Blackwell Scientific Publications.
Kuijt, I. (2000). People and space in early agricultural villages: Exploring daily lives, community size, and architecture in the late Pre-Pottery Neolithic. *Journal of Anthropological Archaeology* 19, 75–102.
Kuijt, I. (2009). Inventing Storage: Evidence for the Earliest Pre-Domestication Granaries 11,000 years ago in the Jordan Valley. *Proceedings of the National Academy of Sciences, U.S.A.* 106, 10966-10970
Kuijt, I. (in press). What do we really know about food storage, surplus, and feasting in pre-agricultural communities? *Current Anthropology*.
Kuijt, I., Goring-Morris, A.N. (2002). Foraging, farming, and social complexity in the Pre-Pottery Neolithic of the southern Levant: a review and synthesis. *Journal of World Prehistory* 16, 361–440.
Ladizinsky, G. (1989). Origin and domestication of the Southwest Asian grain legumes. In D.R. Harris and G.C. Hillman (Eds.), *Foraging and Farming: The Evolution of Plant Exploitation* (pp. 374–389). London: Unwin Hyman.
Legge, A.J. (1996). The beginning of caprine domestication in Southwest Asia. In D.R. Harris (Ed.), *The Origins and Spread of Agriculture and Pastoralism in Eurasia* (pp. 238–263). Washington D. C.: Smithsonian Institution Press.
Larson, G., Dobney, K., Albarella, U., Fang, M., Matisoo-Smith, E., Robins, J., Lowden, S., Finlayson, H., Brand, T., Willerslev, E., Rowley-Conwy, P., Andersson, L., Cooper, A. (2005). Worldwide pylogeography of wild boar reveals multiple centers of pig domestication. *Science* 307, 1618–1621.
Larson, G., Albarella, U., Dobney, K., Rowly-Conwy, P., Schibler, J., Tresset, A., Vigne, J.-D., Edwards, C., Schlumbaum, A., Dinu, A., Balasescu, A., Dolman, G., Tagliacozzo, A., Manaseryan, N., Miravle, P., Van Wijngaarden-Bakker, L., Masseti, M., Bradley, D.G., Cooper, A. (2007). Ancient DNA, pig domestication, and the spread of the Neolithic into Europe. *Proceedings of the National Academy of Sciences U.S.A* 104, 15276–15281.
Layton, R.R, Foley, R. (1992). On subsistence transtitions: Response to Hawkes and O'Connell. *Current Anthropology* 33, 218–219.
Layton, R.R., Foley, R., Williams, E. (1991). The transition between hunting and gathering and the specialized husbandry of resources. *Current Anthropology* 32: 255–274.
Leonard, R.D., Jones, G.T. (1987). Elements of an inclusive evolutionary model of archaeology. *Journal of Anthropological Archaeology* 6, 199–219.
Lev, E., Kislev, M., Bar-Yosef, O. (2005). Mousterian vegetal food in Kebara Cave, Mt. Carmel. *Journal of Archaeological Science* 32, 475–484.
Lev-Yadin, S., Ne'eman, G., Abbo, S., Flaishman, M.A. (2006a). Comment on "Early Domesticated Fig in the Jordan Valley". *Science* 314, 1683a.
Lipo, C.P., O'Brien, M.J., Collard, M., and Shennan, S.J. (2006). *Mapping our ancestors: Phylogentic Approaches in Anthropology and Prehistory*. New Brunswick: Aldine Transaction.
Lyman, R.L., O'Brien, M.J. (1998). The goals of evolutionary archaeology: History and explanation. *Current Anthropology* 39, 615–652.
Lyman, R.L., O'Brien, M.J. (2000). Measuring and explaining change in artifact variation with clade-diversity diagrams. *Journal of Anthropological Archaeology* 19, 39–74.
Lyman, R.L., O'Brien, M.J. (2001). On misconceptions of evolutionary archaeology: Confusing macroevolution and microevolution. *Current Anthropology* 42, 408–409.
MacArthur, R.H., Pianka, E.R. (1966). On optimal use of a patchy environment. *The American Naturalist* 100, 603–609.

Mayr, E. (1942). *Systematics and the Origin of Species*. New York: Columbia University Press.
Mayr, E. (1963). *Animal Species and Evolution*, Cambridge: Harvard University Press.
Meadow, R.H. (1989). Osteological evidence for the process of animal domestication. In J. Clutton-Brock (Ed.), *The Walking Larder: Patterns of Domestication, Pastoralism, and Predation* (pp. 80–90). London:Unwin Hyman.
Mesoudi, A., O'Brien, M.J. (2008). The learning and transmission of hierarchical cultural recipes. *Biological Theory* 3, 63–72.
McClure, S.B., Jochim, M.A., and Barton, C.M. (2006). Human behavioral ecology, domestic animals, and land use during the transition to agriculture in Valencia, eastern Spain. In D.J. Kennett and B. Winterhalder (Eds.), *Behavioral ecology and the transition to agriculture* (pp. 197–216). Berkeley: University of California Press.
Morrell, P.L., Clegg, M.T. (2007). Genetic evidence for a second domestication of barley (*Hordeum vulgare*) east of the Fertile Crescent. *Proceedings of the National Academy of Sciences U.S.A.* 104, 3289–3294.
Moore, A.T.M. (1991). Abu Hureyra I and the antecedents of agriculture on the middle Euphrates. In O. Bar-Yosef and F.R. Valla (Eds.), *The Natufian Culture in the Levant* (pp. 277–294). Ann Arbor: International Monographs in Prehistory.
Moore, A.M.T., Hillman, G.C., and Legge, A.J. (2000). *Village on the Euphrates: from Foraging to Farming at Abu Hureyra*, Oxford: Oxford University Press.
Morentsen, P. (1974). A survey of Prehistoric settlements in Northern Luristan. *Acta Archaeologica* 45, 1–47.
Morentsen, P. (1975). Survey and soundings in the Holailan Valley 1974. In *Proceedings of the Third Annual Symposium on Archaeological Research in Iran* (pp. 1–2). Iranian Centre for Archaeological Research, Tehran.
Munro, N. (2003). *Small game, the Younger Dryas, and the transition to agriculture in the southern Levant. Mitteilungen der Gesellschaft für Urgeschichte* 12, 47–71.
Munro, N. (2004). Zooarchaeological measures of hunting pressure and occupation intensity in the Natufian implications for agricultural origins. *Current Anthropology* 45, S6–S33.
Murray, M.A. (2003). The plant remains. In E. Peltenburg (Ed.), *The colonization and settlement of Cyprus. Investigations at Kissonerga-Mylouthkia, 1976–1996. Lemba Archaeological Project, Cyprus III.1* (pp. 59–71). Studies in Mediterranean Archaeology 70: 4, Sweden: Åströms Förlag,
Nadel, D. (2004). Wild barley harvesting, fishing, and year-round occupation at Ohalo II (19.5 KY, Jordan Valley, Israel). In Le Secrétariat du Congrés (Ed.), *Section 6: The Upper Paleolithic (General Sessions and Posters), Acts of the XIVth UISSP Congress, University of Liege (September 2001)* (pp. 135–143). British Archaeological Reports International Series No. 1240, Oxford: Archaeopress.
Naderi, S., Rezaei, H.-R., Taberlet, P., Zundel, S., Rafat, S.-A., Nagash, H.-R., El-Barody, M.A.A., Ertugrul, O, Pompanon, F., (2007). Large-scale mitochondrial DNA analysis of the domestic goat reveals six haplogroups with high diversity. *Plos One* 2(10): e1012.
Naderi, S., Rezaei, H.R., Pompanon, F., Blum, M.G., Negrini, R., Naghash, H.R., Balkiz, O., Mashkour, M., Gaggiotti, O.E., Ajmone-Marsan, P., Kence, A., Vigne, J.D., Taberlet, P. (2008). The goat domestication process inferred from large-scale mitochondrial DNA analysis of wild and domestic individuals. *Proceedings of the National Academy of Sciences USA* 105, 17659–17664.
Nesbitt, M. (2002). When and where did domesticated cereals first occur in Southwest Asia? In R.T.J. Cappers and S. Bottema (Ed.), *The Dawn of Farming in the Near East* (pp. 113–132). Studies in Early Near Eastern Productions, Subsistence, and Environment 6, Berlin: ex Orient 2.
O'Brien, M.J. (1987). Sedentism, population growth, and resource selection in the Woodland Midwest: A review of coevolutionary developments. *Current Anthropology* 28, 177–197.
O'Brien, M.J. (Ed.), (2008). *Cultural Transmission and Archaeology: Issues and Case Studies*. Washington, D.C.: Society for American Archaeology Press.

O'Brien, M.J., Holland, T.D. (1992). The role of adaptation in archaeological explanation. *Archaeological Method and Theory* 2, 31–79.
O'Brien, M.J., Lyman, R.L. (2000). *Applying Evolutionary Archaeology*, New York: Kluwer Academic/Plenum.
O'Brien, M.J., Lyman, L., Leonard, R.D. (1998). Basic incompatibilities between evolutionary and behavioral archaeology. *American Antiquity* 63, 485–498.
O'Connell, J.F., Hawkes, K., Lupo, K.D., Blurton Jones, N.G. (2002). Male strategies and Plio-Pleistocene archaeology. *Journal of Human Evolution* 43, 831–872.
Odling-Smee, F.J., Laland, K.N., and Feldman, W. (2003). *Niche Construction*, Monographs in Population Biology, No. 37. Princeton: Princeton University Press.
Özbaşaran, M. (2000). The Neolithic Site of Musular – Central Anatolia. *Anatolica* 26, 129–151.
Özbaşaran, M., and Buitenhuis, H. (2002). Proposal for a regional terminology for central Anatolia. In F. Gérard and L. Thissen (Eds.), *The Neolithic of Central Anatolia. Internal Developments and External Relations during the 9th-6th Millennia cal BC* (pp. 67–78). Istanbul: Ege Yayınları.
Özdoğan, M. (1997). Çayönü. *The Oxford Encyclopedia of archaeology in the Near East* 1, 444–446.
Özdoğan, M. (1997). Anatolia from the Last Glacial maximum to the Holocene climatic optimum. Cultural formations and the impact of the environmental setting. *Paléorient* 23(1), 25–38.
Özdoğan, M. (2002). Redefining the Neolithic in Anatolia. A critical overview. In Cappers, R.T.J. Bottema, S., (Eds.), *The Dawn of Farming in the Near East* (pp. 153–158). Studies in Near Eastern Production, Subsistence, and Environment, No. 6, Berlin: Ex Oriente.
Özdoğan, M., Özdoğan, A. (1989). Çayönü, a conspectus of recent work. *Paléorient* 15(1), 65–74.
Ozkan, H., Brandolini, A., Schafer-Pregl, R., Salamini, F. (2002). AFLP analysis of a collection of tetraploid wheats indicates the origin of emmer and hard wheat domestication in southeast Turkey. *Molecular Biology and Evolution* 19, 1797–1801.
Peltenburg, E. (Ed.) (2003). *The Colonization and Settlement of Cyprus. Investigations at Kissonerga-Mylouthkia, 1976–1996. Lemba Archaeological Project, Cyprus III.1*. Studies in Mediterranean Archaeology 70: 4, Sweden: Åströms Förlag.
Peltenburg, E. (2004). Social space in early sedentary communities of Southwest Asia and Cyprus. In Peltenburg, E. and Wasse, A. (Eds.), *Neolithic Revolution. New Perspectives on Southwest Asia in Light of Recent Discoveries in Cyprus* (pp. 71–90). Levant Supplementary Series 1, Oxford:Oxbow.
Perrot, J. (1983). Terminologie et cadre de la préhistoire récents de Palestine. In T.C. Young, Jr., P.E.L. Smith, and P. Mortensen (Eds.), *The Hilly Flanks. Essays on the Prehistory of Southwest Asia* (pp. 113–122). Studies in Ancient Oriental Civilization, No. 36, Oriental Institute, Chicago: University of Chicago.
Peters, J., Helmer, D., von den Driesch, A., Segui, S. (1999). Animal Husbandry in the northern Levant. *Paléorient* 25(2), 27–48.
Peters, J., Von Den Driesch A., and Helmer, D. (2005). The upper Euphrates-Tigris Basin, cradle of agro-pastoralism? In Vigne, J.-D., Peters, J. and Helmer, D. (Eds.), *The First Steps of Animal Domestication*, Oxford: Oxbow Books, pp. 96–124.
Piperno, D.R. (2006). The origins of plant cultivation and domestication in the Neotropics: A behavioral ecology perspective. In D.J. Kennett and B. Winterhalder (Eds.), *Behavioral ecology and the transition to agriculture* (pp. 137–166). Berkeley: University of California Press.
Piperno, D.R., Pearsall, D.M. (1998). *The Origins of Agriculture in the Lowland Neotropics*. San Diego: Academic Press.
Piperno, D.R., Weiss, E., Holst, I., and Nadel, D. (2004). Processing of wild cereal grains in the Upper Paleolithic revealed by starch grain analysis. *Nature* 407, 894–897.
Prentiss, W.C., Chatters, J.C. (2003). Cultural diversification and decimation in the prehistoric record. *Current Anthropology* 44, 33–58.

Redding, R.W. (2005). Breaking the mold, a consideration of variation in the evolution of animal domestication. In J.-D. Vigne, J. Peters and D. Helmer (Eds.), *The First Steps of Animal Domestication* (pp. 41–48). Oxford: Oxbow Books.

Richerson, P.J., Boyd, R. (2005). *Not by Genes Alone: How Culture Transformed Human Evolution*. Chicago: University of Chicago Press.

Richerson, P.J., Boyd, R., Bettinger, R.L. (2001). Was agriculture impossible during the Pleistocene by mandatory during the Holocene? A climate change hypothesis. *American Antiquity* 66, 387–412.

Rindos, D. (1984). *The Origins of Agriculture: An Evolutionary Perspective*. Orlando: Academic Press.

Rindos, D. (1985). Darwinian selection, symbolic variation, and the evolution of culture. *Current Anthropology* 26, 65–77.

Rindos, D. (1987). Comment on O'Brien "Sedentism, population growth, and resource selection in the Woodland Midwest: A review of coevolutionary developments". *Current Anthropology* 28, 192–193.

Rindos, D. (1989a). Diversity, variation, and selection. In R.D. Leonard and G.T. Jones, (Eds.), *Quantifying Diversity in Archaeology* (pp. 13–23). Cambridge: Cambridge University Press.

Rindos, D. (1989b). Undirected variation and the Darwinian explanation of culture change. *Archaeological Method and Theory* 1, 1–45.

Rollefson, G. (2000). Ritual and social structure at Neolithic 'Ain Ghazal. In I. Kuijt (Ed.), *Life in Neolithic Farming Communities. Social Organization, Identity and Differentiation* (pp. 165–190). London: Kluwer Academic/Plenum.

Rollefson, G.O. (2004). A reconsideration of the PPN koine: cultural diversity and centralities. *Neo-Lithics* 1, 46–48

Rollefson, G.O., Gebel, H. -G. (2004). Towards new frameworks: supra-regional concepts in Near Eastern neolithization. *Neo-Lithics* 1, 21–22.

Rosenberg, M. (1990). The mother of invention: Evolutionary theory, territoriality, and the origins of agriculture. *American Anthropologist* 92, 399–415.

Rosenberg, M. (1994a). Pattern, process, and hierarchy in the evolution of culture. *Journal of Anthropological Archaeology* 13, 307–340.

Rosenberg, M. (1994b). Hallan Çemi Tepesi: Some further observations concerning stratigraphy and material culture. *Anatolica* 20, 121–140.

Rosenberg, M. (1998). Cheating at musical chairs: territoriality and sedentism in an evolutionary context. *Current Anthropology* 39, 653–684.

Rosenberg, M., Davis, M. (1992). Hallan Çemi Tepesi, and Early Aceramic Neolithic site in eastern Anatolia: Some preliminary observations concerning material culture. *Anatolica* 18, 1–18.

Rosenberg, M., and Redding, R. (2000). Hallan Çemi and early village organization in eastern Anatolia. In: Kuijt, I. (Ed.). *Life in Neolithic Farming Communities: Social Organization, Identity, and Differentiation* (pp. 39–62). New York: Kluwer Academic /Plenum Press.

Rosenberg, M., Nesbitt, M. Redding, R.W., Peasnall, B.L. (1998). Hallan Çemi. Pig husbandry, and post-Pleistocene adaptations along the Taurus-Zagros arc (Turkey). *Paléorient* 24(1), 25–41.

Ryder, M.L. (1983). *Sheep and Man*. London: Duckworth.

Salamini, F., Öakan, H., Brandolini, A., Schäfer-Pregl, R., Martin, W. (2002). Genetics and geography of wild cereal domestication in the Near East. *Nature Reviews/Genetics* 3, 429–441.

Schmidt, K. (2005). "Ritual Centers" and the Neolithisation of Upper Mesopotamia. *Neo-Lithics* 2/05, 13–21.

Schultz, T., Mueller, U.G., Currie, C.R., and Rehner, S. (2005). Reciprocal illumination: a comparison of agriculture in humans and in fungus-growing ants. In: F. Vega, M. Balckwell, (Eds.), *Ecological and Evolutionary Advances in Insect-Fungal Associations* (pp. 149–190). New York: Oxford University Press.

Seilacher, A. (1972). Fabricational noise in adaptive morphology. *Systematic Zoology* 22, 451–465.

Service, E.R. (1962). *Primitive Social Organization: An Evolutionary Perspective*. New York: Random House.

Simpson, G.G. (1944). *Tempo and Mode in Evolution*. New York: Columbia University Press.
Smith, B.D. (1998). *The Emergence of Agriculture*. New York: W. H. Freeman and Co.
Smith, B.D. (2001). Low level food production. Journal of Archaeological Research 9, 1–43.
Smith, B.D. (2006). Human behavioral ecology and the transition to food production. In D.J. Kennett and B. Winterhalder (Eds.), *Behavioral ecology and the transition to agriculture* (pp. 289–303). Berkeley: University of California Press.
Smith, B.D. (2007a) The ultimate ecosystem engineers. *Science* 315, 1797–1798.
Smith, B.D. (2007b). Niche construction and the behavioral context of plant and animal domestication, *Evolutionary Anthropology* 16, 189–199.
Smith, P.E.L. (1972). Ganjdareh Tepe, *Iran* 10, 165–168.
Smith, P.E.L. (1976). Reflections on four seasons of excavations at Tapeh Ganj Dareh. In Bagherzadeh, F. (Ed.), *Proceedings of the Fourth Annual Symposium on Archaeological Research in Iran* (pp. 11–22). Tehran: Iranian Centre for Archaeological Research.
Solecki, R.L. (1981). *An Early Village Site at Zawi Chemi Shanidar*, Bibliotheca Mesopotamica, No. 13, Malibu: Undena Publications.
Solecki, R.L., and McGovern, T. (1980). Predatory birds and prehistoric man, In S. Diamond (Ed.), *Theory and Practice. Essays Presented to Gene Weltfish* (pp. 79–95). The Hague: Mouton.
Solecki, R.S. (1965). Prehistory in Shanidar valley, Northern Iraq, *Science* 139, 179–193.
Spencer, C.S. (1990). On the tempo and mode of state formation: Neoevolutionism reconsidered, *Journal of Anthropological Archaeology* 9, 1–30.
Spencer, C.S. (1993). Human agency, biased transmission, and the cultural evolution of chiefly authority, *Journal of Anthropological Archaeology* 12, 41–74.
Spencer, C.S. (1997). Evolutionary approaches in archaeology. *Journal of Archaeological Research* 5, 209–264.
Spencer, C.S., and Redmond, E. (2001). Multilevel selection and political evolution in the Valley of Oaxaca, 500-100 B.C. *Journal of Anthropological Archaeology* 20, 195–229.
Stephens, D., and Krebs, J. (1986). *Foraging theory*. Princeton: Princeton Univerisity Press.
Stordeur, D. (2004). New insights and concepts: Two themes of the Neolithic in Syria and southeast Anatolia, *Neolithics* 1, 49–51.
Sudupak, M.A., Akkaya, M.S., Kence, A. (2004). Genetic relationships among perennial and annual*Cicer* species growing in Turkey assessed by AFLP fingerprinting. *Theoretical and Applied Genetics* 108, 937–944.
Tanno, K., Wilcox, G. (2006a). How fast was wild wheat domesticated? *Science* 311, 1886.
Tanno, K., Wilcox, G. (2006b). The origins of cultivation of *Cicer arietinum* L. and *Vicia faba* L.: early finds from tell el-Kerkh, north-west Syria, late 10th millennium B.P. *Vegetation History and Archaeobotany* 15, 197–204.
Tchernov, E. (1991). Biological evidence for human sedentism in Southwest Asia during the Natufian. In O. Bar-Yosef and F.R. Valla, F.R. (Eds.), *The Natufian Culture in the Levant* (pp. 315–340). Ann Arbor: International Monographs in Prehistory.
Tchernov, E., Valla, F.R. (1997). Two new dogs, and other Natufian dogs, from the southern Levant. *Journal of Archaeological Science* 24, 65–95.
Tucker, B. (2006). A future discounting explanation for the persistence of a mixed foraging-horticulture strategy among the Mikea of Madagascar. In D.J. Kennett and B. Winterhalder (Eds.), *Behavioral ecology and the transition to agriculture* (pp. 22–40). Berkeley: University of California Press.
Turnbull, P.F. and Reed, C.A. (1974). *The Fauna from the Terminal Pleistocene of Palegawra Cave, a Zarzian occupation site in the northeastern Iraq*. Fieldiana: Anthropology, vol. 63 (3), Chicago: Field Museum of Natural History.
Ucko, P. (1968). *Anthropomorphic Figurines of Predynastic Egypt and Neolithic Crete with Comparative Material from the Prehistoric Near East and Mainland Greece*. Royal Anthropological Institute Occasional Paper, No 24, London: Andrew Szmidla.
Uerpmann, H.-P. (1978). Metrical analysis of faunal remains from the Middle East. In R.H. Meadow and M.A. Zeder (Eds.), *Approaches to Faunal Analysis in the Middle East* (pp. 41–45). Peabody Museum Bulletin, No 2. Cambridge: Peabody Museum.

Uerpmann, H.-P. (1979). *Probleme der Neolithisierung des Mittelmeeraumes*. Biehefte zum Tübinger Atlas des Vordern Orients, Reihe B, Nr. 28. Wiesbaden: Dr Ludwig Reichert.

Vigne, J.-D., Buitenhuis, H., Davis, S. (1999). Les premiers pas de la domestication animal à l'Oest de l'Euphrate: Chyupre et L'Anatolie centrale. *Paléorient* 25(2), 49–62.

Vigne, J.-D., Carrére, I., Saliége, J.-F., Person, A., Bocherens, H., Guilaine, J., and Briois, F., (2000). Predomestic cattle, sheep, goat, and pig during the late 9th and the 8th millennium cal. B.C. on Cyprus: preliminary results of Shillourokambos (Parekklisha, Limassol). In Mashkour, M. Choyke, A.M., Buitenhuis, H., and Poplin, F. (Eds.), *Archaeozoology of the Near East, IVA* (pp. 83–106). ARC Publication No. 32, Groningen: ARC.

Vigne, J.-D., Carrère I., and Guilaine, J (2003). Unstable status of early domestic ungulates in the Near East: The example of Shillourokambos (Cyprus, IX-VIIIth millennia cal. B.C.). In J. Guilaine et A. Le Brun éds., *Le Néolithique de Chypre* (pp. 239–251). Actes Coll. Int. Nicosie, 17–19 mai 2001, *Bulletin de Correspondance Hellenique*, suppl. 43.

Voigt, M.M. (2000). Çatal Höyük in context: Ritual at early Neolithic sites in Central and Eastern Turkey. In Kuijt, I. (Ed.), *Life in Neolithic Farming Communities. Social Organization, Identity and Differentiation* (pp. 253–294). London: Kluwer Academic/Plenum.

Vrba, E., Eldredge, N. (1984). Individuals, hierarchies, and processes: Towards a more complete evolutionary theory. *Paleobiology* 10, 146–171.

Watkins, T. (1996). Excavations at Pinarbaşi: The early stages. In I. Hodder (Ed.), *On the surface: Çatalhöyük 1993–1995* (pp. 47–57). Cambridge and London: MacDonald Archaeological Institute and the British Institute at Ankara.

Weiss, E., Wetterstrom, W., Nadel, D., Bar-Yosef, O. (2004). The broad spectrum revolution revisited: Evidence from plant remains. *Proceedings of the National Academy of Science, USA* 101, 9551–9555.

Weiss, E., Kislev, M.E., Hartmann, A. (2006). Autonomous cultivation before domestication. *Science* 312, 1608–1610.

Wilcox, G. (2002). Geographical variation in major cereal components and evidence for independent domestication events in Western Asia. In R.T.J. Cappers and S. Bottema (Eds.), *The Dawn of Farming in the Near East* (pp. 133–140). Studies in Near Eastern Production, Subsistence, and Environment, No. 6, Berlin: Ex Oriente.

Wilcox, G. (2005). The distribution, natural habitats, and availability of wild cereals in relation to their domestication in the Near East: Multiple events, multiple centres. *Vegetative History and Archaeobotany* 14, 534–541.

Wilcox, G., Fornite, S., Herveux, L. (2008). Early Holocene cultivation before domestication in northern Syria. *Vegetative History and Archaeobotany.* 17, 313–325.

Winterhalder, B. (1986). Diet choice, risk, and food sharing in a stochastic environment. *Journal of Anthropological Archaeology* 5, 369–392.

Winterhalder, B., Goland, C. (1993). On population, foraging efficiency, and plant domestication. *Current Anthropology* 34:710–715.

Winterhalder, B., and Goland, C. (1997). An evolutionary ecology perspective on diet choice, risk, and plant domestication. In K.J. Gremillion (Ed.), *People, Plants, and Landscapes: Studies in Paleoethnobotany* (pp. 123–160). Tuscaloosa: University of Alabama Press.

Winterhalder, D., and Kennett, D.J. (2006). Behavioral ecology and the transition from hunting and gathering to agriculture. In D.J. Kennett and B. Winterhalder (Eds.), *Behavioral ecology and the transition to agriculture* (pp. 1–21). Berkeley: University of California Press.

Winterhalder, B., Kennett, D.J. (in press). Four neglected concepts with a role to play in explaining the origins of agriculture. *Current Anthropology*.

Winterhalder, B., Smith, E.A. (1981). *Hunter-gatherer foraging strategies: Ethnographic and Archaeological Analyses*. Chicago: University of Chicago Press.

Wright, S. (1931) Evolution in Mendelian populations. *Genetics* 16, 97–159.

Wright, S. (1932). The roles of mutation, inbreeding, crossbreeding and selection in evolution. *Proceedings of the VI International Congress of Genetrics* 1, 356–366.

Zeder, M.A. (1994). After the revolution: Post-Neolithic subsistence strategies in Northern Mesopotamia. *American Anthropologist* 96, 97–126.

Zeder, M.A. (2001). A metrical analysis of a collection of modern goats (*Capra hircus aegargus* and *Capra hircus hircus*) from Iran and Iraq: Implications for the study of caprine domestication. *Journal of Archaeological Science* 28, 61–79.
Zeder, M.A. (2005). New Perspectives on Livestock Domestication in the Fertile Crescent as viewed from the Zagros Mountains. In J.-D. Vigne, J. Peters and D. Helmer (Eds.), *The First Steps of Animal Domestication* (pp. 125–147). Oxford: Oxbow Books.
Zeder, M.A. (2006a). Central questions in the domestication of plants and animals. *Evolutionary Anthropology* 15, 105–117.
Zeder, M.A. (2006b). Archaeological approaches to documenting animal domestication. In M.A. Zeder, E. Emshwiller, B.D. Smith, and D.G. Bradley (Eds.), *Documenting Domestication: New Genetic and Archaeological Paradigms* (pp. 171–180), Berkeley: University of California Press.
Zeder, M.A. (2006c). A Critical Examination of Markers of Initial Domestication in Goats (*Capra hircus*). In M.A. Zeder, E. Emshwiller, B.D. Smith, and D.G. Bradley (Eds.), *Documenting Domestication: New Genetic and Archaeological Paradigms* (pp. 181–208). Berkeley: University of California Press.
Zeder, M.A. (2006d). Archaeozoology in Southwest Asia: A status report based on the eighth meeting of the Archaeozoology of Southwest Asia and Adjacent Areas Working Group, *Paléorient* 32(1), 137–147.
Zeder, M.A. (2008a). Domestication and early agriculture in the Mediterranean Basin: Origins, diffusion, and impact. *Proceedings of the National Academy of Sciences, USA* 105, 11597–11604.
Zeder, M.A. (2008b). Animal domestication in the Zagros: An update and directions for future research. In E. Vila, L. Goucherin, A. Choyke, and H. Buitenhuis (Eds.), *Archaeozoology of the Near East VIII, Proceedings of the eight international symposium on the Archaeozoology of Southwestern Asia and Adjacent areas* (pp. 243–278). Lyon: Travaux de la Maison de l'Orient et de la Méditerranée (TMO).
Zeder, M.A. (2009). The Neolithic macro-(r)evolution: Macroevolutionary theory and the study of culture change. *Journal of Archaeological Research* 17:1–63.
Zeder, M.A. (in press). Pathways to animal domestication. In: P. Gepts (Ed.), *Biodiversity in Agriculture: Domestication, Evolution, & Sustainability*.
Zeder, M.A., Hesse, B. (2000). The initial domestication of goats (Capra hircus) in the Zagros Mountains 10,000 years ago. *Science* 287, 2254–2257.
Zeder, M.A. and Smith, B.D. (in press). A Conversation on Agricultural Origins: Talking Past Each Other in a Crowded Room. *Current Anthropology*.
Zeder, M.A., Bradley, D.G., Emshwiller, E., and Smith, B.D. (Eds.). (2006a). *Documenting Domestication: New Genetic and Archaeological Paradigms*, University of California Press, Berkeley.
Zeder, M.A., Emshwiller, E., Smith, B.D., Bradley, D.G. (2006b). Documenting domestication, the intersection of genetics and archaeology, *Trends in Genetics* 22, 139–155.
Zohary, D. (1992). Domestication of the Neolithic Near Eastern crop assemblage. In Anderson, P.C. (Ed.), *Préhistoire de l'Agriculture* (pp. 81–86). Paris: Editions du CNRS.
Zohary, D. and Hopf, M. (2000). *Domestication of Plants in the Old World*, ed. 3, Oxford: Oxford Science.

Part III
Cultural Diversification, Stasis and Extinction as Macroevolutionary Processes

Chapter 7
A Macroevolutionary Perspective on the Archaeological Record of North America

James C. Chatters

Introduction

The first archaeologists to work in any region of North America from a scientific perspective readily distinguished chronological patterns in artifact assemblages, dwelling forms, and demographic patterning. The cultural sequences they defined from these patterns have in most cases defied dismantling by later generations of investigators. With continued research, subsequent scholars have been able to make chronological refinements within each archaeological tradition, more precisely characterize settlement patterns, technologies, and subsistence practices, and define regional variants on each basic theme, but the overall structure has remained. Terms like Paleo-Indian, Western Pluvial Lakes Tradition, Basketmaker, Middle Archaic (of Eastern North America), Middle Woodland, Plains Village, and Thule all retain much the same meaning they had when coined. In every region, each distinguishable archaeological entity is followed by another whose characteristics are equally distinct. Each, thus, represents an interval of relative behavioral stasis lasting from a few hundreds to a few thousands of years. The stasis can be so strict, in fact, that we are often hard pressed to distinguish chronological subunits within high-level archaeological complexes without assistance from some precise method of chronometry. Subunits, when distinguishable, tend to be stylistic or to represent minor adjustments to environmental fluctuations.

Because these chrono-spatial archaeological units are robust in the face of analysis and persist in time and space, it is reasonable to see them, not as arbitrary constructs from an otherwise continuously varying chaos of independently evolving elements (as some members of the evolutionary archaeology school contend; e.g., Leonard and Jones 1987 O'Brien and Lyman 2000) but as representing real higher-level evolving entities (Chatters and Prentiss 2005). Like the species recognized by paleontologists, each higher-level archaeological unit is made up of multiple

J.C. Chatters (✉)
Applied Paleoscience, AMEC Earth and Environmental, Inc, Bothell, WA; Department of Earth and Environmental Sciences, California State University, Fresno, CA, USA
e-mail: james.chatters@amec.com

A.M. Prentiss et al. (eds.), *Macroevolution in Human Prehistory*,
DOI 10.1007/978-1-4419-0682-3_8,

representatives of its kind, whose distribution is bounded in space and time, with a point of inception, period of florescence (maximum fitness), and point of extinction.

The archaeological record of North America is thus a record of macroevolutionary events. When we take the opportunity to step back from our sometimes bewildering assemblages of lithic debitage, bone fragments, pot sherds, and feature diagrams three facts become evident.

First, stasis is not the exception in human cultural history; the record is dominated by stasis. Change in higher-level cultural entities is a very rare event. To borrow a phrase from Eldredge and Gould (1972), the history of human cultural development is one of punctuated equilibria. Because of this fact, the processes that support stasis are at least as important to understand as—and in part lead us to the processes that drive—the emergence of new higher-level entities. As Prentiss and I have argued elsewhere (Prentiss and Chatters 2003a, b; Chatters and Prentiss 2005), if indirectly, a disruption of this stabilizing process is required, if new forms are to have the *opportunity* to emerge.

Second, change from one higher-level cultural form to the next is usually so rapid that we are unable to distinguish any transition between the two. Transitions occur on the scale of decades rather than centuries; our chronometric methods are simply too imprecise in most cases to allow us to perceive sequences of events on this small time scale.

Third, when we again step a bit closer to the assemblages and see the inner workings of chrono-spatial units, we see that the building blocks of these units tend to have histories that are independent of the histories of the chrono-spatial units themselves. High-level cultural entities are comprised of nested hierarchies of less-complex entities, including subsistence technologies, demographic patterns, and scheduling patterns, and symbolism, each of which is itself comprised of yet-less-complex subunits and so on, until we reach the level of individual cultural traits. Most, if not all, of these lower-level entities did not develop coincidentally with the chrono-spatial unit but have simply been reassembled into the new configuration that the unit represents.

I return to these observations later in this chapter when I present examples of macroevolutionary events in North American prehistory. First, it is important to clarify what is evolving, how its fitness is manifest, how stasis is maintained, and how new forms emerge.

The Evolving Entity

The chrono-spatial units archaeologists have identified were produced by a wide range of human behaviors, but they are dominated by the traces of socioeconomic activities. Each human community collectively holds a set of ideas for supplying food and shelter, along with the materials needed to produce the tools and structures that help fulfill these needs. Prentiss and I have elsewhere termed this set of ideas the Resource Management Strategy or RMS (Chatters and Prentiss 2005; see also Prentiss, this volume; Prentiss et al., this volume)[1]. Nested within the RMS are sub-strategies for subsistence technology, labor management, community positioning and scheduling, resource distribution, consumption, and so on. Each of these substrategies is comprised of less-complex components, including such things as technologies for capturing, processing, or distributing different kinds of foods.

These sets of ideas are heritable in the sense that they can be learned by social transmission through various forms of training, active mimicry, and passive mimicry (Boyd and Richerson 1985, 2005; Richardson and Boyd 1985). Like the interacting assemblages of genes that produce a living organism, the associated sets of ideas, including the complete assemblage of associated sets of ideas that comprise the RMS, are the evolving entities.

Selection and the Manifestation of Fitness

The RMS achieves its physical expression through tactical behaviors, and it is these tactics that are the point of interaction with the RMS's (or sub-strategy's) environment. These tactics leave physical traces as artifacts, patterns of associated artifacts, artificial soil stains, and seasonally patterned plant and animal materials that comprise the archaeological record. As the tactics play out, their success in relation to the tactics of other human individuals or communities is observed and evaluated by both the performing community's members and by the members of other competing communities. If the tactics are ineffective, the RMS, or one or more of its offending component substrategies, may be abandoned (if the practicing community does not fail altogether and cease to exist biologically) in favor of a more effective RMS or RMS component sub-strategy. If effective, the tactics, or more appropriately the ideas that produced them, may persist and ultimately be adopted by the next human generation of the community applying the RMS. If the RMS allows the practicing community to grow through internal reproduction, bud off new communities, and expand its geographical distribution at the expense of its neighbors, it achieves greater fitness than the neighbors' RMS. In this case both the RMS and the population applying it gain fitness, one cultural, the other biological. This increased fitness through growth of the biological population supported by the RMS I will refer to here as *organic* fitness.

An RMS may also achieve greater fitness through what Boyd and Richerson (2005) have called horizontal and oblique transmission. Neighboring individuals or communities aware of the success of another RMS in relation to their own may mimic the successful tactics in whole or in part. In this way the RMS, or its component sub-strategies may gain additional communities, thus increasing its frequency of expression and geographic range. This may be called an improvement in *acquisitive* fitness.

Fitness of the RMS or its component parts is thus manifest in the number of communities expressing the tactics that leave the physical trace we call the archaeological record, regardless of whether an increase in communities results from organic or acquisitive processes. To the expanding RMS, it is of no consequence and may or may not be discernible archaeologically. The RMS becomes extinct when all practicing communities have abandoned it in favor of another (or others) that is (are) perceived to be more effective, or failed biologically.

The reader will note that, unlike some contributors to this volume, I am not attempting to equate cultural evolution with human biological fitness. I believe that to do so confuses the issue. The fitness of an RMS and that of its human practitioners may be increasing or decreasing in tandem, as in the case of the organic fitness of a strategy, but the two are not inextricably coupled. From an archaeological standpoint, the question of whether an RMS expands organically or acquisitively is of historical, not theoretical, interest.

Stabilizing Selection and the Emergence of New Strategies

Each RMS is in competition with others around it but is also subject to continual evaluation by community members, based on its long-term history of effectiveness. If a strategy at whatever level has been effective, even though imperfect, community members are unlikely to be willing to implement a new strategy that is not tried and true. This is especially likely when the competition with other groups is high or environmental conditions are tenuous, and the failure of a new strategy holds the threat of tipping the balance against the experimenting community. In the terms of adaptive landscape theory (Wright 1931, 1932; see Bettinger this volume; Spencer, this volume), and thinking of adaptive peaks as islands surrounded by moats of failure, the community may not reside on the highest or most adapted peak but is unwilling to try a new strategy if it risks the community drowning in the moat. Under conditions of intense competition, environmental continuity, and close communication between human groups, therefore, there will be both strong internal pressure and outside selective pressure for conformity. Under these circumstances, changes are likely to be small scale; any new strategy that is adopted will be one that tracks changes in the community's environment and maintains the community's position at the apex of its island in the face of surrounding pressures. Evolution will, therefore, be micro in scale and stabilizing in function. It is for these reasons that we see prolonged stasis in the historical record. It is also for these reasons that new forms of RMS—featuring new socioeconomic structures or *Baupläne*, to use Rosenberg's (1994) term borrowed from biology, have the best opportunity to emerge either when communities are effectively or geographically isolated from social contact (reducing competition and the pressure to conform), or when the effective population density is reduced, relaxing competitive pressure (Prentiss and Chatters 2003a,b; Chatters and Prentiss 2005). To return to the island metaphor, a new RMS can arise when the moat becomes shallow enough that the community can cross from one adaptive peak to another without the real or perceived risk of drowning. Once one community has successfully made the crossing and demonstrated the new peak to be higher than other already-occupied peaks nearby, other communities may follow with minimal risk, hence the rapidity with which wholesale changes occur in the archaeological record.

As archaeologists, we have developed methods to indirectly observe cultural entities at the scale of technologies, technological complexes, organization systems, and entire RMS on a regional scale by mapping the traces left by their tactical expression (e.g., Chatters 1995; Spencer and Redmond 2002; chapters in this volume). We recognize the emergence of new forms when we see a disjunction in the pattern of these physical manifestations, particularly in the evidence for time–space positioning, demography, consumption, and technology that signal an altered relationship between an interacting set of ancient communities and their resource base. Cultural evolution at the macroscale is what we are best equipped to investigate.

Cultural Macroevolution in North America

The North American archaeological record is rife with macroevolutionary events at a variety of levels within the RMS hierarchy. We see evolutionary changes in the complete RMS as economic revolutions but also witness numerous examples of evolution in the realms of technology, labor organization (sociopolitical organization), community positioning and scheduling, and distribution. In the following pages, I address a few examples of evolution in two levels of strategy: technologies and complete socioeconomic *Baupläne*.

Technologies

In the introduction, I noted that an RMS is a constellation of substrategies which have their own, usually independent, evolutionary histories. Subsistence technologies provide some of the best examples of this independence. Technologies that have developed or been prominent in North American socioeconomic behavior include, but are far from limited to, the fluted projectile, the weighted atlatl-and-dart projectile delivery system, composite and toggling harpoons, acorn processing, cultivation of indigenous domesticates, maize farming, copper metallurgy, net fishing, various food grinding methods, stone boiling, and kayak, dogsled, and equestrian transportation. Here I discuss two technologies, whose histories can be traced and which illustrate the different rates at which fitness can improve. These are the earth oven method of plant food processing and the bow-and-arrow projectile delivery system.

Earth Ovens in Northwest America

During the ethnographic period, the resource management strategies of Pacific Northwest hunter-gatherers were organized on a collector *Bauplan* that included mass harvest of fish, berries, game, and roots in varying proportions (see papers in Suttles 1990; Walker 1998). As Prentiss describes in this volume (also Prentiss and

Chatters 2003a. b; Chatters and Prentiss 2005), this *Bauplan* developed between 4000 and 5000 cal B.P. and expanded rapidly after 4000 cal B.P. to become the dominant regional RMS by around 3500 cal B.P. An important technology in the root-gathering component of this strategy was the earth oven, which was used most often for camas in the Southern Plateau and coastal regions (Thoms 1989, and for balsamroot and avalanche lily on the Canadian Plateau (Peacock 2002).

An earth oven is a means for reducing the complex, indigestible carbohydrates found in some abundant plant species into readily digestible starches and sugars (Thoms 1989, 2008). To oven bake roots and other potential edibles, a shallow, basin-shaped pit, ranging from 0.5 to more than 3 m in diameter, is dug in the ground and lined with stones. A fire is built over and among the stones and fed until the stones are well heated. A layer of damp vegetation, such as grass or reeds, is then laid over the stones, and a mass of the raw plant food is inserted, usually in some form of bag. Another layer of vegetation is laid over the food and the whole covered with earth. A fire is usually then kindled atop the filled pit and burned overnight. Once the fire is dead, the roots are exhumed, ready for consumption. Large quantities of roots can be cooked in a single oven, but the process has a high labor cost. Archaeologically, the earth oven appears as a large, shallow basin with an oxidized lining, fire-broken rock, and charred wood and root tissue.

The earliest earth oven yet documented for the Northwest was a 1.5 m diameter feature found at the Hannavan Creek site in the southern Willamette Valley (Cheatham 1988; O'Neil et al. 2004). Radiocarbon dates on samples of the more than 350 charred camas bulbs placed the use of this oven at around 8500 cal B.P. Another, similar feature was found at the nearby Ralston site and dates to ca 7500 cal B.P. No other well-dated ovens have been found in the region until much later, although high rates of dental caries in human skeletal material from the DeMoss site in the Rocky Mountains of central Idaho may be indirect evidence that people in one remote valley were relying on camas, perhaps cooked in earth ovens, at around 6800 cal B.P. (Pavesic 1985). In the Kalispel Valley of Northeastern Washington, people began occasionally baking camas roots as early as 6300 cal B.P. (Thoms 1989; Andrefsky et al. 2000). After its early appearance, the earth oven continued to be nothing more than an occasional tool for camas processing in the Willamette and Kalispel Valleys; no other examples have been reported.

It was not until after 4800 cal B.P. that the use of earth oven technology exploded in the Willamette Valley. The technology remained important locally from then onward. Despite its importance in the Willamette, however, the earth oven continued to be nothing more than an occasional means to a few sweet meals until after an RMS based on the collector *Bauplan* entered the region at around 3700 cal B.P. (Chatters and Prentiss 2005). Numbers of ovens sharply increased at that time in the Kalispel valley, and the features began to occur in sites along the upper Columbia River (Campbell 1985) and in the Columbia Basin (Daugherty 1952). They also spread shortly thereafter into the Canadian Plateau, becoming a major component of the collector strategy after 2400 cal B.P. (Lepofsky and Peacock 2004; Peacock 2002). It also became an important facility in central California following a shift in subsistence organization after 2500 cal B.P. As part of the collector RMS earth ovens were used to process massive quantities of starch for delayed consumption.

The earth oven is an example of a set of related ideas that remained known to some communities in a region for countless generations but never expanded—increased its fitness—until its use became integral to the functioning of a new overall strategy for acquiring and managing the food supply. It was simply too labor-intensive a technology for regular use in a forager *Bauplan*. Another example of such a "sleeper" technology is the use of the mortar and pestle as a food grinding implement in the Central Valley of California. Mortar and Pestle were in limited use in the productive Clear Lake area from as early as 8000 cal B.P. but not accepted elsewhere or frequently employed until leaching became an important method for processing acorn meal (White et al. 1999). Another example is a series of subsistence technologies developed in southwest Asia in the Epipaleolithic that did not become fully integrated into regional RMS until the Neolithic (Zeder, this volume).

The Bow-and-Arrow

The bow-and-arrow provides a contrasting example. Before 2000 cal B.P., projectile delivery technology in the Americas south of the arctic was limited to the atlatl and dart complex and probably the lance or javelin. The bow-and-arrow, developed in Africa and Southwest Asia as much as 13,000 years ago (Clark 1970) and was widely used during the Mesolithic and Neolithic of Eurasia. Five thousand years ago, Neolithic Paleoeskimo immigrants from northeast Siberia (Hoffecker 2004) introduced the technology to the Western Hemisphere as part of the Arctic Small Tool tradition, which included tiny bifacial end blades for barbed arrow tips. While identification of these implements as arrow points can be contested, the presence of compound bow parts in the descendant pre-Dorset assemblages across the whole of the Canadian arctic confirms the technology's presence after 4500 cal B.P. (Maxwell 1985; McGhee 1996). There it existed until 2800 cal B.P. but was abandoned during Dorset (see also Prentiss and Lenert, this volume). Although bow and arrow continued to be used in the western arctic, it did not cross the sub-arctic wastes into Temperate North American until around 2000 years ago.

The timing of the introduction of the bow and arrow is somewhat problematic, because this technology must be distinguished from the atlatl and dart primarily on the basis of their imperishable tips. Numerous authors have sought means for making this distinction, basing it on one or another measure of size. Arrows are smaller than darts, so they deliver smaller tips. Using the few complete, stone-tipped arrows and darts that are available in museum collections, various researchers have made the distinction between technologies on the basis of weight (Fenenga 1953; Heizer and Baumhoff 1961), neck width (Corliss 1972), basal width (Thomas 1981), or the mathematical relationships among length, width, thickness, and neck width (Shott 1997; Thomas 1978). Often, the technological change is recognized when large projectile tips are replaced by much smaller, thinner types, with the earliest accepted date on a persistent smaller style marking the introduction of the bow.

Blitz (1988) and Nassaney and Pyle (1999) have surveyed the chronology of the bow's introduction across North America. Their data, along with more recent updates, demonstrate that the new technology was accepted across the continent within less than 800 years of its first appearance south of the arctic. Arrow points appear earliest in the southern Plateau, eastern Great Basin, and Northern Great Plains as miniature versions of contemporary dart points. In the southern Plateau, the small Columbia Corner- and Basal-Notched styles appear alongside Quilomene Bar Basal-Notched by between 2200 and 1850 cal B.P. (Chatters 2004). In the eastern Great Basin, the small, corner or basal-notched Rosegate joins the large Elko Corner-Notched as early as 1950 cal B.P. (Aikens and Madsen 1986). On the Plains, the small, side-to-corner-notched Avonlea replaces the larger Besant between 1850 and 1950 cal B.P. (Kehoe 1966; c.f. Vickers 1986). Given the limitations of radiocarbon dating, these introductions were nearly simultaneous. The tendency for the arrow point to be a miniature version of the contemporary dart tip demonstrates that transmission of the idea came through the process of guided variation (Boyd and Richerson 1985). The neophyte archers observed the technology in use among neighbors or visitors to their respective regions and duplicated the components in their own local metaphors. All three of these regions abut the north and central Rocky Mountains, so the people who provided the introduction may have been moving south along this mountain chain, as did later Athabascan-speaking peoples (see Kehoe 1966; Matson and Magne 2007).

The bow and arrow appear at progressively later dates to the south, west, and east of these first introductions. Arrow points identical to those on the Southern Plateau are fully established on the Northwest Coast at 1550 cal B.P. (Chatters et al. 1990) (Determination of precisely when the bow arrived on the coast can be even more problematic than it is in the interior because of an ethnographically known tendency to use hardwoods rather than stone for arrow tips). In the western Great Basin and California, archaeologists do not acknowledge arrival of the bow until after 1500 cal B.P. (Bettinger and Eerkens 1999; Elston 1986; Moratto 1984). They appear as the same, small corner-notched styles found in adjacent regions: the Rosegate in the Basin and Gunter Barbed in California, suggesting that transmission took the form of direct or indirect bias in these cases (Bettinger and Eerkens 1999). By 1350 cal B.P., complete arrows found in dry caves demonstrate that the bow and arrow had been implemented in the Southwest (Morris 1980).

Ideas for the new projectile system spread south and west of the northern Plains at nearly the same rate as they did in the far west. With the exception of a few argued older dates in the Carolinas (Oliver 1985; Sassaman et al. 1990) and the southernmost Great Plains (Vehik 1984), most evidence indicates widespread acceptance of the Bow between 1350 and 1550 cal B.P. (Nassaney and Pyle 1999: 254–255). Stylistic comparisons between southern Plains and lower Mississippi Valley arrow points show that the new technology was introduced into the latter region from the west, whereas evidence elsewhere in the eastern woodlands is indicative of diffusion of the ideas from the north (Nassaney and Pyle 1999).

Throughout most of the continent, the bow and arrow, and the atlatl and dart coexisted for centuries as alternative technologies, with the atlatl persisting in some areas of the Southeast and California until European contact (Ames et al. 1999; Nassaney and Pyle 1999; Shott 1993). The bow was, at least at first, an addition to existing projectile propulsion technology, not a replacement for it. This co-occurrence often confuses the dating of the bow's appearance, making it difficult to distinguish contemporaneity from the mixing of components in stratified chronological sequences. If we accept the co-occurrence of dart and arrow points as evidence for this persistent use of two technologies, this review suggests that we are likely to find that the bow spread across the continent in even less than the 300–500 years.

The bow-and-arrow's status as a projectile co-technology, rather than an entirely new replacement for old technologies, makes its rapid acceptance into societies as divergent as mobile, egalitarian foragers and settled socially ranked horticulturalists readily understandable. All of these societies already possessed projectile propulsion systems capable of casting stone-tipped shafts long distances into enemies or prey animals. Inclusion of one more method for achieving this same end had little chance of disturbing any community's position on its adaptive island. In cases where the bow's characteristics were advantageous, whether for hunting (e.g., Rousseau 2004) or for war (Blitz 1988), it would have given the adopting community a competitive edge over its non-adopting neighbors, improving its position on the adaptive peak. Conversely, it is difficult to envision a risk to a community associated with bow adoption as long as existing atlatl and dart technology remained available.

The bow and atlatl have similar initial production costs. Each requires acquisition of appropriate wood blank and material for wrapping the handle. The atlatl maker must obtain and shape material for a weight, whereas the bow maker must produce a string of suitable length and strength. Once this initial investment has been made, the atlatl user must obtain and straighten a small number of large shafts and foreshafts, and obtain and reduce stone for large points. The arrow maker requires only smaller flake blanks for the tips and shorter shafts, both of which could be more readily obtained than the larger blanks needed for the dart (Frison 1978). Although the energetics of each technology has not been calculated, it is intuitively evident that the cost of each arrow is much less than the cost of a dart.

The bow has certain advantages in application as well (Blitz 1988; Chatters 2004; Christenson 1986; Frison 1978; Hughes 1998; Shott 1993; Nassaney and Pyle 1999). It is more quickly mastered than the atlatl, can be more effectively used with stealth, and can hurl more projectiles at a higher rate and velocity without having to retrieve the shafts before each new salvo. A solitary hunter could, therefore, successfully use a bow, whereas an atlatl might be better used in a cooperative hunt wherein hunters intentionally reveal themselves. That bow-armed hunter/warrior would also require less skill with his implement, since he could take multiple, inexpensive shots at his quarry in a matter of moments. It is often said that the bow has a longer range than an atlatl (e.g., Morse and Morse 1990), but this was probably not the case with the early bows. The early bows would also have had lower impact force than the atlatl (Raymond 1986), making them less effective for larger, thicker-skinned prey or armored foes. Use in cooperative hunting, and superiority for large prey, might explain why the atlatl coexisted with the bow for so many centuries (Nassaney and Pyle 1999).

Ultimately, the complex of ideas for producing and using the bow-and-arrow attained greater fitness than any of the RMS of which it became a component. Once accepted, it probably conferred selective advantage to the biological communities and RMS that first accepted it and, in some cases, altered demographic and community patterns (e.g., Chatters 2004; Rosseau 2004). It also outlasted—was more successful in an evolutionary sense than—nearly all of those first RMS, often by many centuries.

Resource Management Strategies

Prentiss and I (2003a) have presented several examples of the process of RMS diversification in North America, including the emergence of a collector-like *Bauplan* in the Pacific Northwest between 5000 and 4000 cal B.P. during a period of climatic improvement and its expansion at the expense of competing *Baupläne* when resources became more restricted (Prentiss this volume); the emergence of the Numa "processors" (Bettinger and Baumhoff 1982) during the disruption caused by the Little Climatic Anomaly and their subsequent spread across the desert west; and the emergences of multiple alternative foraging strategies and sole survival of intensive bison hunting on the Central Great Plains at about this same time. Elsewhere in this volume, Prentiss and Lenert present the example of Paleoeskimo and Thule expansion across the arctic. These cases include examples of increased organic fitness (Thule, Numa, and the Salish in the Northwest) and others where fitness was acquisitive (acceptance of the collector *Bauplan* of Salish speakers by their Sahaptian neighbors). Numerous other examples exist in our continent's prehistory and are ripe for investigation, including but not limited to:

Expansion of the Clovis technology, whose spread across temperate North American within less than 200 years (Waters and Stafford 2007) was so rapid that it must have entailed both organic growth and acquisition by an existing pre-Clovis population.

Development of numerous post-Clovis local strategies at the end of the Pleistocene, prior to the widespread adoption of RMS following a high-mobility Archaic *Bauplan*.

Emergence of the Middle Archaic and later, woodland, in the American Midwest.

Expansion of the acorn-focused collector *Bauplan* in the Central Valley of California.

Development and expansion of Basketmaker in the Southwest.

Emergence of agrarian Mississippian chiefdoms.

It is this last example that I expand upon in this chapter.

The Mississippian Package

By 650 cal B.P. (ca 1300 A.D.), the southeastern United States was occupied by communities distributed from Wisconsin and Minnesota to the Gulf coast and from Oklahoma on the west to Florida and the Carolinas on the east that expressed resource management strategies that were variants on a single *Bauplan*. Archaeologically, the Mississippian *Bauplan* is represented by intensive maize agriculture using flint hoes, large-scale storage in deep, bell-shaped pits or above-ground granaries, specialized ceramic forms for maize processing, evidence for social ranking, and a hierarchical settlement pattern. Settlements increased in complexity from dispersed farmsteads and small hamlets to small, fortified centers with a single platform mound to large multi-mound centers (Fowler 1978; Nass and Yerkes 1995; Scarry 1996a; Steponaitis 1983). From farmsteads to multi-mound centers, the houses were usually built in a rectangular plan with a single set-post and walls built in shallow trenches. Platform mounds in single-mound centers and the primary mound in multi-mound centers were built in stages, with remains of a large rectangular building beneath the mound and on the mound platform beneath each successive construction level (e.g., Anderson 1994). Mound centers received tribute of commodities and manufactured goods from the smaller communities (Blitz 1993; Emerson 1991; Jackson and Scott 1995). Religious iconography and paraphernalia indicate widespread existence of one or a set of closely similar belief systems, known as the Southern Cult or Southeastern Ceremonial Complex (Waring and Holder 1945) based on communal rites of ancestor worship and world renewal (Blitz and Lorenz 2006; Knight 1986). Mississippian societies are inferred on the basis of mortuary patterns and settlement hierarchies to have been agrarian-based, socioeconomically ranked, hereditary chiefdoms led by a politico-religious elite (Pauketat 2004), as were the historic period societies in the same region, such as the Natchez (Swanton 1911). By European contact, variants on this *Bauplan* were in use by numerous ethno-linguistic groups belonging to multiple language families (papers in Fogelson 2004).

The Mississippian was once thought to have been introduced from Mesoamerica because of its basis in maize agriculture, social complexity, use of platform mounds, and some aspects of its iconography. More recent research has shown it to have been an indigenous development (Byers 2006; Caldwell 1958; Kelly 1990a,b; Pauketat 2004; Smith 1990) During the latter part of the twentieth century, when cultural process was emphasized over cultural historical explanations for socioeconomic change, many researchers saw Mississippian emergence as occurring more or less simultaneously throughout the Southeast, as maize agriculture led to population packing and the consequent need for centralized authority (e.g., Blitz 1988; Kelly 1990b; Schroedl et al. 1990, Steponaitis 1983; Welch 1990). Now, however, it has become increasingly accepted that, as Caldwell (1958) first suggested, the Mississippian *Bauplan* developed in the central Mississippi River Valley, from which it quickly expanded through either organic (Caldwell 1958; Pauketat 2004) or acquisitive (Muller 1986) growth.

Mississippian Precursors

The history of the Mississippian Bauplan, which includes emergence in effective isolation and expansion through organic and acquisitive fitness begins with the development of RMS constructed on what may be called the Woodland *Bauplan* in the mid-continent. Around 2500 cal B.P., toward the end of the demographic circumscription, regionalization, and widespread conflict that characterized the Late Archaic Period, some thus socially isolated community developed an RMS with four key components: dispersed sedentary or semi-sedentary settlements in small hamlets of semi-subterranean key-shaped houses by what were probably extended family groups; intensive harvest of wild mast, game, and fish resources in proximity to each hamlet; limited horticulture of indigenous starchy seeds; and maintenance of community cohesion and probably the redistribution of resources from ecologically diverse family territories through communal mortuary ceremonialism. This new RMS apparently had tremendous selective advantage over existing archaic strategies, because within 300 years it had spread throughout most of temperate eastern North American and moved up the rivers of the Great Plains, nearly to the Rocky Mountains. Along the sea coasts and western Great Plains, horticulture played little or no part, but the overall Bauplan is similar across the entire subcontinent (Jefferies 2004; Johnson 2001).

By 1850 cal B.P., these far-flung communities, undoubtedly derived from a bewildering array of ethnolinguistic groups, interacted extensively—at least indirectly—becoming the great, albeit loosely connected and polyglot, Hopewell interaction sphere. Some ceremonial centers, like Marksville in Louisiana, Mandeville in Georgia, Pinson in Tennessee, and especially Newark in Ohio grew and, judging from their massive earthworks and wealth of grave offerings from throughout the continent, called on a tremendous number of constituents. Populations rose and demographic packing again became evident (Jefferies 2004). In the early centuries A.D., during this time of widespread communication among a dense populace, the force of stabilizing selection on each community's RMS must have been tremendous (see Prentiss and Lenert, this volume, regarding evolutionary forces associated with stasis).

Around 1450 cal B.P., there began a prolonged period of what many archaeologists call cultural decline. During the Late Woodland Period, ceramic traditions indicate that societies became markedly circumscribed (Milner 1999). In the American Bottom region of the central Mississippi (Pauketat 2004) and the middle Tombigbee of Alabama (Blitz and Lorenz 2006), for example, multiple stylistic traditions show that social interaction spheres were reduced to the local scale. Resource stress is evident in the poor health and life expectancy of the population (Peebles 1983); warfare became endemic (Knight and Steponaitis 1998; Welch 1990). Elaborate mortuary ritualism and interregional trade markedly diminished. As societies became increasingly insular, communities became effectively isolated, weakening the forces for stabilizing selection and opening the door for the emergence of new RMS.

The longstanding unity of the Woodland *Bauplan* began to rapidly break down. As Jefferies (2004) describes the situation for the Southeast, "considerable social and political diversity existed" (see also Nassany and Cobb 1991). In northern Alabama, for example, people of the West Jefferson Culture were organized in small groups who moved seasonally, living off nuts, wild game and small-scale gardens of native cultigens and a small amount of maize. The Cofferdam-Gainesville peoples of the nearby Tombigbee River valley had a similar subsistence base and were also socially undifferentiated but lived in large, nucleated base camps (Welch 1990). People of the Effigy Mound Culture in southern Wisconsin and the Mason Phase of East Tennessee continued some of the practices of Early and Middle Woodland, living by a combination of wild foods and indigenous starch crops in widely dispersed villages and hamlets that remained socially integrated through construction of mortuary mounds (Birmingham et al. 1997; Faulkner 2001; Nassany and Cobb 1991). The Coles Creek Culture was the most socioeconomically complex of the Late Woodland cultures. Coles Creek people lived in dispersed hamlets and small villages supported primarily by wild mast, game, and fish in the lowlands of the Arkansas and southern Middle Mississippi rivers (Nassany 1994; Pauketat 2004, Rollingson 1990; Smith 1990). These communities supported small politico-religious centers (Smith 1990) that consisted of platform mounds built around central plazas (Williams and Brain 1983). Although middens are found on the mound slopes, there has been no evidence found of structures (Rollingson 1990). Pauketat (2004) interprets these as communal feasting sites; Kidder (1992) sees evidence of incipient chiefdom organization. Another RMS developed on the American Bottom of the Mississippi River, between the mouths of the Missouri and Ohio.

Emergence of the Mississippian

Between 1350 and 1200 cal B.P., the increasingly insular societies of the American Bottom lived in small, nucleated villages, primarily off gardens of indigenous starches. Like most other Late Woodland Groups they occupied small, key-shaped dwellings surrounded by large numbers of earth ovens and storage pits. After the 1200 cal B.P. introduction of maize, however, there began a series of changes that ultimately led or contributed to the development of the Mississippian RMS (Kelly 1990a, b; Pauketat 2004; Scarry 1996a, b).

Maize, farmed with chert hoes, quickly became a staple in the American Bottom as people developed new ways of storing and processing it. Use of earth ovens sharply declined as deep, bell-shaped storage pits increased. Traditional Woodland ceramic jars and bowls gave way to new forms, including the constricting necked jars that would later be such a hallmark of the Mississippian. Multiple ceramic traditions existed in the valley, each with its own tempers and stylistic elements. Communities were organized in small villages that consisted of small, rectangular, hearthless, slightly subterranean houses built around small courtyards. In the center of the courtyard was either a post or square with a small pit at each corner. Toward the end of the period, a large, rectangular house with a hearth was built on the ground surface. This large house, interpreted as either a headman's residence or ceremonial building, shows evidence of periodic reconstruction.

Change accelerated in the American Bottom between 1050 and 950 cal B.P. Interaction among communities within the Bottom was renewed, as indicated by the appearance of different ceramic traditions in the larger towns. Shell-tempered ceramics were made and, apparently, traded within the region. Ceramics diversified in a variety of flat-bottomed bowls, jars, bottles, and pots. Small hamlets of the preceding period coalesced into larger villages that consisted of multiple house groups, each with its own plaza containing a post or set of ceremonial pits. Only one of the plaza groups contained a large, repeatedly rebuilt house with an internal hearth, which Pauketat (2004) refers to as a "courtyard temple". Kelly (1990b) interprets this pattern as evidence for incipient ranking among the kin groups that comprised the villages, and sees the rebuilding of the courtyard temples as a ceremonial renewal that marked the installation of a new community leader—a precursor to the process later repeated in the construction of platform mounds. Sizes of village groupings varied across the Bottom, with the largest, like Lunsford-Pulcher and Cahokia, having populations upward of 1000 (Pauketat 2004).

By 950 cal B.P., the region's settlement patterns underwent an abrupt rapid transformation. Village dwellers either dispersed into small farmsteads or migrated to increasingly large towns. Small farmstead sites, exemplified by the Carbon Dioxide (Finney 1985; Jackson and Hanenberger 1990) and Range sites (Kelly 1990a) consisted of one or a few rectangular wall-trench buildings built with single set posts and, in some cases, large storage pits within them. These farmsteads were tributary to the towns, where the courtyard temples were now built and renewed on mounds overlooking small plazas, in an apparent duplication of the pattern seen earlier in the Coles Creek Culture. Among these early, larger centers were Lohman, Horseshoe Lake, Macys, and Washausen (Kelly 1990; Rolingson 1990, Scarry 1996; cf Pauketat 2004). Cahokia was by far the largest. Evidence for marked social ranking and the beginnings of so-called Southern Cult motifs can be seen in Cahokia's Mound 74, which contained the burial of a man lying on an eagle-shaped blanket of shell beads, surrounded by sacrificed attendants. Goods, including those used in mortuary ritual, flowed into the mound centers, and populations were drawn into the Bottoms and surrounding upland rim from adjacent regions (Emerson 2000; Pauketat 2004).

The components of the Mississippian—intensive maize horticulture supported by flint hoes and complex cooking and storage technology, hierarchical settlement pattern based on dispersed farmsteads, and mound centers associated with a cult of renewal serving as hubs in a network of tribute and exchange—had coalesced. This RMS then quickly began to expand, both organically and acquisitively.

Mississippian Expansion

Within the first 50–75 years after its full emergence, the Mississippian RMS had scattered throughout much of the region between the Ozark Highland and the Appalachian Mountains. It first appears in a discrete, repetitive package as a discontinuity from earlier, or sometimes contemporaneous, Late Woodland local cultures. Earliest mound centers occur at Zebree and Bay Mounds in Arkansas (Morse and Morse 1990); the Martin Farm site, in East Tennessee (Schroedl et al. 1990), Moundville, Bessemer, and Lubbub in northern Alabama (Bozeman 1982; Knight and Steponaitis 1998; Steponaitis 1983; Welch 1990); Etowah, in northeast Georgia (King 2003), Kinkaid near the Tennessee/ Ohio River confluence (Muller 1986); Angel, in Indiana (Black 1967) and Aztalan, in Wisconsin (Goldstein 1991 Goldstein and Richards 1991;). Expansion also took place southward along the Mississippi River as the Plaquemine Culture. The small, early mound centers are associated with dispersed farmsteads of rectangular houses built in wall trenches by the single set-post design. Beneath each mound are the remains of a large, rectangular building of the same design, with an internal hearth. Such buildings were constructed on each successive surface in the mound. Each mound center was surrounded by a bastioned palisade. Maize replaced wild foods as the staple; shell-tempered ceramics appeared as handled jars, bottles, and other exotic forms, replacing indigenous forms and pastes.

These early settlements, with such repetitive components, discontinuity with local traditions, and bastioned fortifications, have the appearance of colonies emigrated from the Mississippian heartland. At Aztalan, Angel, and somewhat later settlements of Dixon Mounds and Eveland in Central Illinois, this explanation is acknowledged. For other areas, indigenous development is argued, although the connection to earlier Woodland cultures is often strained. It is true that cultural transmission via indirect bias may sometimes explain the high degree of similarity between the new settlements and precursors in the American Bottom. Certainly, neighboring groups or visitors to the newly emerged polities in the Middle Mississippi would have been aware of the success of the new strategy and may have attempted to match it precisely. That success is evident in the improved health and longevity of the participants (e.g., Peebles 1983).

Expansion continued in the period between 850 and 750 cal B.P., but during this period, we see the initial cultural package undergo local modification as a result of the infusion of local ideas or local experimentation. In addition to the central Illinois centers, which are acknowledged colonies from the Cahokia center, expansion continued in the southeast, as colonists from northwest Georgia founded multiple new centers in the Chatahoochee and Savannah River valleys (Blitz and Lorenz 2006; Anderson 1994, 1996). Some of these new centers begin to show local differences, such as the replacement of rectangular buildings with oval or circular ceremonial designs on the Savanna River (Anderson 1994, 1998), the change from rectangular to oval structures in later phases of the Bessemer mound and village (Mistovich 1995), and the abandonment of shell tempering in ceramics on the Chattahoochee following the initial founding of Mississippian centers (Blitz and Lorenz 2006). Cahokia colonies in central Illinois quickly took on characteristics of the indigenous Late Woodland cultures, as local people began to participate in the new strategy (Harn 1994). Ranked, agrarian societies organized as chiefdoms also appeared on the periphery of this Mississippian expansion in Florida, Georgia, Illinois, and Oklahoma, (e.g., Scarry 1998; Blitz and Lorenz 2006; Rogers 1995a, b). By shortly after 750 cal B.P., the Mississippian *Bauplan* reached its apogee.

Macroevolution and the Mississippian

Emergence and expansion of the Mississippian is another example of diversification and decimation, first described as a cultural macroevolutionary process by Prentiss and Chatters (2003a). Late Woodland societies had become insular—socially isolated—during the period between 1450 and 1000 cal B.P., permitting socioeconomic experimentation. In the American Bottom region of the Middle Mississippi, local horticulturalists first adopted intensive maize agriculture, and worked out effective storage and processing methods along with a sociopolitical system based on a concept of world renewal (Byers 2006; Pauketat 2004).

Independent communities combined into villages, wherein one clan assumed the role of maintaining the renewal rites, leading to a stratified society with centralized authority. Ideas about mound building were adopted from the Coles Creek Culture as the communities reorganized their settlement, scattering much of the populace on small farms and centralizing further in a few politico-religious centers. Political organizations were able to organize labor for massive public works and, no doubt, more importantly, for subsistence pursuits and warfare. They also served as a conduit for redistribution of produce within the community and of technologically important goods and materials derived from outside the community

The Mississippian RMS was highly successful at improving the lives of its participants and enabling populations to grow rapidly. These led to increases in both organic and acquisitive fitness of the RMS, enabling its geographic expansion. Mississippian's inception and success may have been influenced, in part, by the selective forces of a warming climate, which allowed maize to be productive at higher latitudes, but this explanation does little to account for its expansion across Tennessee, Mississippi, Alabama, Georgia, and Florida. Whatever the selective forces, most alternative strategies disappeared from the region within 200 years of the new RMS's emergence. By this time regional differences had appeared; a new cultural clade had emerged.

Conclusion

The Mississippian emergence and the histories of projectile propulsion and plant processing technologies discussed in this chapter are but a few of the many examples of macroevolutionary events that have punctuated the history of North America's cultures. If we open our eyes to the reality of macroevolutionary process and the scale on which it occurs, we will become aware of macroevolutionary events marking the divisions within such venerable constructs as the Archaic (Sassaman 2008). To do this, we must think on a different scale than we are prone to do. Americanist Archaeology has for too long been shackled by the Culture Area Concept and the derived idea that historic-era cultural boundaries always delimited cultural developments. As a result of this geographic circumscription of the past, processualist thinkers have been stuck in the idea that change in a culture area, or even region within it, must necessarily be explained as a local transformation. Near-simultaneous changes in resource management systems often seem to be attributed to some mystical *Zeitgeist* rather than to linkage of human minds through observation and emulation. The appearance in adjacent regions of nearly identical RMS is seen not as wholesale expansion of systems or peoples but rather as a set of coincident responses to similar conditions. We see this in much of the discourse about the emergence of Mississippian. It is acceptable, it seems, to view the expansion of small cultural packages, like the bow-and-arrow or earth oven, on a continental scale, but similar movement of larger-scale packages is anathema. If we can discard these notions and assume a transregional macroevolutionary perspective, we will be able to better understand the continent's record as the development, expansion, stasis, and extinction of competing socioeconomic entities that are historically bound and whose fortunes are influenced by changing conditions in their natural, social, and competitive environments.

Notes

1. In Chapter 10, Bettinger questions the use of RMS instead of "adaptive strategy" which he sees as a simpler phrasing. Prentiss and I use RMS because a resource management strategy is not necessarily adaptive. It may be an attempt to adapt to environmental conditions but whether or not it is successful depends on its interaction with that environment. In the case of our modern RMS, global climate changes resulting from our use of fossil fuel resources calls our strategy's adaptedness into question.

References

Aikens, C.M. and Madsen, D.B. (1986). Prehistory of the Eastern Area. In W.L. D'Azevedo (Ed.), *Handbook of North American Indians, Vol 11 Great Basin* (pp. 149–160). Washington D.C.: Smithsonian Institution.

Ames, K.M., Dumond, D.E., Galm, J.R., and Minor, R. (1998). Prehistory of the Southern Plateau. In D.E. Walker (Ed.), *Handbook of North American Indians, Volume 12, Plateau* (pp. 103–119). Washington D.C.: Smithsonian Institution.

Anderson, D.G. (1994). *The Savannah River Chiefdoms*: Political Change in the Late Prehistoric Southeast. Tuscaloosa: University of Alabama Press.

Anderson, D.G. (1996). Chiefly Cycling and Large-Scale Abandonments as Viewed from the Savannah River Basin. In J.F. Scarry (Ed.), *Political Structure and Change in the Prehistoric Southeast* (pp. 150–191). Gainesville: University Press of Florida.

Andrefsky, W. Jr., Burtchard, G.C., Presler, K.M., Samuels, S.R., Sanders, P.H., and Thoms, A.V. (2000). *The Calispel Valley Archaeological Project, Project Final Report.* Center for Northwest Anthropology Project Report 16. Pullman: Washington State University.

Bettinger, R.L. and Baumhoff, M.A. (1982). Great Basin Cultures in Competition. *American Antiquity* 47, 485–503.

Bettinger, R.L. and Eerkens, J. (1999). Point Typologies, Cultural Transmission, and the Spread of Bow and Arrow Technology in the Prehistoric Great Basin. *American Antiquity* 64, 231–242.

Birmingham, R.A., Mason, C.I., and Stoltman, J.B. (1997). *Wisconsin Archaeology*. The Wisconsin Archaeological Society.

Black, G.A. (1967). *Angel Site, an Archaeological, Historical, and Ethnographic Study*. Indianapolis: Indiana Historical Society.

Blitz, J.H. (1988). Adoption of the Bow in Prehistoric North America. *North American Archaeologist* 9:97–122.

Blitz, J.H. (1993). *Ancient Chiefdoms of the Tombigbee*. Tuscaloosa: University of Alabama Press.

Blitz, J.H. and Lorenz, K.G. (2006). *The Chattahoochee Chiefdoms*. Tuscaloosa: University of Alabama Press.

Boyd, R. and Richerson, P.J. (1985). *Culture and the Evolutionary Process*. Chicago: University of Chicago Press.

Boyd, R. and Richerson, P.J. (2005). *The Origin and Evolution of Cultures*. Oxford: Oxford University Press.

Bozeman, T.K. (1982). Moundville Phase Communities in the Black Warrior Valley, Alabama. Ph.d. Dissertation Department of Anthropology, University of California, Santa Barbara.
Byers, M.A. (2006). *Cahokia: A World Renewal Heterarchy*. Gainesville: University Press of Florida.
Caldwell, J.R. (1958). *Trend and Tradition in the Prehistory of the Eastern United States*. American Anthropological Association Memoir 88. Washington, DC: Smithsonian Institution
Campbell, S.K. (Ed.) (1985). *Summary of Results. Chief Joseph Dam Cultural Resources Project.* Office of Public Archaeology, Institute for Environmental Studies, University of Washington, Seattle.
Chatters, J.C. (1995). Population Growth, Climatic Cooling, and the Development of Collector Strategies on the Southern Plateau, Western North America. *Journal of World Prehistory* 9, 341–400.
Chatters, J.C. (2004). Safety in Numbers: The Influence of the Bow and Arrow on Village Formation on the Columbia Plateau. In W.C. Prentiss and I. Kuijt (Eds.), *Complex Hunter-Gatherers: Evolution and Organization of Prehistoric Communities on the Plateau of Northwest North America* (pp. 67–83). Salt Lake City: University of Utah Press.
Chatters, J.C. and Prentiss, W.C. (2005). A Darwinian Macroevolutionary Perspective on the Development of Hunter-Gatherer Systems in Northwestern North America. *World Archaeology* 37, 46–65.
Chatters, J.C., Rhode, D.E. and Hoover, K.A. (1990). Tualdad Altu (45KI59): A Prehistoric Riverine Village in Southern Puget Sound. *Archaeology in Washington* 2, 23–48.
Cheatham, R. (1988). *Late Archaic Settlement Patterns in the Long Tom Sub-Basin, Upper Willamette Valley, Oregon*. University of Oregon Anthropological Papers 39, Eugene, OR: Department of Anthropology, University of Oregon.
Christenson, A.L. (1986). Projectile Point Size and Projectile Aerodynamics: An Exploratory Study. *Plains Anthropologist* 31, 119–128.
Clark, D. (1970). *The Prehistory of Africa*. New York: Praeger Publishers.
Corliss, D.W. (1972). Neck Width of Projectile Points: An Index of Culture Continuity and Change. *Occasional Papers of the Idaho State University Museum 29*, Pocatello.
Daugherty, R.D. (1952). Archaeological Investigations of O'Sullivan Reservoir, Grant County, Washington. *American Antiquity* 17, 374–383.
Eldredge, N. and Gould, S.J. (1972). Punctuated Equilibria: An Alternative to Phyletic Gradualism. In T.J.M. Schopf (Ed.), *Models in Paleobiology* (pp. 82–115). San Francisco: Freeman, Cooper, and Company.
Elston, R.G. (1986). Prehistory of the Western Area. In W.L.D'Azevedo (Ed.), *Handbook of North American Indians Volume 11, Great Basin* (pp. 135–148). Washington D.C.: Smithsonian Institution.
Emerson, T.E. (1991). Some Perspectives on Cahokia and the Northern Mississippian Expansion. In T.E. Emerson and R.B. Lewis (Eds.), *Cahokia and the Hinterlands* (pp. 221–236). Urbana: University of Illinois Press.
Emerson, T.E. (2000). A Reconsideration of the Cultural Processes Linking Cahokia to its Northern Hinterlands During the Period 1000–1200. In S.R. Ahler (Ed.), *Mounds, Modoc, and Mesoamerica. Papers in Honor of Melvin L. Fowler* (pp. 439–467). Illinois State Museum Scientific Paper 281, Illinois: Springfield.
Faulkner, C.H. (2001). Woodland Cultures of the Elk and Duck River Valleys, Tennessee: Continuity and Change. In D.G. Anderson and R.C. Mainfort, Jr. (Eds.), *The Woodland Southeast* (pp. 185–203). Tuscaloosa: University of Alabama Press.
Fenenga, F. (1953). The Weights of Chipped Stone Points: A Clue to their Functions. *Southwest Journal of Anthropology* 9, 309–323.
Finney, F.A. (1985). *The Carbon Dioxide Site*. Urbana: University of Illinois Press, American Bottom Archaeology, FAI-270 Site Reports.
Fogelson, R.D. (Ed.) (2004). *Handbook of North American Indians Volume 14, Southeast*. Washington D.C.: Smithsonian Institution.

Fowler, M.L. (1978). Cahokia and the American Bottom: Settlement Archaeology. In B.D. Smith (Ed.), *Mississippian Settlement Patterns* (pp. 455–478). New York: Academic Press, New York.

Frison, G.C. (1978). *Prehistoric Hunters of the High Plains*. New York: Academic Press.

Goldstein, L.G. (1991). The Implication of Aztalan's Location. In J.B. Stoltman (Ed.), *New Perspectives on Cahokia: Views from the Hinterland* (pp. 209–228), Madison: Prehistory Press.

Goldstein, L.G. and Richards, J. (1991). Ancient Aztalan: The Cultural and Ecological Context of a Late Prehistoric Site in the Midwest. In T.E. Emerson and R.B. Lewis (Eds.), *Cahokia and the Hinterlands* (pp. 193–206). Urbana: University of Illinois Press.

Harn, A.D. (1994).*Variation in Mississippian Settlement Patterns: The Larson Settlement System in the Central Illinois Valley*. Illinois State Museum Reports of Investigations No 50. Illinois: Springfield.

Heizer, R.F. and Baumhoff, M.A. (1961).The Archaeology of Wagon Jack Shelter. In The Archaeology of Two Sites at Eastgate, Churchill County, Nevada. *Anthropological Records* 20(4), 119–138. University of California, Berkeley.

Hoffecker, J.F. (2004). *A Prehistory of the North: Human Settlement of the Higher Latitudes*. New Brunswick: Rutgers University Press.

Hughes, S.S. (1998). Getting to the Point: Evolutionary Change in Prehistoric Weaponry. *Journal of Archaeological Method and Theory* 5, 345–408.

Jackson, D.K. and Hanenberger, N.D. (1990). *Selected Early Mississippian Household Sites in the American Bottom*. Urbana: University of Illinois Press, American Bottom Archaeology, FAI-270 Site Reports.

Jackson, H.E. and Scott, S.L. (1995). Faunal Record of the Southeastern Elite: The implications of Economy, Social Relations, and Ideology. *Southeastern Archaeology* 14,103–119.

Jefferies, R.W. (2004). Regional Cultures, 700 BC-AD 1000. In R.D. Fogelson (Ed.), *Handbook of North American Indians, Volume 14, Southeast* (pp. 115–127). Washington D.C.: Smithsonian Institution.

Johnson, F.E. (2001). Plains Woodland Tradition. In R. DeMallie (Ed.), *Handbook of North American Indians, Volume 13, Plains* (pp. 159–172). Washington D.C.: Smithsonian Institution.

Kehoe, T.F. (1966). The Small Side-Notched Point System of the Northern Plains. *American Antiquity* 31, 827–841.

Kelly, J.E. (1990a). Range Site Community Patterns and the Mississippian Emergence. In B.D. Smith (Ed.), *The Mississippian Emergence* (pp. 67–112). Washington D.C.: Smithsonian Institution Press.

Kelly, J.E. (1990b). The Emergence of Mississippian Culture in the American Bottom Region. In B.D. Smith (Ed.), *The Mississippian Emergence* (pp. 113–152). Washington D.C.: Smithsonian Institution Press.

Kidder, T.R. (1992). Coles Creek Period Social Organization and Evolution in Northeast Louisiana. In A.W. Barker and T.R. Pauketat (Eds.), *Lords of the Southeast: Social Inequality and the Native Elites of Southeastern North America* (pp. 145–162). Archaeological Papers of the American Anthropological Association 3. Washington, DC.

King, A (2003). *Etowah: The Political History of a Chiefdom Capital*. Tuscaloosa: University of Alabama Press, Tuscaloosa.

Knight, V.J. Jr. (1986). The Institutional Organization of Mississippian Religion. *American Antiquity* 51, 675–687.

Knight, V.J. Jr. and Steponaitis, V. (1998). *Archaeology of the Moundville Chiefdom*. Washington D.C.: Smithsonian Institution Press.

Leonard, R.D. and Jones, G.T. (1987). Elements of an Inclusive Evolutionary Theory Model for Archaeology. *Journal of Anthropological Archaeology* 6,199–219.

Lepofsky, D. and Peacock, S.L. (2004). A Question of Intensity: Exploring the Role of Plant Foods in Northern Plateau Prehistory. In W.C. Prentiss and I. Kuijt (Eds.), *Complex Hunter-Gatherers:*

Evolution and Organization of Prehistoric Communities on the Plateau of Northwest North America (pp. 115–139). Salt Lake City: University of Utah Press.
Matson, R.G., and Magne, M.P.R. (2007). *Athapaskan Migrations: The Archaeology of Eagle lake, British Columbia*. Tucson: University of Arizona Press.
Maxwell, M. (1985). *The Prehistory of the Eastern Arctic*. Orlando: Academic Press.
McGhee, R. (1996). *Ancient People of the Arctic*. Vancouver: UBC Press.
Milner, G.R. (1999). Warfare in Prehistoric and Early Historic Eastern North America. *Journal of Archaeological Research* 7, 105–151.
Mistovich, T.S. (1995). Toward and Explanation of Variation in the Moundville Phase Households of the Black Warrior Valley, Alabama. In J.D. Rogers and B.D. Smith (Eds.), *Mississippian Communities and Households* (pp. 156–180). Tuscaloosa: University of Alabama Press.
Moratto, M.J. (1984). *California Archaeology*. New York: Academic Press, New York.
Morris, E.A. (1980). *Basketmaker Caves in the Prayer Rock District, Northeastern Arizona*. Anthropological Papers of the University of Arizona 35. Tucson: University of Arizona Press.
Morse, P and Morse, D.F. (1990). Emergent Mississippian in the Central Mississippi Valley. In B.D. Smith (Ed.), *The Mississippian Emergence* (pp. 153–173). Washington D.C.: Smithsonian Institution Press.
Muller, J. (1986). *Archaeology of the Lower Ohio River Valley*. Orlando: Academic Press.
Nass, R.P. Jr. and Yerkes, R.W. (1995). Social Differentiation in Mississippian and Fort Ancient Societies. In J.D. Rogers and B.D. Smith (Eds.), *Mississippian Communities and Households* (pp. 58–80). Tuscaloosa: University of Alabama Press.
Nassany, M.S. (1994). The Historical and Archaeological Context of Plum Bayou Culture in Central Arkansas. *Southeastern Archaeology* 13, 36–55.
Nassany, M.S. and C.R. Cobb (1991). Patterns and Processes of Late Woodland Development in the Greater Southeastern United States. In M.S. Nassany and C.R. Cobb (Eds.), *Stability, Transformation, and Variation: The Late Woodland Southeast* (pp. 285–322). New York: Plenum Press.
Nassany, M.S. and Pyle, K. (1999). The Adoption of the Bow and Arrow in Eastern North America: A View from Central Arkansas. *American Antiquity* 64, 243–364.
Oliver, B.L. (1985). Tradition and Typology: Basic Elements of the Carolina Projectile Point Sequence. In R.S. Dickens Jr. and H.T. Ward (Eds.), *Structure and Process in Southeastern Archaeology* (pp. 195–211). Tuscaloosa: University of Alabama Press.
O'Brien, M.J., and Lyman, R.L. (2000). *Applying Evolutionary Archaeology*. New York: Kluwer Academic/Plenum.
O'Neill, B.L., Thomas, C., and D.E. Friedel (2004). *The Holocene Geoarchaeological Record of the Upper Willamette Valley: The Long Tom and Chalker Sites*. University of Oregon Anthropological Papers 61. Eugene:Oregon.
Pauketat, T.R. (2004). *Ancient Cahokia and the Mississippian*. Cambridge: Cambridge University Press.
Pavesic, M.D. (1985). Cache Blades and Turkey Tails: Piecing Together the Western Idaho Archaic Burial Complex. In M.G. Ples, J.C. Woods, and M.G. Pavesic (Eds.), *Stone Tool Analysis: Essays in Honor of Don E. Crabtree* (pp. 55–89). Albuquerque: University of New Mexico.
Peacock, S.L. (2002). Perusing the Pits: The Evidence of Prehistoric Geophyte Processing on the Canadian Plateau. In S.L.R. Mason and H.G. Hather (Eds.), *Hunter-Gatherer Ethnobotany: Perspectives from the Northern Temperate Zone* (pp. 44–61). Institute of Archaeology Occasional Publication, Archetype Publications, University College of London.
Peebles, C.S. (1983). Summary and Conclusions: Continuity and Change in a Small Mississippian Community. In C.S. Peebles (ed.), *Excavations in the Lubbub Creek Archaeological Locality* (pp. 253–277). Vol 1. Prehistoric Agricultural Communities in West Central Alabama. Report Submitted to the USACE, Mobile District.
Prentiss, W.C. and Chatters, J.C. (2003a). Cultural Diversification and Decimation in the Prehistoric Record. *Current Anthropology*, 44, 33–58.
Prentiss, W.C. and Chatters, J.C. (2003b). The Evolution of Collector Systems in the Pacific Northwest Region of North America. *Senri Ethnological Studies* 63, 49–82

Richerson, P.J. and Boyd, R. (1985). *Culture and the Evolutionary Process*. Chicago: University of Chicago Press.

Raymond, A. (1986) Experiments in the Function and Performance of the Weighted Atlatl. *World Archaeology* 18, 153–177.

Rogers, J.D. (1995a). Dispersed Communities and Integrated Households: A Perspective from Spiro and the Arkansas Basin. In J.D. Rogers and B.D. Smith (Eds.), *Mississippian Communities and Households* (pp. 53–68). Tuscaloosa: University of Alabama Press.

Rogers, J.D. (1995b). Markers of Social Integration: The Development of Centraqlized Authority in the Spiro Region. In J.F. Scarry (Ed.), *Political Structure and Change in the Prehistoric Southeastern United States* (pp. 12–52). Gainesville: University Press of Florida.

Rolingson, M.A. (1990). The Toltec Mounds Site: A Ceremonial Center in the Arkansas River Lowland. In B.D. Smith (Ed.), *The Mississippian Emergence* (pp. 27–49). Washington D.C.: Smithsonian Institution Press.

Rosenberg, M. (1994). Pattern, Process, and Hierarchy in the Evolution of Culture. *Journal of Anthropological Archaeology* 13,307–340.

Rousseau, M.K. (2004). A culture historic synthesis and changes in human mobility, sedentism, subsistence, settlement, and population on the Canadian Plateau from 7000 to 200 BP. In W.C. Prentiss and I. Kuijt (Eds.), *Complex Hunter-Gatherers: Evolution and Organization of Prehistoric Communities on the Plateau of Northwestern North America* (pp. 3–16). Salt Lake City: University of Utah Press.

Sassaman, K.E. (2008). The New Archaic, It Ain't What it Used To Be. *SAA Archaeological Record* 8, 6–8.

Sassaman, K.E. and Anderson, D.G. (1990). *Native American Prehistory in the Middle Savannah River Valley: A Synthesis of Archaeological Investigations on the Savannah River Site, Aiken and Barnwell Counties, South Carolina*. Savannah River Archaeological Research Papers 1. South Carolina Institute of Archaeology and Anthropology, University of South Carolina, Columbia.

Scarry, J.F. (1996a). The Nature of Mississippian Societies. In J. F. Scarry (Ed.), *Political Structure and Change in the Prehistoric Southeastern United States* (pp. 12–52). Gainesville: University Press of Florida.

Scarry, J. F. (1996b). Stability and Change in the Apalachee Chiefdom. In J. F. Scarry (Ed.), *Political Structure and Change in the Prehistoric Southeastern United States* (pp. 192–227). Gainesville: University Press of Florida.

Schroedl, G.F., Boyd, C.C. and Davis, R.P.S. (1990). Explaining Mississippian Origins in East Tennessee. In B.D. Smith (Ed.)*The Mississippian Emergence* (PP. 175–196). Washington D.C. Smithsonian Institution.

Shott, M.J. (1993). Spears, Darts and Arrows: Late Woodland Hunting Techniques in the Upper Ohio Valley. *American Antiquity* 58, 425–443.

Shott, M.J. (1997). Stones and Shafts Redux: The Metric Dimensions of Chipped Stone Dart and Arrow Points. *American Antiquity* 62, 86–101.

Smith, B.D. (1995). The Analysis of Single Household Mississippian Settlements. In J.D. Rogers and B.D. Smith (Eds.), *Mississippian Communities and Households* (pp. 224–249). Tuscaloosa: University of Alabama Press.

Smith, B.D. (1990). Introduction: Research on the Origins of Mississippian Chiefdoms in Eastern North America. In B.D. Smith (Ed.), *The Mississippian Emergence* (pp. 1–8). Washington D.C.: Smithsonian Institution Press.

Spencer, C.S. and Redmond, E.M. (2002). Militarism, Resistance, and Early State Development in Oaxaca, Mexico. *Social Evolution & History* 2(1), 25–70.

Steponaitis, V. (1983). *Ceramics, Chronology, and Community Patterns: An Archaeological Study of Moundville*. New York: Academic Press.

Suttles, W. (Ed.) (1990). *Handbook of North American Indians, Vol 7, Northwest Coast*. Washington D.C.: Smithsonian Institution Press.

Swanton, J.R. (1911). *Indian Tribes of the Lower Mississippi Valley and Adjacent Coast of the Gulf of Mexico*. Bulletin 43, Bureau of American Ethnology, Smithsonian Institution, Washington DC.

Thomas, D.H. (1978). Arrowheads and Dart Points: How the Stones Got the Shaft. *American Antiquity* 43, 461–472.
Thomas, D.H. (1981). How to Classify the Projectile Points from Monitor Valley, Nevada. *Journal of California and Great Basin Anthropology* 3, 7–43.
Thoms, Alston V. (1989). *Northern Roots of Hunter-Gatherer Intensification: Camas and the Pacific Northwest.* PhD Dissertation, University Microfilms, Ann Arbor.
Thoms, A. (2008). The Fire Stones Carry: Ethnographic Records and Archaeological Expectations for Hot Rock Cookery in Western North America. *Journal of Anthropological Archaeology* 27, 443–460.
Vehik, S.C. (1984). The Woodland Occupations. In R.E. Bell (Ed.), *Prehistory of Oklahoma* (pp. 175–187). Orlando: Academic Press.
Vickers, R. (1986). *Alberta Plains Prehistory, a Review*. Occasional Paper 27, Archaeological Survey of Alberta, Edmonton.
Walker, D.E. Jr. (Ed.). (1998). *Handbook of North American Indians, Volume 12, Plateau*. Washington D.C.: Smithsonian Institution Press.
Waring, A.J. and Holder, P. (1945). A Prehistoric Ceremonial Complex in the Southeastern United States. *American Anthropologist*, new series, 47,1–34.
Waters, M.R. and Stafford, T.W. (2007). Redefining the Age of Clovis: Implications for the Peopling of the Americas. *Science* 315,1122 – 1126.
Welch, P.D. (1990). Mississippian Emergence in West Central Alabama. In B.D. Smith (Ed.), *The Mississippian Emergence*. Washington D.C.: Smithsonian Institution Press.
White, G., Frederickson, D., Meyer, J., Rosenthal, J., Waters, M., and West, G.J. (1999). Final Report of the Anderson Flat Project, Lower Lake, Lake County, California. California Department of Transportation, Eureka.
Williams, S. and Brain, J.P. (1983).*Excavations at the Lake George Site, Yazoo County, Mississippi, 1958–1960*. Papers of the Peabody Museum of Archaeology and Ethnology, 74, Harvard University, Cambridge.
Wright, S. (1931).Evolution in Mendelian Populations. *Genetics* 16, 97–159.
Wright, S. (1932). The Roles of Mutation, Inbreeding, Crossbreeding and Selection in Evolution. *Proceedings of the VI International Congress of Genetrics* 1, 356–366.

Chapter 8
Cultural Stasis and Change in Northern North America: A Macroevolutionary Perspective

Anna Marie Prentiss and Michael Lenert

Introduction

To successfully build a macroevolutionary theory of culture change, we, like paleobiologists (e.g., Eldredge 1985, 1989; Gould 2002; Stanley 1979), must carefully address a number of critical topics including scale and hierarchy, emergence processes, stasis, extinction, and long-term evolutionary trends. While archaeologists have long recognized the stable persistence of cultural patterns (or cultural stasis) in the archaeological record, they have rarely examined this phenomenon from an evolutionary perspective. Rare exceptions include Prentiss and Chatters (2003) and Richerson et al. (2001). In this chapter, we seek to build on these studies to open a discussion of long-term cultural stasis as a process.

As a tangential problem, we also examine factors that allow these tenacious cultural entities to expand geographically, sometimes outcompeting and replacing other indigenous strategies. Many researchers have been interested in understanding the archaeological record of major population movements such as the Neolithic expansion across Europe (Colledge et al. 2004; Rowly-Conwy 2004; Shennan 1993; 2002), the Polynesian dispersal in the Pacific (Kirch 2000), the Numic spread in western North America (Bettinger and Baumhoff 1982), and the Paleoeskimo and Thule movements across the American Arctic (McGhee 1996). Consequently, there is substantially more developed literature on this issue (see also Rockman, this volume) to which we add our thoughts.

Ultimately, we hope that this discussion will also shed light on processes of change. As has been recognized by paleobiologists for years (e.g., Stanley 1992), we cannot fully understand processes behind the emergence of new cultural or biological variants without understanding processes of stasis. Cultural characters, whether artifacts or socioeconomic strategies, seem at times to become "locked" (Gould 2002) into particular design structures or *Baupläne*. Breaking the lock may be fundamental to change. The archaeological record of the pre-Dorset, Dorset, and

A. Prentiss (✉)
Department of Anthropology, The University of Montana, Missoula, MT 59812, USA
e-mail: anna.prentiss@umontana.edu

A.M. Prentiss et al. (eds.), *Macroevolution in Human Prehistory*,
DOI 10.1007/978-1-4419-0682-3_9,

Thule phases in North America's Arctic region dramatically illustrates some core macroevolutionary principles associated with these processes of stasis and change.

Why Stasis in the Archaeological Record?

Our focus in this chapter is on stasis at scales of socioeconomic strategies, recognizing that this process could be recognized on multiple scales. In order to more fully develop our theoretical ideas, we first explore some basic cultural macroevolutionary concepts associated with the structure of cultural entities and their implications for understanding cultural evolution. We follow with a more in-depth consideration of stasis as an evolutionary process.

Macroevolutionary Units

In order to explore the persistence of socioeconomic strategies over long time spans, we rely on the concept of a Resource Management Strategy (compare with Bettinger's [this volume] complex adaptive strategies) as a macroevolutionary unit of evolution (Chatters, this volume; Chatters and Prentiss 2005; Prentiss, this volume). The Resource Management Strategy (RMS) provides the organizational logic (e.g., Boyd and Richerson 1985; Boyd et al. 1997; Prentiss, this volume; Prentiss et al. this volume) that structures and integrates human subsistence and settlement behavior and related aspects of social organization. It is heritable in that it can be taught to new generations and it can be transmitted between peers and groups (admittedly, it is a complex process [e.g., Bettinger 2003]). Behavioral or phenotypic manifestations of the RMS are termed applied tactics. Because tactics involve economic practices (hunting, fishing, food production, seasonal mobility, food sharing, etc.), they have the potential to also affect reproductive fitness within human populations. Because the tactical manifestation of an RMS will always impact the local ecosystems, we can also refer to it as a cultural niche (Laland et al. 1996, 2000) and even view it as a "peak" within an adaptive landscape (per Wright 1932; see Bettinger, this volume; Spencer, this volume).

Since Resource Management Strategies (RMS) evolve as packaged units, they resemble in some ways the evolution of higher phyletic units in biology such as species (Stanley 1979). While RMS are not exactly the same thing as species (Prentiss et al., this volume), they do exhibit formal variation in structural designs or *Baupläne* (Prentiss and Chatters 2003). The *Bauplan* concept can also be useful for understanding the evolutionary process (e.g., Gould and Lewontin 1979; McGhee 1999), in that it allows us to recognize that change at lower levels (e.g., memetic level) can often have far-reaching impacts elsewhere within sets of integrated tactics (or systems). If this is the case then not all change is necessarily the target of selection but may be some form of accommodation to change elsewhere and may or may not be adaptive (Prentiss, this volume; Prentiss et al. 2007).

The archaeological record consistently supports the contention that RMS are stable cultural entities that emerge, persist, evolve, and go extinct. We recognize signatures of human behavior that seem to persist for long time spans (Prentiss and Chatters 2003; Richerson et al. 2001) and also move across significant geographic space (e.g., Bettinger and Baumhoff 1982; Rockman, this volume). This pattern of persistence among evolutionary entities has been termed stasis (Eldredge and Gould 1972; Gould and Eldredge 1977). Although change does occur within periods of stasis, it is typically on lower scales (for example technologies or specific tactics) and is not always directional in the long term. Following Gould (2002), we could say that these entities are "locked."

Cultural Stasis on the Macroevolutionary Scale

To understand stasis, we must examine the factors that promote stability while curtailing variation. Paleobiologists have developed a number of models that might have analogical utility for understanding cultural phenomena. The microevolutionary model of stabilizing selection asserts positive reinforcement for maintenance of the same design if that RMS is at peak adaptation. There is reason to believe that the adaptation need not be entirely optimal since the humanly created niche will substantially aid in buffering the harder impacts of natural selection (Laland et al. 2000, 2001; see also Kuijt and Prentiss this volume). For example, an adaptation could be maintained through habitat tracking (e.g., Eldredge 1995, 1999), whereby the social group moves, bringing along its RMS and cultural niche. In another example, accommodations by various populations to local conditions could create a pattern of temporally shifting variation (e.g., Lieberman and Dudgeon 1996) around a common RMS. In this case, none would be necessarily at peak adaptation, but local adaptive performance would be adequate to prevent breakdown and loss of that design, and stability would emerge as a compromise (Gould 2002).

The internal structure of the strategy and its implied cultural niche will have an impact on the longevity of the cultural pattern. The prevailing cultural transmission system may also be important. As argued by Cavalli-Sforza and Feldman (1981) and Boyd and Richerson (1985), when learning occurs substantially by vertical cultural transmission (as in the parent to offspring pattern so typical of hunter-gatherers), opportunities for creativity or significant error (Eerkens and Lipo 2003) may be reduced, resulting in patterns of cultural conservatism (MacDonald 1999). Further, some cultural prohibitions (maintained in this conservative context) could act as selective factors, also stifling outside influence (Boyd and Richerson 1985).

There may be very good reasons to stifle within-group variation, particularly if success or even survival is predicated on careful scheduling and teamwork throughout the year to acquire, process, and distribute resources and to meet critical social obligations. While human niches will always be broader and more flexible in terms of habitat than those associated with other organisms due to our propensity for culture (Laland et al. 1996, 2000, 2001), some strategies and associated cultural niches will be more specialized than others entailing a more restricted range of habitats available for colonization. This pattern of greater specialization could favor extreme stasis under stable environmental conditions via something akin to the adaptive peak model. Habitat tracking, as well, could be critical for long-term survival and consequently cultural stasis within this scenario. Inflexibility in ability to use new habitats could also lead to more frequent opportunities for extinction within a changing environment (e.g., Janssen et al. 2003). In contrast, other strategies and their niches will be structured in such a way that many kinds of habitat could be co-opted by their users with only minor technological or organizational shifts. Habitat tracking would be less important to survival, since the cultural niche is more likely to be functional in a wider range of habitats. Rather, survival and cultural stasis might be predicated on adaptive flexibility favoring stasis as the mean of a range of local variants on that basic design. Local variation could be created via movement into new habitats that could also entail competition with other populations operating different niches. Kuijt and Prentiss (this volume) provide examples of these processes illustrating specialization and a shorter evolutionary history associated with the Natufian culture of the Upper Pleistocene Near East, and flexibility, expansion, and longer term macroevolutionary stasis with the Neolithic culture of the early-to-middle Holocene Near East and Europe.

Despite the processes that promote stability, change still occurs. Processes of stasis and change can be understood using Wright (1931, 1932; see also Bettinger, this volume; Spencer, this volume) shifting balance or adaptive landscape theory (SBT). Within SBT the framework, local group adaptations (as reflected in Resource Management Strategies) are viewed as peaks on a topographically variable landscape (Joshi 1999). Evolutionary change is envisioned as crossings between alternative peaks. Shallow valleys between peaks imply little maladaptation; deep chasms are more risky and have significant potential for extinction. Wright originally envisioned peak crossings as initiated by genetic drift causing local populations to move from their peaks into that maladaptive zone. If lucky, they encountered and ascended a new peak; otherwise they faced potential extinction. Contemporary theorists offer a wide variety of scenarios as to how peak crossing might occur. Environmental change could draw an adapted population rapidly into a maladaptive zone as could a simple process of maladaptive behavior. In contrast, it also appears likely that selection could drive crossings between peaks, sometimes permitting targeted characters to "carry" other correlated traits (Bettinger, this volume; Price et al. 1993). There is substantial agreement that major peak shifts are facilitated by small populations in relatively isolated contexts (Coyne et al. 1997).

With these ideas in mind we propose that, on a macroevolutionary scale, special conditions may often be required to open the lock that prevents significant change (Prentiss and Chatters 2003). First, whether adaptive entities "sit" on that metaphorical adaptive peak or remain at a slightly lower level (as in the accommodation model), resource conditions will often change and remain predictably different for long enough to permit exploratory cultural changes (as noted by Prentiss, this volume, this could occur by conscious design or unintentional drift).

This could mean that the natural environment offers new opportunities, a major innovation alters energy budgets within a given group leading to the precipitation of broader organizational change, or that unintentional rearrangements in subsistence schedules or other activities precipitate a series of unforeseen changes in human population–resource relationships (Bettinger, this volume; Kuijt and Prentiss, this volume; Prentiss, this volume). Second, environments must remain productive enough that initial steps toward change are not immediately halted by social prohibitions or outright economic disasters (Prentiss and Chatters 2003). Third, groups are expected to be generally small in which new patterns of culturally endogamous behavior can quickly emerge and be rewarded (Boyd and Richerson 1985; Laland et al. 2000). Finally, there must be some form of geographic and/or social isolation from dominant cultural patterns to permit variation to emerge and not have it swamped by a cultural version of gene flow. This latter factor may also affect the persistence of smaller scale RMS when exposed to expanding larger entities.

Arctic Prehistory

The evolutionary origin of the pre-Dorset/Dorset or Paleoeskimo culture is not altogether clear though the Middle Neolithic Bel'kachinsk culture of eastern Siberia is a likely ancestor (Hoffecker 2003). Over its developmental history, some technological elements originally present in pre-Dorset were refined and others lost, giving rise to a new form identified as Dorset. Variation between pre-Dorset and Dorset RMS suggests two extended periods of significant stasis at the strategic level. Populations associated with the Dorset cultural signature were ultimately replaced by Thule Inuit populations

Pre-Dorset

The pre-Dorset RMS was a highly successful Eastern Arctic adaptation that persisted roughly 4000–2500 years ago. Based on similarities in ecological adaptations and artifacts, the pre-Dorset appears rooted in the Western Paleoeskimo, Arctic Small Tool Tradition. These founding populations migrated from the Western Arctic, rapidly expanding across the Eastern Arctic, and settled in the Barren Grounds (Arctic and Subarctic interior west of Hudson bay), Arctic Archipelago, Hudson Bay, coastlines of the Hudson Strait, Newfoundland, and northern Greenland giving rise to a variety of local phases including Independence I, Independence II, Sarqaq, and Groswater Dorset (Maxwell 1985; McGhee 1976).

Archaeological materials attest to hunting adaptations in coastal and interior settings. Technological hallmarks in coastal sites include harpoon heads, and other artifacts of seal and walrus bone and ivory, illustrating a focus on marine resources. The chronology of harpoon head types shows that over time, open socket, single line hole, barbed heads lost their barbs and shrank in size but retained open sockets (Maxwell 1976). Some may have been armed with chipped stone points. Sites in the interior document the importance of hunting terrestrial resources in pre-Dorset times. Populations acquired game by hunting with bow and arrow and lance. Faunal remains from residential site middens include caribou, musk ox, polar bear, and birds.

Pre-Dorset peoples lived in seasonal shelters. Small rings of boulders or simply dense oval patterns of artifacts and small rocks mark locations of summer tent shelters (Dekin 1976). Although there is some disagreement over what winter season structures may have looked like (see McGhee 1976), some believe they are evidenced by small oval areas with raised stone axial features containing hearth and storage bins. During the pre-Dorset phase social units were likely organized at the scale of small family groups that changed residence frequently and exploited resources on an encounter basis.

Pre-Dorset archaeological records exhibit general continuity, although site assemblages from outside the Foxe Basin and Baffin Island core area do show some variation. The west-central Arctic (e.g., Victoria Island and surrounding locales) assemblages evince an affinity to those of Alaskan Paleoeskimos (Denbigh Flint Complex); southwestern Hudson Bay assemblages contain standard pre-Dorset lithics plus heavy-duty tools like mattocks, picks, and gouges; the northern Labrador assemblages are akin to those of the Sarqaq of the western Greenland coast; and Independence I of the High Arctic (e.g., northern Greenland, Ellesmere Island, Queen Elizabeth Islands) reflects a stronger focus on terrestrial game than those to the south (Maxwell 1972; McGhee 1996).

The Transitional period separating the pre-Dorset and Dorset periods dates back to 2800–2500 years ago, postdating the late peak of the Neoglacial climatic period (Wright et al. 1993). New unstable and warmer weather conditions likely resulted in unpredictable variation in snow and ice with negative impacts on terrestrial fauna (Dekin 1975). These conditions may also have affected distributions of sea mammals, particularly in warmer years. Warmer conditions may have also decreased access to sea mammals by human hunters in some areas. One by-product of this period for human groups may have been increased rates of residential mobility and exploration of new foraging opportunities in wider foraging ranges. Greater mobility and movement into more distant landscapes could have had the effect of severing social links between pre-Dorset groups, thereby giving rise to greater regional variation in material culture and adaptive behavior. This is potentially supported by evidence from several regions. Paleoeskimos reappear in the High Arctic of Greenland and Devon and Ellesmere Islands as the Independence II culture (Knuth 1967, 1968; McGhee 1981). A similar cultural pattern is also evident in the Groswater Dorset phase of Labrador and Newfoundland (Fitzhugh 1972, 1976; Renouf 2005). While it is not clear whether Independence II and Groswater Dorset left obvious direct descendents (e.g., Fitzhugh 1980), late pre-Dorset sites from Igloolik and Lake Harbour area of southern Baffin Island do suggest that during a period of population reduction, Paleoeskimos in this area underwent a dramatic process of *in situ* change from the pre-Dorset to Dorset pattern that included not only diversification of and intensification of some food resources but also the seemingly strange loss of technologies like the bow and arrow and bow-drill (Maxwell 1973, 1985; Meldgaard 1960, 1962).

McGhee (1996) outlines a scenario at the pre-Dorset to Dorset transition similar to the one offered by Henrich (2004) for loss of technological skill in Tasmania. He suggests that periods of devastating famine, as may have happened in the Arctic at ca. 2500–2800 cal. B.P. could have removed knowledgeable tool makers from pre-Dorset populations. Catastrophic loss of knowledge could not be made up by transmission from the outside due to extreme isolation in the Eastern Arctic.

While the McGhee and Henrich models imply region-wide gradual microevolutionary change driven in part by acts of nature implying something akin to genetic drift, we suggest that the emergence of Dorset could also be viewed from a macroevolutionary perspective. The pre-Dorset to Dorset transition is clearly not a gradual process that occurred simultaneously region-wide. Rather, it appears to have been characterized by late pre-Dorset cultural diversification, subsequent regional decimation, and replacement by one homogeneous pattern, Early Dorset, perhaps arising in the Foxe Basin area. If this is the case it provides yet another example of the cultural diversification-decimation macroevolutionary process increasingly recognized in other parts of the world (Chatters, this volume; Kuijt and Prentiss, this volume; Prentiss and Chatters 2003).

Dorset

The Dorset phase postdates 2800 cal. B.P. and is known for a number of distinctive cultural characteristics. Technologically, Dorset assemblages include closed socket toggling harpoons (some slotted for chipped stone points) of a variety of sizes, bone sled shoes, crampons for ice walking, snow knives for constructing shelters, and occasionally, small kayaks. We also find an increase in the variety of raw materials used for tool making (Nagy 2000). Evidence for bow-drills, throwing boards, and bows and arrows disappears (McGhee 1996). Some of these changes in technology support a contention for greater emphasis on winter sea mammal hunting (Maxwell 1972; cf. Murray 1996, 1999). Faunal assemblages contain abundant seal remains as well as lesser amounts of walrus, narwhal, and beluga. Communal hunting of caribou is suggested by associations of Dorset artifacts with *inuksuit* alignments (Maxwell 1972, 1985). Fish weirs appear to have been widely used, but associated technology was simple and dominated by leisters and bilaterally barbed spears. As in the previous pre-Dorset, open socket lances continue to be used to hunt terrestrial resources in the Dorset period. Accumulations and storage of food surpluses for later use appears in the form of caches in some areas (Murray 1999; Nagy 2000).

Winter dwellings in some places consisted of rectangular structures with central platforms and probably housed multiple families, reinforcing the mode of periodic communal hunts (Maxwell 1972). Snow houses or igloos were often used as well (McGhee 1996). Ephemeral summer tent camps consist of small oval or large rectangular rings of small stones and boulders. Like the previous pre-Dorset populations, the Dorset were also probably regionally scattered, egalitarian bands who aggregated in larger groups at certain times of the year to participate in social networking, alliance building, as well as group hunting endeavors (Damkjar 2005; Murray 1999).

Homogenous cultural traits, synchronous style shifts, and contemporaneous changes in artifact forms demonstrate intensive interaction among populations over vast distances (in the core area—Foxe Basin, Hudson Strait, Baffin Island region) (Friesen 2000; Maxwell 1985). Maxwell (1985) claims stylistic shifts had no bearing on increased socioeconomic efficiency. During Dorset (particularly Late and Terminal Dorset) times, we see periodic repopulations of peripheral areas including northern Devon Island (located south of Ellesmere Island), Ellesmere Island, and western Greenland. These "fringe" areas demonstrate less chronological continuity and evince intermittent occupations in hyper-isolated locations (McGhee 1981). The gaps or occupational hiatuses associated with these places may represent population withdrawals into localized, advantageous ecological niches or even the extinction of some Dorset groups. Isolated Newfoundland and Labrador Dorset groups also appear to vary from the core area. In fact, archaeological records in these regions are more similar to the Greenland Dorset than to the core. West of the core, the Victoria and Banks Island Dorset variants are similar to the core but contain additional items like ice chisels, barbed dart heads, and projectile tips with side and end blades.

Arctic archaeologists have offered a range of opinions regarding the basic socioeconomic structure of Dorset societies. Maxwell (1980, 1985) argues that organizational changes between pre-Dorset and Dorset were more quantitative than qualitative, with an increased focus on marine mammals and a reduced emphasis on terrestrial game, despite the appearance of caribou drive systems employing *inuksuit*. The implication is that Dorset people intensified aspects of their economy like breathing hole and ice lead sea mammal hunting, while dropping other parts (e.g., stalking of terrestrial mammals).

Alternatively, Nagy (2000) argues that over the span of pre-Dorset occupation, populations became increasingly familiar with their landscape and environment, and in turn they were better acquainted with the best hunting locations in relation to the seasonal availability of those resources. Thus, Nagy claims pronounced change involving a shift from pre-Dorset forager to Dorset collector strategies (collector per Binford 1980). She bases her argument on five excavated sites from Ivujivik in the High Arctic whose faunas are dominated by seals. She also argues that there is evidence for caching behavior. While the seal-dominated faunas could also be interpreted as a reflection of a more specialized wintertime foraging economy as might be expected of Igloolik or Netsilik seal-hunting groups (Balicki 1984), caching of seal meat could reflect some degree of storage similar to a collector-like system (Binford 1980).

Murray (1996, 1999; see also Friesen 2007) presents the strongest case for Dorset collector-like behavior drawing data from sites in the Igloolik area. Inter-site variability support settlement complexity indicative of collector systems. Caribou faunas appear to reflect intensive harvest elsewhere and long distance transport.

Indeed, use of drive-line systems for mass kills would make little sense if a large amount of the food was not intended for storage as occurs among modern Eskimos (e.g., Binford 1978). Large walrus faunas may also reflect large-scale harvest for storage purposes. Finally, the Igloolik sites have frequent large cache features, strongly indicative of large-scale storage.

It would appear that Dorset likely had a greater degree of organizational flexibility than earlier pre-Dorset despite constraints associated with loss of key technologies. Where resources permitted, as in the Foxe Basin with its year-round walrus population and nearby caribou herds, Dorset groups operated a collector strategy. Elsewhere, in more marginal places, subsistence and mobility was perhaps more of an intensified (e.g., serial specialist) forager-like strategy. However, we should also realize that lack of good sleds and open boats would have prevented a significant amount of flexibility concerning predation and transportation tactics. Open water hunting of seals, walrus, and whales as well as encounter-oriented hunting of caribou and musk ox using bows and arrows was substantially out of the question.

Debate continues over the Terminal Dorset period. In general, most Arctic researchers argue for acculturation and overlap of Dorset and Thule occupations ca. 800–1000 cal. B.P. Scholars who subscribe to this hypothesis generally rely on three main lines of evidence: artifacts in contexts that suggest direct contact between the two; evidence for acculturation, wherein Thule may have adopted some Dorset harpoon styles; and overlapping radiocarbon dates for both Dorset and Thule occupations. The general assumption is that there were at least occasional brief encounters and ephemeral interactions between the two groups (see Friesen 2000, 2005; McGhee 1996).

Alternatively, Park (2000, see also 1993) challenges contentions for cultural contemporaneity by demonstrating that radiocarbon dates for Dorset and Thule do not overlap in time and, in fact, appear to fall in bimodal distribution, suggesting Dorset was going extinct as the Thule entered the Eastern Arctic (cf. Friesen 1999, 2005; McGhee 1997). Park (2000) and others (Fitzhugh 1994; Tuck and Fitzhugh 1986) posit that Dorset repopulations of the fringe regions after 2300 cal. BP were unlikely to have been very large but instead consisted of small scattered groups. Park (2000) argues that these highly dispersed small populations would have required intergroup social interactions for biological reasons as well as for the kind of social networking typically required by foragers in such an environment. Given the distance between these small populations, it was probably difficult to maintain social interactions across the Arctic, and Park (2000) claims that this led to the loss of social networking that may have spurred the collapse of these far-flung Dorset populations prior to the arrival of the Thule groups some time after ca. 1000 cal. B.P.

Thule

The Thule phase reflects a cultural strategy originally derived from another Siberian ancestor (likely the Late Neolithic Ymyakhtakh culture), best described as a winter village collector RMS/cultural niche. The Thule manifestation of this RMS evolved in northwestern Alaska ca. 1200–800 cal. BP in the setting of highly resource-productive capes and islands of the Bering Strait and northern coastal Alaska (Mason, this volume). Due to the productive nature of their surroundings, some low mobility populations in these areas began to explore new socioeconomic options that included whaling and the development of surplus-oriented economies in localized places such as St. Lawrence Island and East Cape, Siberia (Mason 1998). Although many details have yet to be worked out (Mason, this volume), it is clear that the Thule socio-economy or niche was eventually born and characterized by high population densities, winter village sedentism, extensive logistical mobility, whaling, storage, social inequality, and warfare. As the Thule people spread east out of the Alaskan Arctic after 1000 cal. B.P. they were well-equipped, flexibly adapted, and capable of employing foraging strategies, in addition to collecting if necessary to meet the requirements of survival in new contexts.

Thule probably underwent some changes to its socio-economy, associated with different environmental and social conditions in the Eastern Arctic (Friesen 2000; Savelle 2001). Nonetheless, it remained qualitatively different from that of Dorset. While Thule subsistence was relatively broad, bowhead whale hunting was a primary subsistence focus, wherever possible. Many of the same terrestrial and marine food options favored by the Dorset people were also hunted.

Early Eastern Thule populations were often more sedentary than Dorset. They occupied substantial semi-subterranean dwellings, sometimes stratigraphically cutting into former Dorset occupations, and engaged in extensive logistical mobility. Like Dorset, Thule appears to have aggregated in large winter villages, but unlike Dorset, the Thule communities appear to have been more permanent. Additionally, large Thule populations may have aggregated in the warm season as well, to conduct trade and communally hunt bowhead whales (Savelle 1987). They also appear to have occupied productive resource locales, which effectively barred access to Dorset populations. Whether Thule employed violence/warfare for defense of these locales is not well understood archaeologically. Friesen (2000) argues that open water hunting, one of many Thule subsistence tactics, would have been less spatially restricted as it was previously in the Western Arctic, and therefore it may have been less practical or necessary to defend access to favorable hunting locations. Later, as climatic conditions cooled, Thule people became increasingly mobile, though still occasionally aggregating in groups of about 100 persons, probably to participate in exchange and communal hunts (e.g., Savelle 2001).

The technology of the Thule people was far more complex, diverse, and in some ways efficient than that of the Dorset people. A few of the artifact hallmarks included the bow and arrow, dog sled, harpoon floats, large and very effective harpoon heads, bow drills, and elaborate watercraft such as kayaks and umiaks (Maxwell 1985). Thule also maintained a lance system for early fall and spring hunting of caribou, and specialized equipment for hunting birds and fishing. Shifts in subsistence pursuits occurred over the life of the Thule phase. After about 600 cal. B.P., for example, whaling decreased, and seal hunting became more important as the sea ice expanded with the onset of cooler climate. At this time, serial specialist foraging (e.g., Binford 1980) becomes more evident in the central Arctic, as bands mapped on to spatially dispersed clusters of food resources.

Socially, the Thule people may have been more aggressive than Dorset. The Dorset people employed what we might consider an 'open' territorial system based on widespread interaction and exchange, which is demonstrated well in the Dorset record by region-wide stylistic continuity. Eastern Thule predecessors, the Western Thule, operated under a more 'closed' system based on tightly bounded and defended territories that prevented the same degree of interaction and the flow of ideas between regional groups (Friesen 2000). However, after migrating eastward, the degree of 'closed-ness' is unclear. Friesen (2000) notes that Thule material culture fails to support strongly defined 'closed' social units. Despite initial variation among early eastern Thule assemblages, Thule technology becomes overwhelmingly homogeneous across the region. This suggests that inter-regional interaction was maintained over a vast area and supports contentions for a more 'open' social system in the east than the 'closed' system employed back in Alaska.

Thule also differed from Dorset in the nature of their exchange economy. Whereas Dorset exchange focused on alliance-network building and maintenance, the Thule mode of exchange often emphasized individual success, whether purely economic as in the exchange of better quality raw materials for tools, or social, associated with the acquisition of status-conferring items made from amber, copper, or iron. Such prestige or wealth items may have played crucial roles as sources of power for elites or individuals aspiring to be elites. The Dorset record remains largely silent in respect of social differentiation among its populations (but see Friesen 2007; Murray 1999).

In summary, the Eastern Thule socio-economy involved flexible mobility and subsistence strategies with complex technologies that could successfully maintain populations regardless of resource density or predictability. We also see hints of status differentiation and formal leadership, specialized task groups who sought foodstuffs, and knowledge of war tactics that initially evolved in the Western Arctic under different conditions. Simply put, the Eastern Thule winter village RMS/cultural niche evolved under conditions of intense competition for control of optimal positions on the Western Arctic landscape. Consequently the Thule people were well prepared to outcompete and displace the less complexly organized and economically more specialized Dorset peoples of the Eastern Arctic (Friesen 2000).

Extinction of Dorset

The Dorset RMS represented a cultural niche that relied upon cold conditions and access to extensive sea ice. Consequently it is not surprising that the Dorset population peaked (as measured by numbers of dated sites) during the colder period of 1600–1300 cal. B.P. and subsequently declined with the rapid warming of the Medieval Climatic Optimum (ca. 1200–650 cal. B.P.) that followed. As conditions grew more stressful, Dorset peoples moved into many previously more marginal areas but maintained long distance networks facilitated by apparently more frequent social gatherings at "long-houses" (Damkjar 2005). While it is reasonable to suggest that Late Dorset expansion is linked to resource conditions (McGhee 1996;

Park 1993), one archaeologist has suggested a social explanation linked to stress caused by rising social inequality in the Foxe Basin area (Friesen 2007).

We argue that the lock tightened, putting an overspecialized Dorset on the pathway to eventual near-oblivion due to habitat loss associated with climatic warming and competition with a more flexible and technologically powerful RMS associated with the Thule Inuit peoples. Perhaps reflective of Paleoeskimo cultural conservatism, recent research suggests that one socially isolated group, known as the Sadlermiut, persisted on Southampton Island into the early twentieth century with much of their original Dorset lifeway intact (Hayes et al. 2005; Maxwell 1985).

Advanced technology and organizational capacities permitted Thule groups to flexibly adapt to warmer or colder conditions. Persistence of the pre-Dorset forager and later Dorset collector-like RMS had been facilitated by specialization to Neoglacial cold conditions. Reductions in sea ice during the warm periods of ca. 2800–2500 and 1300–650 cal. B.P. was catastrophic to its niche and led to stress at ca. 2500 cal. B.P. and near extinction some time after 1000 cal. B.P. In contrast, stability in the Thule RMS was capable of operating in the widest range of Arctic environments, thereby reducing the need for the kind of habitat tracking required for Dorset and pre-Dorset. Despite impacts by Europeans, many elements of the Thule RMS remain in operation today in communities across the North American Arctic.

Discussion

The Arctic provides a particularly good place to look at processes of cultural evolution. Assuming that "stasis is data" (Gould 2002), studying processes of stasis is equally important to developing explanations for change. SBT provides a useful metaphor for understanding the persistence and eventual extinction of the Arctic Small Tool tradition cultures as manifested archaeologically by the pre-Dorset and Dorset phases of the Eastern Arctic.

Pre-Dorset populations operated a generally mobile forager-like strategy. A warming climate at ca. 2800–2500 cal. B.P. may have triggered descents from original adaptive peaks among late pre-Dorset groups. It seems likely that some groups did not survive this period, whereas at least one crossed an adaptive trough, triggering the shift from pre-Dorset foragers to Dorset collectors. Once present, the Dorset peak was undoubtedly higher, narrower, and surrounded by a far deeper chasm given its isolation and loss of key technological resources (e.g., bow and arrow, throwing board, bow drill). This meant that groups possessing the Dorset RMS were well positioned as long as resource conditions were favorable. And this appears to have been the case since the strategy spread rapidly throughout the region and persisted with little change for nearly two millenia. However, such a specialized strategy was probably not well positioned to survive any rapid descents into a maladaptive zone (e.g., Rockman, this volume).

Hunter-gatherers in more temperate contexts tend to be able to survive climate change and other impacts with greater ease than many more complex societies. They do it primarily by maintaining social and economic flexibility even if it means operating at less than peak (or optimal) levels. It would appear that Dorset, despite some organizational flexibility (i.e., ability to shift between a serial specialist and more collector-like strategy), had substantially lost the technological capacity to react like many other hunter-gatherers. Serious isolation left groups without opportunities for transmission of outside technologies that could have been transformative (e.g., Bettinger 1999). Unlike more temperate contexts, the arctic was far too unforgiving for significant internal socioeconomic flexibility. Further, internal pressures from millennia of culture history were undoubtedly strong providing a sort of inertia preventing major internal developments (e.g., Janssen et al. 2003). Friesen (2007), for example, argues that Dorset peoples maintained strict cultural protocols aimed at maintaining egalitarian social relations. Finally, if isolated pockets of Dorset people remained at the advent of the Thule expansion, they were in no position to compete with the more powerful and technologically flexible newcomers. Thus, when climate change and social competition drew the Dorset socioeconomic strategy from its adaptive peak, the chasm between them proved too deep and Dorset moved toward extinction.

Conclusion

Culture historians have, for decades, told us that cultures appear, persist, and eventually disappear. While it has been easy to write off these conclusions as products of a faulty theoretical paradigm (culture history) or an imperfect archaeological record, our assessment of Arctic data suggests that there can be a high degree of validity to those old ideas. Indeed, we suggest that while previous generations lacked the theoretical apparatus for adequate explanation, their observations have proven largely correct.

In cases like the ones we have presented, the designation of cultural phases (pre-Dorset, Dorset, and Thule), also mark RMS that existed in processes of stasis, at least at the most complex integrative level (see Chatters, this volume). As we have pointed out, this did not mean that evolution did not continue to operate on other scales. Technological changes continued within periods of stasis at the RMS level (e.g., Prentiss and Clarke 2008). The challenge to archaeologists becomes not just why pre-Dorset or Dorset emerged, but why they and their ancestors persisted for so long without significant organizational change. We have tried to point to some potential pathways toward explanation that we admit will require more research.

A macroevolutionary perspective permits novel insights into cultural and environmental processes. In the case study presented here, we argue that the pre-Dorset to Dorset transition is not an anomalous case but rather an example of a process recognized in other areas. Isolation can play a major role in the emergence of new cultural entities. As Chatters (this volume) points out, isolation may be purely social, as in the Late Woodland case, or physical, as was more likely in our arctic example. The Paleoeskimo example also suggests that resource conditions may affect stasis and change. It is no surprise to us that only one major cultural transition occurred across a span of over 3000 years. The combination of cultural factors favoring conservatism and a harsh environment were likely enough to curtail significant cultural variation. Significantly, collector behavior appears to have developed in the one area (Foxe Basin) of the eastern arctic with the right potential combination of resources to support a somewhat more sedentary and storage-oriented economy.

It was this same combination of factors that likely favored a similar transition on the Northwest Coast at ca. 5000 cal. B.P. An implication is that the winter-village collector RMS eventually associated with the Thule culture could have had a similar beginning in Siberia and/or Alaska.

References

Bettinger, R.L. (1999). What Happened in the Medithermal? In C. Beck (Ed.), *Models for the Millenium: Great Basin Anthropology Today* (pp. 62–74). Salt Lake City: University of Utah Press.

Bettinger, R.L. (2003). Comment on "Cultural Diversification and Decimation in the Prehistoric Record" by W.C. Prentiss and J.C. Chatters. *Current Anthropology* 44, 33–58.

Bettinger, R.L. and Baumhoff, M.A. (1982).The Numic Spread: Great Basin Cultures in Competition. *American Antiquity* 47, 485–503.

Binford, L.R. (1978). *Nunamiut Ethnoarchaeology*. New York: Academic Press.

Binford, L.R. (1980). Willow Smoke and Dog's Tails: Hunter-Gatherer Settlement Systems and Archaeological Site Formation. *American Antiquity* 45, 4–20.

Boyd, R., Mulder, M.B., Durham, W.H., and Richerson, P.J. (1997). Cultural Phylogenies Possible? In P. Weingart, S.D. Mitchell, P.J. Richerson, and S. Maasen (Eds.), *Human by Nature: Between Biology and the Social Sciences* (pp. 355–384). Mahwah: Lawrence Erlbaum Associates, Publishers.

Boyd, R. and Richerson, P.J. (1985). *Culture and the Evolutionary Process*. Chicago: The University of Chicago Press.

Boyd, R. and Richerson, P.J. (1992). How Microevolutionary Processes Give Rise to History. In M. Nitecki and D.V. Nitecki (Eds.), *History and Evolution* (pp. 179–209). New York: State University of New York Press.

Cavalli-Sforza, L.L. and Feldman, M.W. (1981). Cultural transmission and Evolution: A Quantitative Approach. Princeton: Princeton University Press.

Chatters, J.C. and Prentiss, W.C. (2005). A Darwinian Macroevolutionary Perspective on the Development of Hunter-Gatherer Systems in Northwestern North America. *World Archaeology* 37, 46–65.

Colledge, S., Connolly, J., and Shennan, S. (2004). Archaeobotanical Evidence for the Spread of Farming in the Eastern Mediterranean. *Current Anthropology* 45, S35-S59.

Coyne, J.A., Barton, N.H., and Turrelli, M. (1997). A Critique of Sewall Wright's Shifting Balance Theory of Evolution. *Evolution* 51, 643–993.

Damkjar, E. (2005). Late Dorset Longhouses: A Look Inside. In P.D. Sutherland (Ed.), *Contributions to the Study of The Dorset Paleo-Eskimos* (pp. 147–165). Ottawa: Mercury Series Archaeology Paper 167, Canadian Museum of Civilization.

Dekin, A. (1975). *Models of Pre-Dorset Culture: Towards an Explicit Methodology*. Unpublished Ph.D. dissertation, Michigan State University, East Lansing.

Dekin, A. (1976). The Arctic Small-Tool Horizon: A Behavioral Model of the Dispersal of Human Population into an Unoccupied Niche. In M. Maxwell (Ed.), *Eastern Arctic Prehistory:*

Paleoeskimo Problems (pp.156–163). Memoirs of the Society for American Archaeology No. 31.

Eerkens, J.W. and Lipo, C.P. (2003). Cultural Transmission, Copying Errors, and the Generation of Variation in Material Culture and the Archaeological Record. *Journal of Anthropological Archaeology* 24, 316–334.

Eldredge, N. (1985). *Unfinished Synthesis*. New York: Oxford University Press.

Eldredge, N. (1995). *Reinventing Darwin: The Great Debate at the High Table of Evolutionary Theory*. New York: John Wiley.

Eldredge, N. (1999). *The Pattern of Evolution*. New York: W.H. Freeman.

Eldredge, N. and Gould, S.J. (1972). Punctuated Equilibria: An Alternative to Phyletic Gradualism. In T.J.M. Schopf (Ed.), *Models in Paleobiology* (pp. 82–115). San Francisco: Freeman, Cooper, and Company.

Fitzhugh, W. (1972). *Environmental Archaeology and Cultural Systems at Hamilton Inlet, Labrador: A Survey of the Central Labrador Coast from 3000 B.C. to the Present*. Washington, D.C.: Smithsonian Contributions to Anthropology No. 16. Smithsonian Institution.

Fitzhugh, W. (1976). Environmental Factors in the Evolution of Dorset Culture. In M. Maxwell (Ed.), *Eastern ArcticPrehistory: Paleoeskimo Problems* (pp.139–149). Memoirs of the Society for American Archaeology No. 31, Washington, DC.

Fitzhugh, W. (1980). Preliminary Report on the Torngat Archaeological Project. *Arctic* 33, 585–606.

Fitzhugh, W. (1994). Staffe Island 1 and the Northern Labrador Dorset-Thule Succession. In D. Morrison and J. Pilon (Eds.), *Threads of Arctic Prehistory: Papers in Honour of William E. Taylor, Jr.* (pp.239–268). Ottawa: Mercury Series Paper, No. 149. Canadian Museum of Civilization, Archaeological Survey of Canada.

Friesen, T. M. (2000). The Role of Social Factors in Dorset-Thule Interaction. In M. Appelt, J. Bergland, and H.C. Gullov (Eds.), *Identities and Cultural Contacts in the Arctic: Proceedings from a Conference at the Danish National Museum Copenhagen, November 30 to December 2 1999* (pp.206–220). Copenhagen: Danish National Museum and Polar Center.

Friesen, T. M. (2005). Contemporaneity of Dorset and Thule Cultures in the North American Arctic: New Radiocarbon Dates from Victoria Island, Nunavut. *Current Anthropology* 45, 685–691.

Friesen, T.M. (2007). Hearth Rows, Hierarchies, and Arctic Hunter-Gatherers: the Construction of Equality in the Late Dorset Period. *World Archaeology* 39, 194–214.

Gould, S.J. (2002). *The Structure of Evolutionary Theory*. Cambridge: Belknap Press.

Gould, S.J. and Eldredge, N. (1977). Punctuated Equilibria: The Tempo and Mode of Evolution Reconsidered. *Paleobiology* 3, 115–151.

Gould, S.J. and Lewontin, R.C. (1979). The Spandrels of San Marco and the Panglossian Paradigm: A Critique of the Panglossian Programme. *Proceedings of the Royal Society of London B* 205: 581–598.

Hayes, M.G., Coltrain, J.B., and O'Rourke, D.H. (2005). Molecular Archaeology of the Dorset, Thule, and Sadlermiut: Ancestor-Descendent Relationships in Eastern North American Arctic Prehistory. In P.D. Sutherland (Ed.) *Contributions to the Study of The Dorset Paleo-Eskimos* (pp. 11–32). Ottawa: Mercury Series Archaeology Paper 167, Canadian Museum of Civilization.

Henrich, J. (2004). Demography and Cultural Evolution: How Adaptive Cultural Processes can Produce Maladaptive Losses—The Tasmanian Case. *American Antiquity* 69, 197–214.

Hoffecker, J.F. (2003). *A Prehistory of the North: Human Settlement of the Higher Latitudes*. New Brunswick: Rutgers University Press.

Janssen, M.A., Kohler, T.A., and Schefer, M. (2003). Sunk-Cost Effects and Vulnerability to Collapse in Ancient Societies. *Current Anthropology* 44, 722–727.

Joshi, A. (1999). The Shifting Balance Theory of Evolution. *Resonance* December, 66–75.

Kirch, P.V. (2000). *On the Road of the Winds: An Archaeological History of the Pacific Islands Before European Contact*. Berkeley: University of California Press.

Knuth, E. (1967). The Ruins of the Musk Ox Way. *Folk* 8–9, 191–219.
Knuth, E. (1968). The Independence II Bone Artifacts and Dorset Evidence in North Greenland. *Folk* 10, 61–80
Laland, K.N., Odling-Smee, J., and Feldman, M.W. (1996). The Evolutionary Consequences of Niche Construction: A Theoretical Investigation Using Two-Locus Theory. *Journal of Evolutionary Biology* 9, 293–316.
Laland, K.N., Odling-Smee, J., and Feldman, M.W. (2000). Niche Construction, Biological Evolution and Cultural Change. *Behavioral and Brain Sciences* 23, 131–175.
Laland, K.N., Odling-Smee, J., and Feldman, M.W. (2001). Cultural Niche Construction and Human Evolution. *Journal of Evolutionary Biology* 14, 22–33.
Lieberman, B.S. and Dudgeon, S. (1996). An Evaluation of Stabilizing Selection as a Mechanism for Stasis. *Palaeogeography, Palaeoclimatology, Palaeoecology* 127, 229–238.
Mason, O.K. (1998). The Contest Between the Ipiutak, Old Bering Sea, and Birnirk Polities and the Origin of Whaling During the First Millenium A.D. Along Bering Strait. *Journal of Anthropological Archaeology* 17, 240–325.
Maxwell, M. (1972). Pre-Dorset and Dorset History of Canada. In D. Damas (ed.), *Handbook of North American Indians, Volume 5, Arctic* (pp.359–368). Washington: Smithsonian Institution Press.
Maxwell, M. (1973). *Archaeology of the Lake Harbour District, Baffin Island.* Ottawa: National Museum of Man Mercury Series Archaeological Survey of Canada Paper 6. Ottawa.
Maxwell, M. (1976). Pre-Dorset and Dorset Artifacts: The View from Lake Harbour. In M. Maxwell (Ed.), *Eastern Arctic Prehistory: Paleoeskimo Problems* (pp.58–78). Memoirs of the Society for American Archaeology No. 31.
Maxwell, M. (1980). Archaeology of the Arctic and Subarctic Zones. In *Annual Review of Anthropology* 9, 161–185.
Maxwell, M. (1985). *The Prehistory of the Eastern Arctic*. Orlando: Academic Press.
MacDonald, D.H. (1999). Subsistence, Sex, and Cultural Transmission in Folsom Culture. *Journal of Anthropological Archaeology* 17, 217–239.
Meldgaard, J. (1960). Prehistoric Sequences in the Eastern Arctic as Elucidated by Stratified Sites at Igloolik. In A.F.C. Wallace (Ed.), *Selected papers of the 5th International Congress of Anthropological and Ethnological Sciences, 1956* (pp. 588–595). Philadelphia: University of Pennsylvania Press.
Meldgaard, J. (1962).On the Formative Period of the Dorset Culture. In J.M. Campbell (Ed.), *Prehistoric Relations between the Arctic and Temperate Zones of North America* (pp. 92–95). Montreal: Arctic Institute of North America Technical Paper 11.
McGhee, G.R. Jr. (1999). *Theoretical Morphology: The Concept and Its Applications*. New York: Columbia University Press.
McGhee, R. (1976). Paleoeskimo Occupations of Central and High Arctic Canada. In M. Maxwell (ed.), *Eastern Arctic Prehistory: Paleoeskimo Problems* (pp.15–39). Memoirs of the Society for American Archaeology No. 31.
McGhee, R.J. (1981). *The Dorset Occupations in the Vicinity of Port Refuge, High Arctic Canada.* National Museum of Man, Mercury Series, Archaeological Survey of Canada Paper 105. Ottawa: National Museums of Canada.
McGhee, R. (1996). *Ancient People of the Arctic*. Vancouver: University of British Columbia Press.
Murray, M. (1996). *Economic Change in the Palaeoeskimo Prehistory of the Foxe Basin, Northwest Territories*. Unpublished Ph.D. dissertation, Department of Anthropology, McMaster University, Hamilton.
Murray, M. (1999). Local Heroes. The Long-Term Effects of Short-Term Prosperity – An Example from the Canadian Arctic. *World Archaeology* 30(3), 466–483.
Nagy, M. (2000). From Pre-Dorset Foragers to Dorset Collectors: Palaeo-eskimo Cultural Change in Ivujivik, Eastern Canadian Arctic. In M. Appelt, J. Bergland, and H.C. Gullov (Eds.), *Identities and Cultural Contacts in the Arctic: Proceedings from a Conference at the Danish National Museum Copenhagen, November 30 to December 2 1999* (pp.143–148). Copenhagen: Danish National Museum and Polar Center.

Park, R. (1993). The Dorset-Thule Succession in Arctic North America: Assessing Claims for Culture Contact. *American Antiquity* 58, 203–234.
Park, R. (2000). The Dorset-Thule Succession Revisited. In M. Appelt, J. Bergland, and H.C. Gullov (Eds.), *Identities and Cultural Contacts in the Arctic: Proceedings from a Conference at the Danish National Museum Copenhagen, November 30 to December 2 1999* (pp.192–205). Copenhagen: Danish National Museum and Polar Center.
Prentiss, A.M. and Clarke, D.S. (2008). Lithic Technological Organization in an Evolutionary Framework: Examples from North America's Pacific Northwest Region. In W. Andrefsky (Ed.), *Lithic Technology: Measures of Production, Use and Curation* (pp. 257–285). Cambridge: Cambridge University Press.
Prentiss, A.M., Lyons, N., Harris, L.E., Burns, M.R.P., and Godin, T.M. (2007). The Emergence of Status Inequality in Intermediate Scale Societies: A Demographic and Socio-Economic History of the Keatley Creek Site, British Columbia. *Journal of Anthropological Archaeology* 26, 299–327.
Prentiss, W.C. and Chatters, J.C. (2003). Cultural Diversification and Decimation in the Prehistoric Record. *Current Anthropology* 44, 33–58.
Price, T., Turelli, M., and Slatkin, M. (1993). Peak Shifts produced by Correlated Response to Selection. *Evolution* 47, 280–290.
Renouf, M.A.P. (1993). Palaeoeskimo Seal Hunters at Port-Au-Choix, Northwestern Newfoundland. *Newfoundland Studies* 9, 185–212.
Renouf, M.A.P. (2005). Phillip's Garden West: A Newfoundland Groswater Variant. In P.D. Sutherland (ed.), *Contributions to the Study of the Dorset Palaeo-Eskimos* (pp. 57–80). Ottawa: Mercury Series, Archaeology Paper 167, Canadian Museum of Civilization.
Richerson, P.J., Boyd, R., and Bettinger, R.L. (2001). Agriculture Impossible During the Pleistocene but Mandatory During the Holocene? A Climate Change Hypothesis. *American Antiquity* 66, 387–412.
Rowly-Conwy, P. (2004). How the West was Lost: A Reconsideration of Agricultural Origins in Britain, Ireland and Southern Scandinavia. *Current Anthropology* 45, S83-S113.
Savelle, J. (1987). *Collectors and Foragers: Subsistence-Settlement Change in the Central Canadian Arctic, AD 1000–1960.* Oxford: BAR International Series No. 358.
Savelle, J. (2001). Logistical Organization, Social Complexity, and the Collapse of Prehistoric Thule Whaling Societies in the Central Canadian Arctic. In B. Fitzhugh and J. Habu (Eds.), *Beyond Foraging and Collecting: Evolutionary Change in Hunter-Gatherer Settlement Systems* (pp. 73–90). New York: Kluwer/Plenum.
Shennan, S. (1993). Settlement and Society in Central Europe 3500–1500 BC. *Journal of World Prehistory* 7, 121–161.
Shennan, S. (2002). *Genes, Memes and Human History: Darwinian Archaeology and Cultural Evolution.* New York: Thames and Hudson.
Stanley, S.M. (1979). *Macroevolution: Pattern and Process.* San Francisco: W.H. Freeman.
Stanley, S.M. (1992). An Ecological Theory for the Origin of *Homo*. *Paleobiology* 18, 237–257.
Tuck, J. and Fitzhugh, W. (1986). Palaeo-Eskimo Traditions of Newfoundland and Labrador: A Re-Appraisal. In *Palaeo-Eskimo Cultures in Newfoundland, Labrador, and Ungava* (pp.161–167). St. John's: Reports in Archaeology No. 1. Memorial University of Newfoundland.
Wright, H.E., Kutzbach, J.E., Webb III, T., Ruddiman, W.F., Street-Perrot, F.A., and Bartlein, J.F. (1993). *Global Climates Since the Last Glacial Maximum.* Minneapolis: University of Minnesota Press.
Wright, S. (1931). *Evolution in Mendelian Populations. Genetics* 16, 97-159.
Wright, S. (1932). Roles of Mutation, Inbreeding, Crossbreeding and Selection in Evolution. *Proceedings of the VI International Congress of Genetics* 1, 356–366.

Chapter 9
Niche Construction, Macroevolution, and the Late Epipaleolithic of the Near East

Ian Kuijt and Anna Marie Prentiss

Introduction

Since the time of George Gaylord Simpson (1944) macroevolutionists have sought to understand variation in the tempo and mode of evolution. Evolutionary tempo and mode have also been a central issue to archaeologists seeking explanation of the great events (e.g., emergence of agriculture, rise of the state) in cultural evolution. In this chapter, we explore the use of a macroevolutionary perspective incorporating niche construction theory to offer new ideas about the tempo, development, and collapse of cultural entities associated with the Near Eastern Epipaleolithic period, the precursor to the later Neolithic developments. The archaeological record of the terminal Pleistocene and Early Holocene in the Near East provides many potentially useful examples of macroevolutionary processes. A brief look at the chronology serves to highlight the potential of this approach to provide novel solutions to some old problems. We provide brief comments on early developments associated with the Epipaleolithic cultures of the Near East in general, with a more detailed focus on the geographical region of the Southern Levant.

Theoretical Perspectives

Macroevolution and Culture

From a Darwinian evolutionary perspective culture can be described as an inheritance system, whereby basic units, whether called memes, traits, or other imaginative names, are transmitted between individuals, accepted, sometimes modified, and used to guide different aspects of behavior (Henrich and McElreath 2003). Boyd et al. (1997) and Holden and Shennan (2005) have suggested that elements of cultures could be transmitted as "packages," since some traits (e.g., tools, foraging tactics, inheritance rules) are actually made of many integrated elements. On a

I. Kuijt (✉)
Department of Anthropology, University of Notre Dame, Notre Dame, IN 46556-5639, USA
e-mail: ikuijt@nd.edu

A.M. Prentiss et al. (eds.), *Macroevolution in Human Prehistory*,
DOI 10.1007/978-1-4419-0682-3_10,

higher scale of complexity, hierarchically integrated systems (of information) could have a "core" defining organizational logic much the way grammar structures language (Boyd et al. 1997). The implication is that these ideas could code for complex sets of organizational behavior as might define a socioeconomic system employed by a given population. Chatters and Prentiss (2005; Prentiss and Chatters 2003) define the most complex socioeconomic entities as Resource Management Strategies (RMS) suggesting that their manifestation in a material and behavioral sense can be called tactics (compare to Bettinger's [this volume] complex adaptive strategies). Within this formulation, RMS can be manifested in behavioral tactics that will vary depending upon local contingencies and individual decision making. This implies that even within a regionally or temporally stable RMS, there will always be variation at the tactical level and correspondingly in the archaeological record.

While there is not enough space in this chapter to fully explore the interesting issue of intentionality (see Rosenberg, this volume; Zeder, this volume), we can suggest that intent plays a significant role in cultural evolution. As modeled by Boyd and Richerson (1985), individuals continuously make choices concerning acceptance of new cultural traits. They also make decisions regarding how those traits are manifested behaviorally and even how they might be transmitted to others. Although disputed by some (Lyman and O'Brien 1998; Rindos 1985), many evolutionists (e.g., Boyd and Richerson 1985) view cultural evolution as at least a somewhat Lamarckian process. From this standpoint, cultural transmission processes generally play a stronger role in cultural microevolutionary process than do the actions of natural selection, which require much longer time spans to really have an impact. A selection-like process may play a more significant role on the macroevolutionary scale associated with the differential persistence of more complex cultural packages that affect the fitness of entire human groups (Prentiss, this volume).

There are many examples from the global archaeological record that illustrate how a complex but stable package such as an RMS could be manifested in different ways over time and space as shaped by local decisions and contingencies. Recent research by Prentiss and Clarke (2008) suggests that a forager-like (e.g., Binford 1980) RMS possessed by Upper Paleolithic hunters in Siberia (Diuktai culture) traveled into interior Alaska (Denali complex) little changed and then shifted down the Northwest Coast (Northwest Microblade tradition and Old Cordilleran culture) where diets shifted to an increasingly marine orientation. Despite the radical change in environments and food targets (mammoths, bison and caribou to fish, seals, deer, and elk) the basic structure of settlement, food processing and distribution, and technology changed little. Another example is the famous Numic Spread whereby southeastern California Numic speakers transported their "processor" (Bettinger and Baumhoff 1982) strategy from the Mojave Desert to the Northern Rocky Mountains with a variety of local subsistence shifts yet stability in the basic strategy focused on broad spectrum foraging and intensified processing activities.

If Resource Management Strategies provide the information code for the technologies and social arrangements that permit acquisition, processing and distribution of resources then RMS must be recognized as high level entities within an evolutionary hierarchy viewed from the genealogical perspective (Eldredge 1985, this volume; Rosenberg, this volume). RMS could persist for extended periods if action in the ecological realm (e.g., Eldredge 1985) provides an adequate payoff to the user population. Thus, if the RMS codes for economic strategies offer adequate cost-benefits to the user population, then we could expect that population to hold on to that strategy, potentially even when exposed to others in nearby areas. Other factors may also condition the persistence of RMS linked to the inherent structure of the actions required by the RMS. As argued by Chatters and Prentiss (2005), annual scheduling probably plays a major role in preventing major change since most cultures, especially hunter-gatherers and incipient food producers, depend upon careful geographic positioning and labor allocation linked to local and seasonally specific resource needs. Under these conditions, selection, or cultural proxies of selection, could act to reduce opportunities for emergence and persistence of new cultural patterns (Chatters, this volume; Prentiss and Lenert, this volume). This may help explain why in some contexts we witness in the archaeological and ethnographic records the presence of two or more groups living adjacent to one another yet not altering their basic approaches to food getting or social arrangements to a significant degree (e.g., Hitchcock and Ebert 1982).

Collectively this implies that radically new RMS should not emerge very often due to the strength of internal and external pressures. Indeed, it should be rare in contexts of high population interaction, since this cultural analogue of gene flow would serve to smother any significant new variants during the germination phase. Harsh natural environments (e.g., the desert areas of Saudi Arabia) may also play a role due to the potential for harsh punishment meted out for mistakes in annual subsistence pursuits and social arrangements. Thus, truly new RMS are expected to be rare, to develop in small groups relatively isolated from groups associated with parent culture, and to be associated with relatively productive resource conditions (Chatters, this volume; Prentiss, this volume; Prentiss and Chatters 2003).

Niche Construction and Culture

Drawing on a biological perspective, Day, et al., (2003) argue that niche construction occurs when organisms directly construct or otherwise modify aspects of the environments including nests, burrows, and surrounding landscapes. Niche construction may play an important role in evolution, as it implies that organisms play a more active role in the development of their selective environment than previously thought (Laland, et al., 1996, 2000; Odling-Smee et al., 1996). While all plant and animal species seem to participate in niche construction, most broadly defined, *Homo sapiens* still provides the best examples in that we use the widest range of artifacts, develop the most complex socioeconomic organizations, and affect our habitats the most severely of any species. We create a habitat for ourselves in the form of houses and other shelters. We develop tools from natural materials used to enhance our diets, we organize socially to accomplish tasks requiring extensive and complex communication systems, and consequently we severely alter the natural spaces around us. Thus, the so-called natural world is actually made up of humanly managed habitats permitting us to exist within a series of constructed cultural niches (Laland, et al., 2000).

One theoretical implication of this work is that we become adapted not just to the natural environment but to the cultural niches we help to create. This permits substantial flexibility to move between habitats without significantly changing the niche. This helps us understand why human groups can move across multiple habitats while changing few fundamentals of their socioeconomic structure (see Rockman, this volume). The record of human history is rife with examples that include the spread of Upper Paleolithic foragers out of Africa and around much of the world (Bar-Yosef 1998), the Clovis dispersal across interior North America (Haynes 2001; Kelly and Todd 1988), and the dispersal of Neolithic Asians across the Pacific Basin (Kirch 2000).

Logically, if we are most fundamentally adapted to our culturally constructed niches, and if those niches can move through many kinds of natural habitats, then from an organizational perspective cultures should always remain in stasis patterns between their periods of emergence and decline. This does not mean that no change will occur—only that the basic constructed niche will not substantially change. The archaeological record provides good examples to support such an idea. For instance, the longest periods of stability are evident in the least complex societies where adaptive flexibility in the form of mobility buffered against niche-threatening impacts (Chatters and Prentiss 2005; Prentiss and Lenert, this volume). However, some well-positioned complex societies (e.g., the Egyptian civilization) have also lasted long periods (Wenke 1991).

A final implication concerns the origin(s) of newly constructed niches. Since niche construction is the by-product of an evolutionary process that has the potential to intensely rearrange organismic behavior patterns and habitats, it can only be modeled on the scale of single geographically or socially constrained (not swamped by gene flow) populations (Laland, et al., 2001). Given the socioeconomic tenacity by which a constructed niche constrains change, it is likely that major cultural niche change within a human group could only come quickly in small and more isolated populations. Logically, it seems unlikely that gradual change across larger populations could give rise to major changes. The implication is that punctuated equilibrium (Eldredge and Gould 1972) should be frequently recognized in the fossil and archaeological record where niche construction is an important process (Laland, et al., 2000:141).

Integrating Evolutionary and Ecological Views of Culture

Niche construction theory provides insight into the workings of the evolutionary process. We suggest that it is particularly effective in helping to imagine the cultural evolutionary process on a macroevolutionary scale. If we view Resource Management Strategies (RMS) as the genealogical packages for human behavior associated with cultural niche construction, then the constructed cultural niche is a simultaneous reflection of the phenotypic actions of a population with a common RMS and its impact on the natural environment. Thus, niche construction theory helps us to understand the ecological half of the broader evolutionary equation.

Probably the most important lesson concerns our conceptualizations of the natural environment and the role of selection in cultural evolution. Niche construction theory makes it very clear that when anthropologists talk about resource conditions, what they are really describing are culturally constructed niches. This has specific implications for understanding macroevolutionary cultural processes of emergence, stasis, radiation, and extinction.

Chatters and Prentiss (2005) predicate their model for the emergence of new RMS on good resource conditions that might permit a new arrangement to fluoresce enough to become fixed as a group's modus operandi. Niche construction theory implies that the new RMS do not necessarily develop at the behest of the environment but in unison with the environment. This does not mean that an independent factor like climate change cannot play a role in this process as it clearly does. What it does suggest is that cultural developments may play an equal role. As argued by Laland, et al., (2001), new niche constructions represent solutions to problems stemming from previous constructions, and they do not require major environmental changes.

Tempo and Niche Construction in the Near East Late Epipaleolithic Period

Having explored a macroevolutionary framework drawing upon niche construction theory, we can now turn to the critical question of how we are to understand the development and collapse of the Near Eastern Epipaleolithic RMS and the subsequent rise of those associated with the earliest Neolithic. The Epipaleolithic of the Near East is divided into a number of temporal and cultural divisions (Table 9.1). Recent synthetic research (Bar-Yosef 2003; Byrd 2005) has outlined the economic and cultural diversity within these cultural periods. The evolutionary origin of the earliest Epipaleolithic pattern, the period known as the Kebaran, remains unclear. This is partially due to the limited number of archaeological sites known from this period, as well as the historical focus among prehistoric archaeologists on other periods. Despite these limitations, however, it is still possible to develop a good understanding of the emergence and collapse of Epipaleolithic cultures in general and Natufian cultures in specific. In many ways, these data act as a series of windows into a critical, yet poorly understood, evolutionary process. It is our hope that this discussion will serve to present ideas for colleagues to reflect upon and with which to explore the implications of the theoretical ideas presented here. While debate exists as to the relative mobility of groups in the early and middle Epipaleolithic, there is generalized agreement that these groups were hunter-gatherers[1] that employed a range of adaptive strategies. At around 15,000 years ago, with the start of the Early Natufian, we find our first evidence for emerging sedentary, and/or semisedentary collector-like hunter-gatherers (collectors per Binford 1980).

Table 9.1 Summary of Relevant Near Eastern Culture Historical Sequence (Drawing on Bar-Yosef 2003; Byrd 2005; Goring-Morris and Belfer-Cohen 1997; Kuijt and Goring-Morris 2002; Munro 2004)

Cultural period	Calibrated years before present (uncalibrated age estimates)	Phase	Possible sub-phases
Early Epipaleolithic	23,000–18,000 B.P. (19,400–15,000 B.P.)	Kebaran and Qalkan	na
Middle Epipaleolithic	18,000–14,900/14,600 B.P. (15,000–12,500 b.p.)	Geometric Kebaran, Mushabian in Levant	na
Late Epipaleolithic	14,900/14,600–12,000/11,700 B.P. (12,500–10,250/10,100 b.p.)	Natufian (in the southern Levant)	Early, Late (Harifian in Negev)
Early Neolithic	12,000/11,700–10,650 B.P. (10,250/10,100–9400 B.P.)	Pre-pottery Neolithic A (PPNA)	Khiamian, Sultanian
	10,650–8400 B.P. (9400–7600 B.P.)	Pre-pottery Neolithic B (PPNB)	Early, Middle, Late

Emerging and Coexisting Niches

The late Epipaleolithic of the Near East is characterized by the emergence of multiple coexisting successful cultural niches (Table 9.1, Fig. 9.1). As is outlined by Bar-Yosef (2003: Fig 5.15), several contemporary socioeconomic strategies emphasizing different habitats developed during the Natufian. Bar-Yosef argues that as an outgrowth of the Middle Epipaleolithic, Early, and Late Natufian people were initially focused on a core area of the Mediterranean Jordan Valley. This "Natufian Homeland," to use his term, was an area of the Near East that had good access to water and subsistence resources and served as an ecotone between other neighboring areas. It is within this core area, moreover, that the earliest semisedentary communities emerged.

By the Late Natufian phase, people apparently expanded out of this core area and into other habitats. It seems unlikely that this regional expansion was linked to some form of population packing, with people moving in from neighboring ecological zones. Instead, following Bar-Yosef (2003) and Özdögan (1998), we argue that this process reflects developments within local cultures featuring new use of other less-preferred habitats. There was, in short, local adaptation to changing conditions and the subsequent creation of multiple new niche opportunities.

This process of socioeconomic diversification was linked to small-scale inter-regional developments. This resulted in a patchwork mosaic of local adaptations. Drawing upon other sources, Bar-Yosef (2003) argues that by the end of the Late Natufian, this resulted in the coexistence of different peoples, and with different economic adaptations, focused in distinct physiographic areas (See Fig. 9.1). According to Bar-Yosef, this probably included the existence of Late Natufian collectors

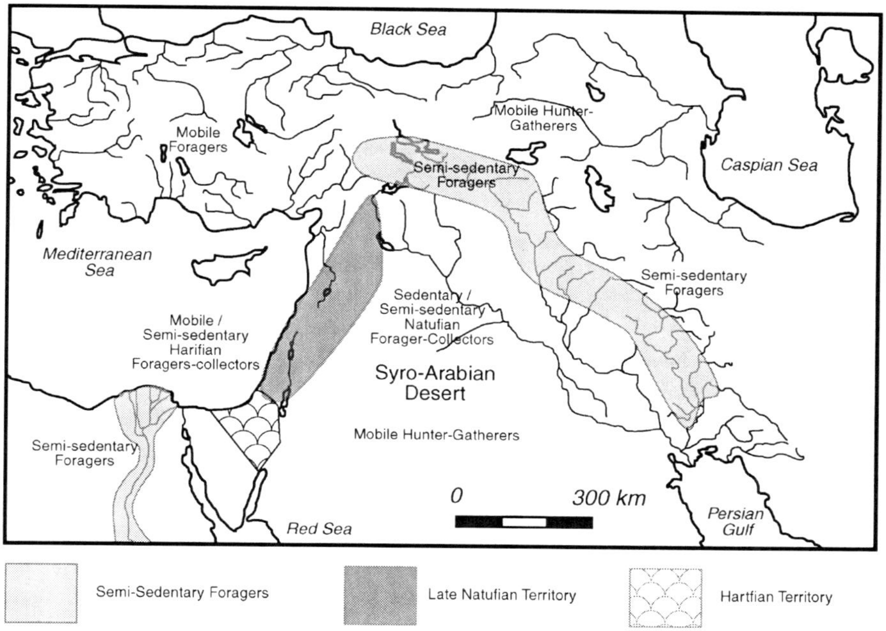

Fig. 9.1 Coexisting cultural and ecological niches in the late Epipaleolithic (Based on Bar-Yosef 2003: Fig. 5.15)

with food storage and social inequality living around the Jordan Valley, possibly more forager-like hunter-gatherer groups adapted to desert environments, and other semisedentary hunter-gatherers in other areas. Rising sea levels and poor preservation levels make it difficult to identify late Epipaleolithic adaptations to coastal settings and resources, but it is entirely possible that various coastal forager-farmer-fisher groups also emerged in these areas. Probably these RMS first emerged on a small scale in one or more places but then grew quickly. They varied in several ways, including the following: (1) production of food, (2) social organization as it impacts food distribution/rights to resources, (3) corresponding ideological changes that contributed toward social cohesion, (4) some technological elements, and (5) architectural changes.

In the wake of the Early Natufian-improved environmental conditions, it can be argued that the Levant provided an opportunity waiting to happen. Multiple new and different RMS emerged from the homogeneous cultural heritage of the Middle Epipaleolithic. Given commonalities in technology, it is clear that most of these groups were interconnected. Thus, we can posit a number of potential ancestor-descendant relationships between Middle and Late Epipaleolithic cultural complexes. At some point one or more of these groups, or some other hunter-gatherer group on the margins, developed a more intensive approach to the cultivation and collection of wild plant resources. While followed by a period of apparent subsistence stress and associated restricted development in the Late Natufian (Fig. 9.2), these developments provided the foundation for later cultural developments and economic adaptations of the Pre-Pottery Neolithic A period (Zeder, this volume).

Emergence Events: Developments During the Epipaleolithic

One of the interesting aspects of the late Epipaleolithic adaptations is that some of these were not only distinctly unique but that they were not seen in the preceding period, and in some cases, in the following periods (Fig. 9.2). In the Early and Middle Epipaleolithic period, RMS were largely focused on mobile foraging (Byrd 2005; Goring-Morris 1987). While there is some evidence for decreased residential mobility, such at Ohalo II, this was probably seasonal and limited to relatively short periods of time. With the start of the early Natufian around the Jordan Valley, we witness the emergence of large residential structures, the construction of an entirely new suite of features inside and outside of the structures, and for the first time, the systematic placement of burials within settlements.

Among others, Bar-Yosef (2003), Goring-Morris and Belfer-Cohen (1997) and Byrd (2005) illustrate that the Middle Epipaleolithic period focused on a mobile hunter-gathering lifeway. With the appearance of improved warm and wet climatic conditions around 14,600 years ago, we find our earliest appearance of Early Natufian sedentary/semisedentary collectors. This was a distinctly different Resource Management strategy. The period of 14,600 to ca. 12,800 was characterized by an increase in the number of unique economic developments. Early Natufian people moved into the rich valleys of the Levantine Corridor where new richer resource configuration permitted longer stays and opportunities for reorganized mobility schedules and foraging practices. As is illustrated in Fig. 9.2, this reflects a rapid shift to semisedentary life with at least some logistical mobility and food storage.

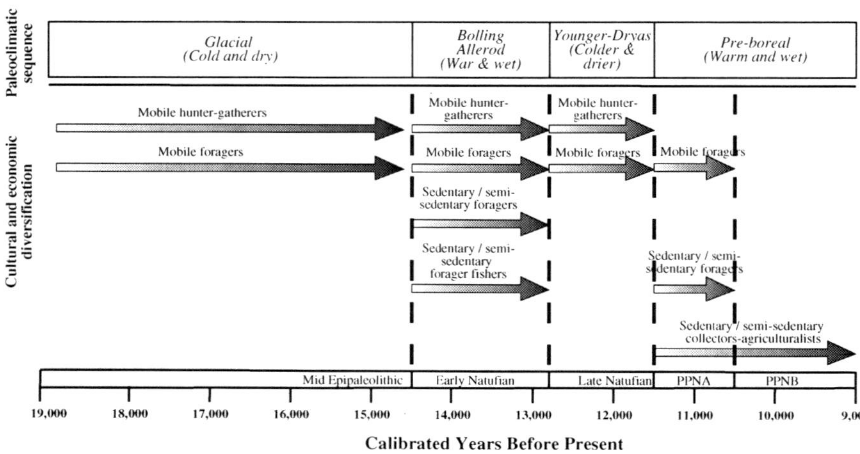

Fig. 9.2 Cultural entities, chronology, and economic cultural diversity in the southern Levant (based on Bar Yosef 2000, 2003; Byrd 2005)

The transition from the early and middle Epipaleolithic and from the Geometric Kebaran to the Early Natufian provides evidence for a major transition in settlement scheduling, food harvest and storage, social relations, technology, and architecture. Collectively these clearly show reconfigured cultural niches and the emergence of multiple coexisting RMS.

When discussing the Epipaleolithic as an evolutionary foundation for the origins of agriculture, some researchers overlook the Early-to-Late Natufian transition. This transition, however, illustrates the linkages between restricted resource conditions and foraging opportunities. With the start of the Younger Dryas at around 12,800 years ago, there was a dramatic increase in residential mobility. Compared to the Early Natufian, Late Natufian people constructed fewer examples of residential architecture, smaller buildings, produced shallow cultural deposits, and developed new mortuary practices reflecting at least in part, different systems of social organization (Belfer-Cohen 1995; Bar-Yosef 2003; Kuijt 1996). Early Natufian people relied on some form of food storage, as well as rearranged subsistence cycle based around it. While not always easy to identify, it is clear that Late Natufian people abandoned some of these practices and adopted more frequent residential group mobility and a generalized more forager-like (per Binford 1980) hunting and gathering adaptation (Munro 2004). Moreover, comparative studies of Early and Late Natufian mortuary practices (Belfer-Cohen 1991; Byrd and Monahan 1995; Kuijt 1996) illustrate that this reflects a rather rapid rearrangement of social relations.

It is interesting to note that this pattern is similar to many North American examples where more sedentary hunter-gatherers experienced subsistence stress and then reverted to higher mobility regimes. A strong case is seen with Amsden's (1977) work with the historic Nunamiut, who, when the caribou crashed in about 1895–1900 (AD), substantially left the Brooks Range for the Coast. The small remaining group then shifted back to earlier adaptations that focused on more frequent residential moves and reduced storage. This also appears to have been what happened on the Columbia Plateau at ca. 4000 B.P. with the collapse of the Pithouse I culture (Chatters 1995; Prentiss and Chatters 2003). In brief, it appears that this major economic disaster was followed by a pattern of group breakup and higher mobility. As is outlined by several authors (Bar-Yosef 2003; Belfer-Cohen and Bar-Yosef 2000; Byrd 2005), in the Late Natufian we witness a similar pattern with a fragmentation of the original model. At the core of this pattern, we see a severely interrupted strategy causing social breakups and a regional population seeking to make a living while preserving what is feasible.

The Epipaleolithic of the Near East: Stasis and Tempos of Change

Several archaeological studies (e.g., Byrd 2006; Zeder this volume) have drawn increased attention to the tempos of culture change during the Epipaleolithic and Neolithic periods. As we have noted earlier, and as illustrated in Fig. 9.3, there is a clear connection between major paleo-climatic changes and the timing and tempos of culture change. Oxygen isotope data from the GRIP ice core

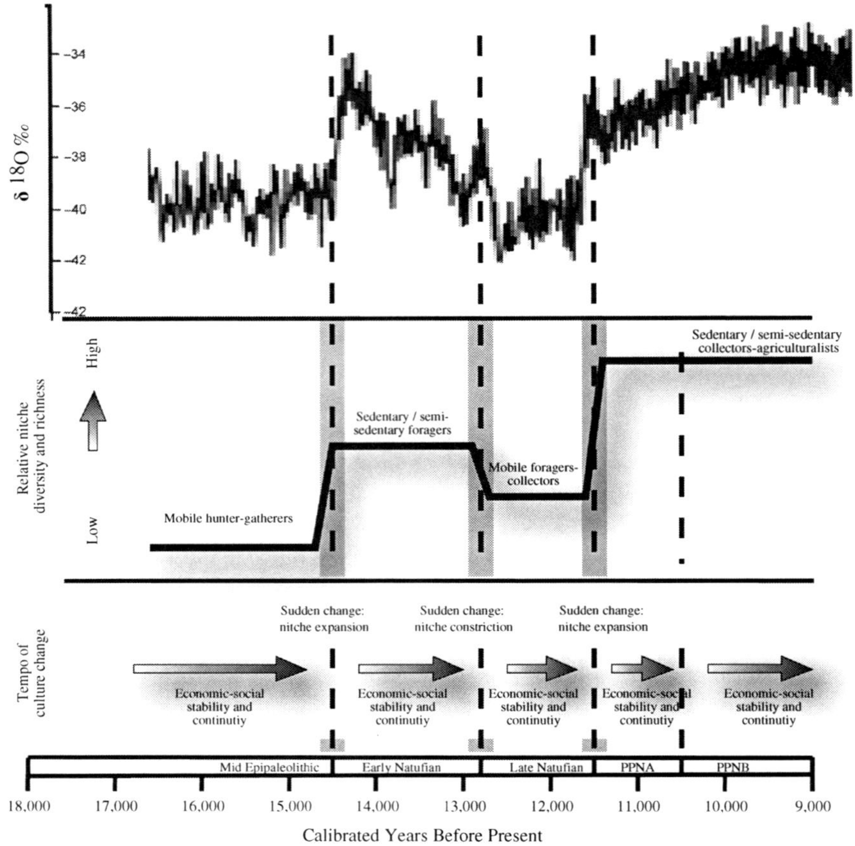

Fig. 9.3 Tempos of culture change, niche diversity/richness (based on Byrd 2005; Prentiss and Chatters 2003)

(Ditlevsen, et al., 1996; see also Richerson, et al., 2001) clearly illustrate a strong relationship between Upper Pleistocene cultural macroevolution and global climate change. Analysis of radiocarbon data by Byrd (2006) indicates that the transition from the Middle Epipaleolithic to Early Natufian was rapid. Once established, Early Natufian practices became fixed, and we find evidence for a period of relative cultural stability for the next 1,500 years coinciding with the period of generally warmer and wetter conditions preceding the Younger Dryas. The rich architectural evidence from these sites, as well as data indicating some scale of food storage, supports the argument that this Resource Management Strategy was quite successful along the Jordan Valley. Mortuary practices, chipped stone tool technology, and systems of constructing buildings remained largely unchanged during this time (Henry 1989). Once in place, groups appear to have maintained the underlying economic strategy, social structure, and supporting belief system (Byrd 1992; Kuijt 1996).

With the dramatic temperature downturn of the Younger Dryas (Fig. 9.3), Early Natufian groups abandoned these tried and true lifeways, reverting to more a mobile settlement system as a means of dealing with reduced subsistence resource productivity. We believe that this transition was connected to a reduction in opportunities for successful use of cultural niches (Fig. 9.3). Many settlements were abandoned, and if not totally abandoned, they became characterized by smaller buildings, fewer features, and generally smaller quantities of deposited cultural materials. Burials become less frequent. The relative paucity of Late Natufian archaeological materials, especially for the final stages of the Late Natufian, has been interpreted by researchers as reflecting increased residential mobility, smaller group size, and a general reversion to more nomadic small-scale communities (e.g., Bar-Yosef 2003; Byrd 2006; Munro 2004). This was at least partially connected to the development of dry and cold conditions, which probably required Late Natufian people to utilize lower ranked food resources.

At the start of the Pre-Pottery Neolithic A (PPNA) period, around 11,500 years ago, paleo-climatic conditions again changed drastically (Fig. 9.3). The wetter and warmer climate now dramatically improved growing conditions for plants. This climatic shift is echoed by the remarkable contemporaneous transition from the more mobile Late Natufian settlement and subsistence practices to the PPNA lifeways with much reduced residential mobility, construction of building with greater investment of labor with mud floors, walls, and storage features (Kuijt and Goring-Morris 2002). While it is clear that some community members still traveled between settlements, social and economic practices were now focused on dealing with growing communities of forager-collectors-cultivators. Economic strategies continued to focus on the hunting of wild game and collecting of wild plants (Bar-Yosef 2003; Byrd 2006; Munro 2004). At the same time, however, people were now actively involved in either the direct manipulation of plants, or alternatively, the environment in which plants were growing. Either way, this highlights a diversification of targeted subsistence resources. It also reflects a significant expansion in niche richness, some of which was triggered by paleo-climatic changes.

Drawing on other studies (including Bar-Yosef 2003; Byrd 2006; Zeder this volume) this analysis highlights that while change occurred during individual periods, the most dramatic and significant changes were punctuated, rapid, and followed by periods of relative cultural stasis (Fig. 9.3). We believe, moreover, that these were linked to relative niche richness and diversity and temporally linked to climatic changes.

Cracked Niches and RMS Extinction

Extinction occurs frequently in the record of past life whether organismic or cultural in form. Indeed from some perspectives, rates, contexts, and forms of extinctions may play important roles in rates and patterns of macroevolution (Gould 2002; Jablonski 1985). Different from organic evolution, extinction in this context does not necessarily mean loss of human life but simply loss of the RMS. Niche construction theorists suggest that a cultural niche can be "cracked" or suffer "niche destruction" (Laland, et al., 2000:145). One pathway is through a self-generated process as documented so frequently in the archaeological record (Kirch 1997; Redman 1998; Rollefson 1996; Simons 2000), whereby people over-harvest resources, foul their landscape with pollutants, or trigger catastrophic loss of productive land through maladaptive agricultural practices. Some argue that economic crises may generate new cultural options for success (Fitzhugh 2001; Rosenberg 1990). However, evidence is sparse for such events, leaving others (e.g., Janssen et al. 2001; Prentiss and Chatters 2003) to argue that bad times are more often associated with conservative behavior and low rates of innovation associated with those most critical subsistence technologies. Laland, et al., (2000) offer other reasons from the standpoint of niche construction theory why people do not just simply "adapt" when times get bad. First, the human group may not recognize the source of the problem and thus may not be prepared to make the "fix." Second, the "fix" may require technology well beyond their abilities (see also Tainter 1988). Third, the source of the problem may be associated with indirect feedback from the cultural niche that is not readily recognizable. Finally, radical changes in the face of trouble can lead to maladaptive "runaway" processes (see also Boyd and Richerson 1985) whereby so-called solutions lead to even greater maladies.

It is from this perspective that we view the end of the early Natufian as an example of a cracked niche. A number of researchers (e.g., Bar-Yosef 2003; Belfer-Cohen and Bar-Yosef 2000; Byrd 2005; Goring-Morris and Belfer-Cohen 1997) outline the late Epipaleolithic as a period of significant subsistence and settlement change but within the context of observable cultural and technological continuity. In the face of overall continuity in lithic technology, and very generally in terms of architecture, we find evidence for the abandonment of settlement and subsistence systems impacting, in particular, residential stability and food storage. Similarly, mortuary practices suggest a drastic change in how these communities were organized, with less evidence for social differentiation (Belfer-Cohen 1995; Byrd and Monahan 1995; Kuijt 1996). As illustrated in Fig. 9.3, the end of the early Natufian overlaps with the transition to the Younger Dryas. The combination of internal factors, such as population growth, as well as external factors, including global paleo-climatic changes, would have created the context for new social, economic, and subsistence stresses.

Drawing upon data on ratios of high-to-low-ranked game in the late Epipaleolithic, Munro (2004) argues that the transition to the Late Natufian reflects a reduction in site-occupation intensity and increased population mobility. As with other researchers, she argues that this transition indicates that Late Natufian communities developed more cost-effective demographic solutions to environmental stress. We agree with this view and argue that in the face of the combination on internal and external forces, Late Natufian people abandoned earlier settlement and subsistence tactics that had originally included semisedentism, scheduled harvest and storage of grain surpluses, and likely, variability in food distribution pathways affected by social status differentiation. We believe that this collapse was not related to human habitat disturbance as much as changes in local resources and distributions. The Early Natufian RMS appear to have broken down; the later groups now implemented rearranged foraging schedules, new mobility tactics, altered food harvest and consumption patterns, and an apparent return to egalitarianism. The "cracking" of Natufian was different from the declines associated with the Late Pre-Pottery Neolithic. For the former, it represented dissolution of the core socioeconomic organizing principles. For the latter, core economic principles were substantially preserved despite many altered social arrangements (Köhler-Rollefson and Rollefson 1990; Rollefson 1998; Simmons 2000).

Discussion

Triggers to Niche Construction

What then might trigger a major new niche construction (or niche reorganization as described by Prentiss and Chatters 2003)? As noted by Prentiss (this volume), one possibility is a major innovation in technology that has immediate implications for accessing new resources or improving existing approaches. Similarly, Mason (1998, this volume) argues that changes in harpoon and boat technology permitted a few isolated groups in and around the Bering Strait to harvest whales and subsequently to expand and eventually control the entire North American Arctic region. Emergent cultural niches may have come about more often as groups combined prior existing technologies with new resource opportunities to put together novel new packages featuring reorganized subsistence schedules and patterns of labor organization and, consequently, new cultural niches. However, socio-ecological conditions favoring such events may not have come very often. Our Near Eastern data indicate that it only occurred twice during a period of approximately 5,000 years, coincident with rapid climatic warming in each case.

These results closely parallel findings outlined elsewhere in this volume and in the archaeological literature. In prehistoric sequences around the world, global climatic events triggered simultaneous dramatic change in the organization of indigenous societies. Early Holocene warming favored rapid replacement of Paleoindian by Archaic adaptations (e.g., Chatters, et al., 2010) simultaneous with the collapse of Pre-Pottery Neolithic B towns (Kuijt and Goring-Morris 2002) and the expansion of agriculturalists across Europe (Shennan 1993). Early Neoglacial climate changes affected diversification and decimation in the Pacific Northwest giving rise to the region's classic collector-based socio-economies (Chatters, this volume; Prentiss, this volume; Prentiss and Chatters 2003). Simultaneously, Paleoeskimos emerged in the Bering Strait and quickly spread across the Eastern Arctic (Prentiss and Lenert, this volume). The Harappan civilization rose and fell in the Indus valley (Kennoyer 1998) as did the complex towns of the Middle Jomon culture (Habu 2004). Radical temperature and moisture fluctuations of the Medieval Warm period impacted the rise and fall of hunter-gatherer villages in North America's Pacific Northwest (Prentiss, et al., 2007), Arctic (Mason, this volume), and southern California (Arnold 2001), the Norse settlement of the North Atlantic region (Dugmore, et al., 2007;

McGovern 1991), and the rise and fall of agricultural polities in North America's eastern and southwestern regions (e.g., Chatters, this volume).

The pattern of periodic macroevolutionary culture change is reminiscent enough of Vrba's (1985) "turnover pulse" hypothesis (see also Brett and Bairds' [1995] coordinated stasis hypothesis) for us to suggest that there may have been a cycle of periodic "turnovers" in world prehistory coordinated with climate swings much like those described by Vrba for biological species that substantially impacted the ultimate course of human history. If so, it implies that despite our ability to buffer change through cultural practices, we, as a species, have not been immune to major fluctuations in global climate. Indeed, the development of today's cultural institutions may have been heavily contingent on the effects of those short-lived periods.

Cultural Extinction and Cracked Niches

Human adaptation to specific niches is complex, and the abandonment of specific niches, such as with *cracked niches*, can be modeled via abrupt changes in subsistence behavior as manifested in niche expansion and contraction. Following Bar-Yosef (2003) and Byrd (2006), we argue that macroevolutionary changes in the late Epipaleolithic were linked to shifts in local and regional paleo-environmental conditions. This appears, moreover, to be traced to conditions with climatic downturns as well as improvements.

If cultural niches can operate across multiple habitats then it implies high potential for preservation of their associated RMS over time and space. A closely integrated relationship between the structure of human cultural activities and the configuration of subsistence resources virtually guarantees that stasis at the RMS level will characterize the long-term histories of cultures (e.g., Prentiss and Lenert, this volume). This does not mean that gradual evolution did not occur—just that its occurrence would be highly unlikely on this highest and most complexly integrated scale (Chatters, this volume). Indeed, a wide range of evolutionary changes occurred during the long spans associated with stasis at the higher levels. One means by which stability (and thus stasis) may have been maintained is through lower scale cultural evolution, as groups adapted elements of technologies to new resources, encouraged the development of new resources, or reorganized at smaller or larger demographic scales with associated shifts in social relations (e.g., Prentiss and Clarke 2008). In the Late Epipaleolithic, there is a general stability in lithic technology, even when significant economic, mobility, and social changes were emerging. While there are differences in the size of chipped stone tools, such as lunates, the general tools technology is more or less stable from the more sedentary early Natufians to the more mobile Late Natufian groups (Bar-Yosef 1998). There is, in sum, evidence of longer term stability within the context of lower scale techno-economic structures.

Another theoretical implication of this study is that constructed niches can eventually be broken potentially leading to extinctions in biological and cultural systems.

There is good reason to believe that in some cases culture would not offer an adequate buffer to niche cracking (see Laland, et al., 2000:145). Cultural niches logically could also encounter troubles potentially driving human populations to new areas with or without their original niches. If an old niche was in some sense transported with its users, then we might also see rapid perturbations to the new habitat as likely occurred during many Neolithic expansions (e.g., Kirch 1997).

Niche construction theorists argue that if organisms are adapted to constructed niches, then that will act as a buffer against the most severe aspects of natural selection, consequently giving rise to wider gene pools within a given population (Laland, et al., 2000). One by-product of niche cracking and dispersal of local populations could be sudden pulses of evolutionary change drawing from genetic diversity inherent in the dispersing population but also affecting unrelated species as the dispersing groups impact new habitat, thereby altering other selection pressures on local species (e.g., Vrba 1982). It is not currently clear whether a process like this could occur within human cultural contexts. Something like this kind of process is imagined by Rosenberg (1994), who suggests that crystallization of new cultural systems will best occur in the wake of cultural extinctions. On the other hand, others suggest major breakdowns do not necessarily lead to sudden expansions in the range of cultural patterns on a given landscape (e.g., Prentiss and Chatters 2003). One possibility is that some cultural diversity could emerge with the breakup of dense settlements containing subcultural diversity that leads to population dispersal but does not include total loss of the cultural niche. This also implies that some adaptations were focused on, and limited to, specific habitats that were constrained. This might help us understand why the desert adaptation of the Harifian died out, but other more generalized adaptations carried on.

Notes

1. There is a lack of consensus among Near Eastern Archaeologists in the definition and use of the terms collectors, foragers, or farming. With the exception of a few detailed analyses of late Epipaleolithic subsistence practices and regional economy (e.g., Munro 2004), researchers have largely focused on the process of plant and animal domestication end to approach this. There is, in short, a clear disjunction between ethnographic modeling of settlement and subsistence systems (e.g., Binford 1980) and archaeological data sets from the Epipaleolithic. In this chapter we draw upon the primary work of others (e.g., Bar-Yosef 2003; Byrd 2005) in our discussion of the coexistence of different economic and subsistence adaptations but also recognize that additional work is necessary to define the full range of economic adaptations for these periods.

Acknowledgments This study has been directly and indirectly supported by The National Endowment for the Humanities and the Department of Anthropology at the University of Notre Dame. Discussions with and critical editorial comments by G. Rollefson, T. Kohler, M. Chesson,

S. Shennan, and several anonymous reviewers have been crucial to the development of this essay. While not agreeing with some of the concepts and interpretations presented in this manuscript, the constructive criticism, willingness to share information, and patience of all of these individuals has immeasurably improved the clarity and organization of this work. In the end, however, they are not to be held accountable for the results.

References

Amsden, C. (1977). A Quantitative Analysis of Nunamiut Settlement Dynamics: 1898 to 1969. Ph.D. dissertation, Department of Anthropology, University of New Mexico.

Arnold, J.E. (ed.) (2001). *The Origins of a Pacific Coast Chiefdom: The Chumash of the Channel Islands*. Salt Lake City: University of Utah Press.

Bar-Yosef, O. (1998). On the Nature of Transitions: The Middle to Upper Paleolithic and the Neolithic Revolution. *Cambridge Archaeological Journal* 8, 141–163.

Bar-Yosef, O. (1998). The Natufian Culture in the Levant, Threshold to the Origins of Agriculture. *Evolutionary Anthropology* 6, 159–177.

Bar-Yosef, O. (2000). The Impact of Radiocarbon Dating on Old World Archaelology: Past Achievements and Future Expectations. *Radiocarbon* 42, 1-17.

Bar-Yosef, O. (2003). The Natufian. A Complex Society of Foragers. In B. Fitzhugh and J Habu (Eds.), *Beyond Foraging and Collecting. Evolutionary Change in Hunter -Gatherer Settlement Systems* (pp. 91–149). New York: Kluwer Academic/Plenum Publishers.

Belfer-Cohen, A. (1995). Rethinking Social Stratification in the Natufian Culture: The Evidence from Burials. In S. Campbell and A. Green (Eds.), *The Archaeology of Death in the Ancient Near East* (pp. 9–16). Oxford: Oxbow Monograph 51.

Belfer-Cohen, A. and Bar-Yosef, O. (2000). Early Sedentism in the Near East: A bumpy road to Village Life. In I. Kuijt (Ed.), *Life in Neolithic Farming Communities: Social Organization, Identity, and Differentiation* (pp. 19–37). New York: Kluwer Academic/Plenum Press.

Bettinger, R.L. and Baumhoff, M.A. (1982). The Numic Spread: Great Basin Cultures in Competition. *American Antiquity* 47, 485–503.

Binford, L.R. (1980). Willow Smoke and Dog's Tails: Hunter-Gatherer Settlement Systems and Archaeological Site Formation. *American Antiquity* 45, 4–20.

Boyd, R., Mulder, M.B., Durham, W.H. and Richerson, P.J. (1997). Cultural Phylogenies Possible? In P. Weingart, S.D. Mitchell, P.J. Richerson, and S. Maasen (Eds.), *Human by Nature: Between Biology and the Social Sciences* (pp. 355–384). Mahwah: Lawrence Erlbaum Associates, Publishers.

Boyd, R. and Richerson, P.J. (1985). *Culture and the Evolutionary Process*. Chicago: University of Chicago Press.

Brett, C.E. and Baird, G.C. (1995). Coordinated Stasis and Evolutionary Ecology of Silurian to Middle Devonian Faunas of the Appalachian Basin. In Erwin, D.H. and Anstey R.L. (Eds.), *New Approaches to Speciation in the Fossil Record* (pp. 285–317). New York: Columbia University Press.

Byrd, B. F. (1992). The dispersal of food production across the Levant. In Gebaur, A.B. and Price T.D. (Eds.), *Trancitions to Agriculture in Prehistory* (pp. 49-61). Madison: Prehistory Press.

Byrd, B. F. (2005). Reassessing the Emergence of Village Life in the Near East. *Journal of Archaeological Research* 13, 231–290.

Byrd, B. F., and Monahan, C. M. (1995). Death, Mortuary Ritual, and Natufian Social Structure. *Journal of Anthropological Archaeology* 14, 251–287.

Chatters, J.C. (1995). Population Growth, Climatic Cooling, and the Development of Collector Strategies On the Southern Plateau, Western North America. *Journal of World Prehistory* 9, 341–400.

Chatters, J.C., Hackenberger, S., Lenz, B., Prentiss, A.M., and Thomas, J-L. (2010). The Paleoindian to Archaic Transition in the Pacific Northwest: *In Situ* Development or Ethnic Replacement. In C.B. Bousman and B.J. Vierra (Eds.), *On the Brink: Transformations in Human Organization and Adaptation at the Pleistocene-Holocene Boundary in North America*. College Station: Texas A&M Press (In press).

Chatters, J.C. and Prentiss, W.C. (2005). A Darwinian Macro-Evolutionary Perspective on the Development of Hunter-Gatherer Systems in Northwestern North America. *World Archaeology* 37, 45–65.

Day, R.L., Laland, K.N., and Odling-Smee, J. (2003). Rethinking Adaptation: The Niche Construction Perspective. *Perspectives in Biology and Medicine* 46, 80–95.

Ditlevsen, P.D., Svenmark, H., and Johnsen, S. (1996). Contrasting Atmospheric and Climate Dynamics of the Last Glacial and Holocene Periods. *Nature* 379, 810–812.

Dugmore, A.J., Keller, C., and McGovern, T.H. (2007). Norse Greenland Settlement: Reflections on Climate Change, Trade, and the Contrasting fates of Human Settlements in the North Atlantic Islands. *Arctic Anthropology* 44, 12–36.

Eldredge, N. (1985). *Unfinished synthesis: Biological hierarchies and modern evolutionary thought*. New York: Oxford University Press.

Eldredge, N. and Gould, S.J. (1972). Punctuated Equilibria: An Alternative to Phyletic Gradualism. In T.J.M. Schopf (Ed.), *Models in Paleobiology* (pp. 82–115). San Francisco: Freeman, Cooper, and Company.

Fitzhugh, B. (2001). Risk and Invention in Human Technological Evolution. *Journal of Anthropological Archaeology* 20, 125–167.

Goring-Morris, A. N. (1987). *At the Edge: Terminal Pleistocene Hunter-Gatherers in the Negev and Sinai*. Oxford: British Archaeological Reports, International Series 361.

Goring-Morris, N. and Belfer-Cohen, A. (1997). The Articulation of Cultural Processes and Late Quaternary Environmental Changes in Cisjordan. *Paléorient* 23, 71–93.

Gould, S.J. (2002). *The Structure of Evolutionary Theory*. Cambridge: Belknap Press.

Grosman, L. (2003). Preserving Cultural Traditions in a Period of Instability: The Late Natufian of the Hilly Zone. *Current Anthropology* 44, 571–580.

Habu, J. (2004). *Ancient Jomon of Japan*. Cambridge: Cambridge University Press.

Haynes, G. (2001). *The Early Settlement of North America: The Clovis Era*. Cambridge: Cambridge University Press.

Henrich, J. and McElreath, R. (2003). The Evolution of Cultural Evolution. *Evolutionary Anthropology* 12, 123–135.

Henry, D.O. (1989). *From Foraging to Agriculture: The Levant at the End of the Ice Age*. Philadelphia: University of Pennsylvania Press.

Hitchcock, R.K. and Ebert, J.I. (1982). Foraging and Food Production among Kalahari Hunter/Gatherers. In J.D. Clark and S.A. Brandt (Eds.), *From Hunters to Farmers* (pp. 328–348). Berkeley: University of California Press.

Holden, C.J. and Shennan, S. (2005). Introduction to Part 1: How Tree-Like is Cultural Evolution? In R. Mace, C.J. Holden, and S. Shennan (Eds.), *The Evolution of Cultural Diversity: A Phylogenetic Approach* (pp. 11–29). London: UCL Press.

Jablonski, D. (1985). Background and Mass Extinctions: The Alteration of Macroevolutionary Regimes. *Science* 231, 129–133.

Janssen, M.A., Kohler, T.A. and Schefer, M. (2001). Sunk-Cost Effects and Vulnerability to Collapse in Ancient Societies. *Current Anthropology* 44, 722–727.

Kelly, R.L. and Todd, L.C. (1988). Coming into the Country: Early Paleoindian Hunting and Mobility. *American Antiquity* 53, 231–44.

Kennoyer, J.M. (1998). *Ancient Cities of the Indus Valley Civilization*. Karachi: Oxford University Press.

Kirch, P.V. (1997). Microcosmic Histories: Island perspectives on "Global" Change. *American Anthropologist* 99, 30–42.

Kirch, P.V. (2000). On *the Road of the Winds: An Archaeological History of the Pacific Islands Before European Contact*. Berkeley: University of California Press.

Kuijt, I. (1996). Negotiating Equality through Ritual: A Consideration of Late Natufian and Pre-Pottery Neolithic A Period Mortuary Practices. *Journal of Anthropological Archaeology* 15, 313–336.
Kuijt, I., and Goring-Morris, N. (2002). Foraging, Farming, and Social Complexity in the Pre-Pottery Neolithic of the Southern Levant: A Review and Synthesis. *Journal of World Prehistory* 16, 361–440.
Köhler-Rollefson, I., and Rollefson, G. (1990). The Impact of Neolithic Man on the Environment: The Case of 'Ain Ghazal, Jordan. In S. Bottema, G. Entjes-Nieborg and W. Van Zeist (Eds.), *Man's role in Shaping of the Eastern Mediterranean Landscape* (pp. 3–14). Rotterdam: A Balkema.
Laland, K.N., Odling-Smee, J., and Feldman, M.W. (1996). The Evolutionary Consequences of Niche Construction: A Theoretical Investigation Using Two-Locus Theory. *Journal of Evolutionary Biology* 9, 293–316.
Laland, K.N., Odling-Smee, J., and Feldman, M.W. (2000). Niche Construction, Biological Evolution and Cultural Change. *Behavioral and Brain Sciences* 23, 131–175
Laland, K.N., Odling-Smee, J., and Feldman, M.W. (2001). Cultural Niche Construction and Human Evolution. *Journal of Evolutionary Biology* 14, 22–33.
Lyman, R.L. and M.J. O'Brien (1998). The Goals of Evolutionary Archaeology: History and Explanation. *Current Anthropology* 39, 615–652.
Mason, O. (1998). The Contest Betweeen the Ipiutak, Old Bering Sea, and Birnirk Polities and the Origin of Whaling during the First Millenium A.D. along Bering Strait. *Journal of Anthropological Archaeology* 17, 240–325.
McGovern, T.H. (1991). Climate Correlation and Causation in Norse Greenland. *Arctic Anthropology* 28, 77–100.
Munro, N. (2004). Zooarchaeological Measure of Hunting Pressure and Occupation Intensity in the Natufian. *Current Anthropology* 45 (supplement issue), S7–S33.
Odling-Smee, F.J., Laland, K.N., and Feldman, M.W. (1996). Niche Construction. *American Naturalist* 147, 641–648.
Özdögan, M. (1998). Anatolia from the Last Glacial Maximum to the Holocene Climatic Optimum: Cultural Formations and the Impact of the Environmental Setting. *Paléorient* 23 , 25–38.
Prentiss, W.C. and Chatters, J.C. (2003). Cultural Diversification and Decimation in the Prehistoric Record. *Current Anthropology* 44, 33–58.
Prentiss, A.M. and Clarke, D.S. (2008). Lithic Technological Organization in an Evolutionary Framework: Examples from North America's Pacific Northwest Region. In W. Andrefsky (Ed.), *Lithic Technology: Measures of Production, Use and Curation* (pp. 257–285). Cambridge: Cambridge University Press.
Prentiss, A.M., Lyons, N., Harris, L.E., Burns, M.R.P., and Godin, T.M. (2007). The Emergence of Status Inequality in Intermediate Scale Societies: A Demographic and Socio-Economic History of the Keatley Creek Site, British Columbia. *Journal of Anthropological Archaeology* 26, 299–327.
Redman, C. (1998). *Human Impact on Ancient Environments*. Tucson: University of Arizona Press.
Richerson, P.J., Boyd, R. and Bettinger, R.L. (2001). Agriculture Impossible During the Pleistocene but Mandatory During the Holocene? A Climate Change Hypothesis. *American Antiquity* 66, 387–412.
Rindos, D. (1985). Darwinian Selection, Symbolic Variation, and the Evolution of Culture. *Current Anthropology* 26, 65–88.
Rollefson, G. O. (1996). The Neolithic Devolution: Ecological Impact and Cultural Compensation at 'Ain Ghazal, Jordan. In J. S. Seger (Ed.), *Retrieving the Past: Essays on Archaeological Research and Methodology in Honor of Gus. W. Van Beek* (pp. 219–229). Winona Lake: Eisenbrauns.
Rosenberg, M. (1990). The Mother of Invention: Evolutionary Theory, Territoriality, and the Origins of Agriculture. *American Anthropologist* 92, 399–415.

Rosenberg, M. (1994). Pattern, Process, and Hierarchy in the Evolution of Culture. *Journal of Anthropological Archaeology* 13, 307–340.
Shennan, S. (1993). Settlement and Society in Central Europe 3500–1500 B.C. *Journal of World Prehistory* 7: 121–161.
Simmons, A. (2000). Villages on the Edge: Regional Settlement Change and the End of the Levantine Pre-pottery Neolithic. In I. Kuijt (Ed.), *Life in Neolithic Farming Communities: Social Organization, Identity, and Differentiation* (pp. 211–230). New York: Kluwer/Plenum Press.
Simpson, G.G. (1944). *The Tempo and Mode in Evolution*. New York: Columbia University Press.
Tainter, J.A. (1985). *The Collapse of Complex Societies*. Cambridge: Cambridge University Press.
Tainter, J.A. (2006). Archaeology of Overshoot and Collapse. *Annual Review of Anthropology* 35, 4.1–4.16.
Vrba, E.S. (1982). Macroevolutionary Trends: New Perspectives on the Roles of Adaptation and Incidental Effect. *Science* 221, 387–389.
Vrba, E.S. (1985). Environment and Evolution: Alternative Causes of temporal Distribution of Evolutionary Events. *South African Journal of Science* 81, 229–236.
Wenke, R.J. (1991). The Evolution of Early Egyptian Civilization: Issues and Evidence. *Journal of World Prehistory* 5, 279–329.

Part IV
Macroevolutionary Theory in Archaeology

Chapter 10
Macroevolutionary Theory and Archaeology: Is There a Big Picture?

Robert L. Bettinger

Introduction

As the reader will have gathered from the contributions to this volume, the subject of macroevolutionary theory is complex, and when approaching complex subject matters, definitions are always a good place to start. The common understanding has macroevolution acting on large scales, above the level of the individual, over very long periods of time and many generations, as opposed to microevolution, which acts on small scales, at the level of the individual, over one to a few generations. For many biologists the underlying processes are the same, the outward difference between macro- and microevolution being purely an artifact of scale. Short-term microevolutionary change is relatively fast, but much of it is random, and therefore offsetting, over the long periods of macroevolution, at which scale change proceeds more slowly (Gingerich 1983). I take this to be the stance of those evolutionary archaeologists who portray individual human behavior, decision making, and innovation as generating random variation on which natural selection then acts, directing the course of evolution in the long run. Here small-scale events and processes are critical in generating the needed variation, but their individual outcomes are unpredictable and essentially random, depending so much on myriad historically contingent details. The force of evolution only becomes apparent when change is viewed at the macroscale, after natural selection has done its work (monkeys + typewriters + time = Romeo and Juliet).

Evolutionary human ecologists argue pretty much the opposite—that evolutionary forces are most clearly appreciated at the microscale and that natural selection is not directly involved. They cast individual human behavior and decision making as phenotypical variation that *short-circuits*, or substitutes for, natural selection, whose force is expressed indirectly, in the evolved human capacity for behavioral flexibility to cope with difficult problems and changing circumstances. Here, selection has favored the evolution of individuals with the capacity to change behavior in

R.L. Bettinger (✉)
Department of Anthropology, University of California, Davis, CA 95616, USA
e-mail: rlbettinger@ucdavis.edu

A.M. Prentiss et al. (eds.), *Macroevolution in Human Prehistory*,
DOI 10.1007/978-1-4419-0682-3_11, © Springer Science+Business Media, LLC 2009

accordance with changing circumstances, acting faster than natural selection can act but broadly in accord with its expectations. Because phenotypes are not directly heritable and so cannot evolve in the Darwinian sense, evolutionary human ecologists hesitate to label this individual level phenotypical change *microevolution*. Change in culturally transmitted traits (e.g., replacement of dogsleds by snowmobiles) and behaviors dependent on those traits (e.g., foraging itineraries), however, is clearly (micro) evolution, because transmission is involved. With that minor caveat, evolutionary human ecology is almost entirely given to microevolution as it is more strictly defined by neo-Darwinian theorists, referring to processes that take place within a species and fast enough to be observed during the career of an individual scientist or team of scientists pursuing a long-term research project, as contrasted with macroevolution, which refers to processes that occur at and above the species level over intervals far too long for any individual to witness. On this view, the only macroevolutionary events of interest to evolutionary human ecologists would be the speciation events leading to the emergence of anatomically modern Homo sapiens as a separate taxon. Everything since would be the product of microevolutionary processes that are mainly the result of human decision making, major components of which (e.g., human ingenuity) may not be evolving much at all.

There is much to be said for this tack. Against competing theories that offer equally compelling explanations of human behavior, evolutionary ecological accounts enjoy inherent advantages in parsimony, furnishing fairly simple, and eminently testable, predictions about individual behavior under equilibrium conditions. The microscale perspective of evolutionary human ecology rapidly loses this parsimony advantage, however, when asked to explain even moderately complex social arrangements and seems wholly overmatched when brought to bear on a handful of landmark cultural achievements (e.g., evolution of state-level societies) that profoundly changed the evolutionary trajectory of humankind. The question here is not whether microscale decision making and behavior change operate in these settings (they clearly do). It is whether these processes alone can account for the settings in which they operate—the big picture so to speak. This holds even when the big picture is simple hunter-gathers. Ability to predict hunter-gatherer prey choice, for example, requires that one know prey return rates, which are strongly dependent on available technology, the presence of which is not explained but treated as a given. Shifting the locus of decision making up one level, to technology, does not solve this problem; the foraging theory can predict when one technology will be selected over another (e.g., Bettinger et al. 2006) but not the presence of either to begin with. Here again the big picture (presence of the two technologies) is a given, either assumed or drawn from historical fact (e.g., the ethnographic or archaeological record). Evolutionary ecologists do not dispute that this big picture context is important, and that things like the rise of agriculture deserve special attention (e.g., Kennett and Winterhalder 2006), but their methodological individualism remains focused on microscale behavior at equilibrium, mostly leaving the big picture for others to sort out.

Evolutionary archaeologists seem more interested in the macroscale, where many think natural selection has sorted out the random chaos of microevolutionary variation. This can be done by tracking the replicative success of individual artifact forms, which they view as part of the extended human phenotype (e.g., Leonard and Jones 1987). They deny, however, that the big picture is shaped by distinctive macro-processes other than long-term evolutionary averaging, and perhaps occasionally punctuated equilibrium, or that it has an intelligible pattern or trajectory (e.g., progress). Accordingly, they reject special explanations for things like the rise of agriculture or state-level societies which in their view are not separable phenomena but parts of a larger evolutionary continuum. Here, singling out putatively "landmark" human achievements for special explanation is a fairyland exercise in teleological essentialism, where one assumes what happened in human prehistory had to happen and that when you look at human history these landmark events all make perfect sense.

Evolutionary archaeologists think it possible to account for specific phenomenon, say, the rise of the Zapotec state but argue that such accountings cannot possibly be parsimonious, because one must trace a very large number of very long and involved sequences of small-scale microevolutionary events connected mainly by historical chance. They think such cultural phylogenies are attainable using methods developed in genetics and follow geneticists in denying that individual pedigrees (e.g., of the Mesoamerican state) are useful in understanding others (e.g., of the Near Eastern state) beyond the point at which the two share a common ancestor. In granting the possibility of group selection in the special case of complex societies, evolutionary archaeologists do allow that there might be more to the big picture than just cultural genealogies, but in the end most of them side with evolutionary ecologists in arguing that evolutionary processes at work within groups, microevolutionary processes by definition, are vastly more important.

Discussion

The review above is certainly not comprehensive, arguably not even fair, because I am less interested in characterizing evolutionary archaeology and evolutionary human ecology than I am in defining macroevolution in a way not twisted to suit the theoretical predilections of either. What would be the point? Neither school makes enough room for the kind of macroevolution that I have in mind. In my version, macroevolution is not just the big scale sum of small-scale events. Nor is it recapitulated in small-scale events that indirectly express a big-scale force. In my view, cultural macroevolution requires a fundamentally different perspective. The unit of interest is frequently not the individual but a group of individuals, this owing to an evolutionary setting in which the selective differences between groups are greater than those between individuals within those groups, promoting group selection in stage-like leaps in a trajectory that is progressive in character (Richerson et al. 2005). If this sounds dangerously familiar, it should.

My version of cultural macroevolution entails evolutionary units, categories, and trajectories by and large compatible with the progressive social evolutionary framework that dominated anthropology and archaeology until the relatively recent rise of neo-Darwinian theory. The evolutionary accounts of Lubbock, Spencer, Tylor and Morgan, as well as plain old- fashioned culture histories (see Chatters this volume), were couched in terms of collectives (cultures, e.g., the Kwakiutl) that fell into a range of forms, types, or stages (e.g., chiefdoms) that could be ordered in a coherent sequence to represent a basic evolutionary trajectory (progress). Today, of course, all this is passé. The pioneers of neo-Darwinian anthropology made such easy work of progressive social evolutionary theory and traditional culture history that scholars now quite consciously avoid the big picture framework common to both, frequently for the worse. As in John Godfrey Saxe's poem about the blind men and the elephant, without the big picture, the details make no sense—or less than they should. It is remarkable, that this late twentieth and early twenty-first century chauvinism so lightly dismisses the insights of scholarly giants like Morgan. The problem with the progressive social evolutionists was not their vision of the big picture; if we define progress as increase in social complexity, or population size or density, then they got that basically right—surely Darwin thought so: "And as natural selection works solely by and for the good of each being, all corporeal and mental endowments will tend to progress toward perfection" (Darwin 1859: 489; cf. Ghiselin 1995: 1032), and so "the more recent forms must, on my theory, be higher than the more ancients; for each new species is formed by having had some advantage in the struggle for life over other and preceding forms" (Darwin 1859: 337; cf. Richards 2005). Where the progressive social evolutionists failed was explaining the big picture, figuring this was pretty much self-evident: Progress was its own explanation, an innate, perfection-seeking, evolutionary force (Bettinger 1991). While partly empirical and about the big picture (Leonard and Jones 1987), the neo-Darwinian dismantling of progressive social evolutionary theory was mainly logical, centering on this matter of explanation, showing how simple materialist forces like natural selection acting on individuals (or selection-shaped decision making by individuals) could explain important population-level consequences more cleanly and elegantly than a shadowy cosmic force like progress ever could. Neo-Darwinian anthropologists pushed this little picture perspective to its logical extreme. That selection on (or by) individuals could explain "many things" became "most things" and in the limit "everything," leaving no need and little room for the big picture explanation. Never mind that key contributors to the Modern Synthesis, notably Sewall Wright (Wright 1931) and George Gaylord Simpson (Simpson 1944), were convinced that many evolutionary problems required a big picture perspective *and* evolutionary forces other than just natural selection. The case for cultural macroevolution builds from the groundwork they developed using more recent models developed specifically to handle the special case of culture and cultural transmission (Boyd and Richerson 1985; 2005; Richerson and Boyd 2005).

Fitness Landscapes and Adaptive Peaks

Among the volume contributions, Spencer's view of macroevolution is perhaps closest to mine (Bettinger 1978; 1980: Fig. 5.2), because it is framed in terms of the fitness landscape—a concept first laid out by Sewall Wright (1931; 1932) in connection with his shifting balance theory. Spencer sketches the basic idea, but it is worth reviewing.

Wright envisioned evolution playing out on adaptive topographies (fitness landscapes) whose peaks represented "better" evolutionary solutions (higher fitness) and valleys represented "worse" evolutionary solutions (poorer fitness). Wright would later follow Simpson (1944) in casting these solutions as combinations of continuously varying phenotypical traits (e.g., beak length, neck length, stomach capacity, etc.). This makes intuitive sense: Square pegs go well with square holes, round pegs with round holes but not square pegs with round holes or the other way around. Square pegs and square holes define one peak, round pegs and round holes another, and mixed combinations the valley between them. In exactly the same way, and for the same reasons, the familiar contrast between r-selected and K-selected species implies two quite distinct adaptive peaks (small body size, short generation time, high fecundity vs. large body size, long generation time, low fecundity).

The landscape framework helps clarify many otherwise fuzzy macroevolutionary concepts the reader frequently encounters in this volume. Peaks, for example, may be thought of as *emergent properties* of the adaptive landscape, felicitous evolutionary solutions that only become evident (i.e., emerge) when specific individual traits take on just the right value in just the right combination. The trait combinations that define an individual peak thus constitute a discrete evolutionary recipe, blueprint, or as many here have termed it, *Bauplan*; a mountain range made up of many such peaks is an evolutionary *Baupläne*, a family of qualitatively distinct alternative solutions to the same basic problem, something akin to Clark's (1968) techno-complex. The fitness peak enforces the association between individual *Bauplan* traits, constraining them from varying independently, making it difficult to understand the expression of individual traits without reference to the adaptive complex they serve. For example, as Chatters (this volume) notes, mid-latitude hunter-gatherer subsistence-settlement systems are functionally integrated in a way that resists piecemeal change (e.g., a subsistence change may require social, technological, and settlement change) (Bettinger 1994: 54).

Wright argued that many landscapes would have multiple local optima (fitness peaks), some much higher than others. The problem was that natural selection acting alone would drive a newly introduced species to the nearest peak, where it would remain stranded; natural selection would prevent it from moving to adjacent higher peaks owing to the intervening low-fitness valleys. As in the old Maine saying, with natural selection alone, "You can't get there from here." These populations are something like blind mountain climbers, able to sense only whether their last steps moved them up or down. They can readily find nearby local peaks, but not necessarily the highest ones on the landscape, which they cannot see and would require moving down to get to the foot of a higher peak (e.g., Prentiss and Lenert, this volume). In Wright's shifting balance solution, mutation and genetic drift could move moderately sized, moderately isolated, sub-populations randomly away from the local peak. Gene flow would prevent many of these sub-populations from going extinct, allowing some to drift to the foot of another, higher, peak, at which point natural selection would again take over, moving that sub-population up that peak. The higher fitness of this sub-population would then lead to higher densities, out-migration, and thus higher than average interbreeding with, and genetic influence on, other sub-populations, moving the species as a whole to the higher fitness peak. Evolution is thus the result of this delicately shifting balance between natural selection, mutation, drift, and gene flow (Joshi 1999). Simpson's (1944) *quantum evolution* expands on this idea.

While the shifting balance theory itself remains controversial (e.g., Coyne et al. 1997, 2000; Goodnight and Wade 2000), the metaphor of the adaptive landscape remains enormously influential in evolutionary theory, as Spencer notes (cf., Eldredge 1989). The concept, of course is inherently macroevolutionary; the fitness landscape IS the big picture. Its configuration largely determines the character of evolutionary trajectories. If the landscape is smoothly contoured with a single peak, species will move more or less directly to that peak. Once there, mutation- and drift-like forces will shift it about randomly, but natural selection will keep it close to the peak. On the other hand, if the landscape is uneven and deeply fissured, with many local optima, a population introduced to the landscape will find the closest local peak and linger there until some combination of forces cause it to leap rapidly to another, higher peak. Then evolution will proceed discontinuously in a series of stages or steps (e.g., Price et al. 1993), from lower to higher peaks in a progressive trajectory.

From this perspective, microevolution is roughly what happens in relation to individual local peaks or optima, where natural selection dominates—the last phase of the shifting balance process. If the landscape overall has only one peak, macroevolution will be the large-scale result of these garden-variety microevolutionary processes; the population will reach evolutionary equilibrium (the peak) and then drift randomly in a pattern whose long-term (macroevolutionary) average will be close to the peak itself. If the landscape is fissured, however, there will also be occasional, rapid qualitative change, as the population moves from lower to higher peaks; macroevolutionary change will be stage-like and progressive—quite different from microevolutionary change. The view that human cultural evolution is gradual, continuous, and mainly the result of natural selection implies the first condition: a smooth, single-peaked macroevolutionary landscape. Unless this peak is quite gently sloped (meaning selection is weak), or changes its location frequently due to environment, equilibrium will be achieved quite rapidly. This does not seem to describe the empirical trajectory of cultural evolution.

As Kuijt and Prentiss (this volume) demonstrate, culture histories often show repeating sequences of stasis—change—stasis, suggesting the kind of peak-leaping that characterizes macroevolutionary change on complexly contoured adaptive landscapes. Granting that environmental change, by creating new niches, can produce similar stasis—change sequences, and was likely a major determinant of cultural evolution in the wildly unstable Pleistocene (Richerson et al. 2005), environmental change cannot explain the vastly greater culture change that has occurred since, during the relatively quiescent Holocene, with no evolutionary equilibrium yet in sight. Theory, too, suggests that peak-leaping ought to be the dominant mode of cultural evolution because humans are uniquely social, a propensity giving rise to culture evolutionary landscapes with many local optima.

The bulk of human behavior is acquired by social transmission and has payoffs that hinge on the result of social interaction. This is important because when the payoffs to individuals depend on the behavior of other individuals, even the simplest kinds of interaction are likely to have more than one feasible solution. Many such situations can be viewed as problems in coordination. Take driving. It matters less whether drivers in two-way traffic keep to the right-hand side or left-hand side of the road than that they all agree to do one or the other. As illustrated in Fig. 10.1, the payoff to driving on the right or left increases directly with the frequency of other drivers who do the same. In this case, payoffs are exactly the frequencies of the two respective strategies. When everyone stays to the right (frequency right = 1), the payoff for driving on the right is 1; and when no one stays to the right (frequency right = 0), the payoff for driving on the right is 0. Once either convention becomes common (frequency >.50), its payoff advantage will quickly make it universal.

The two peaks (all drive left, all drive right) on this two-dimensional landscape are of equal fitness, but many coordination problems, and particularly social interactions that involve reciprocity or cooperation, are likely to have many peaks of quite different fitness.

A classic example is delayed reciprocity, where individuals are presented the opportunity to provide a benefit ***b*** to a partner at some cost ***c*** to themselves, knowing that partner will be presented the opportunity to reciprocate in kind in the future. This differs from the driving problem, which is purely a matter of self-interest and therefore self-enforcing; it never pays to "fool" your partner by driving on the left when your partner expects you to drive on the right. With delayed reciprocity, however, cheating pays; cheaters get a one-time, cost-free benefit if they accept a benefit but do not reciprocate. If interactions are repeated, however, cheating becomes less

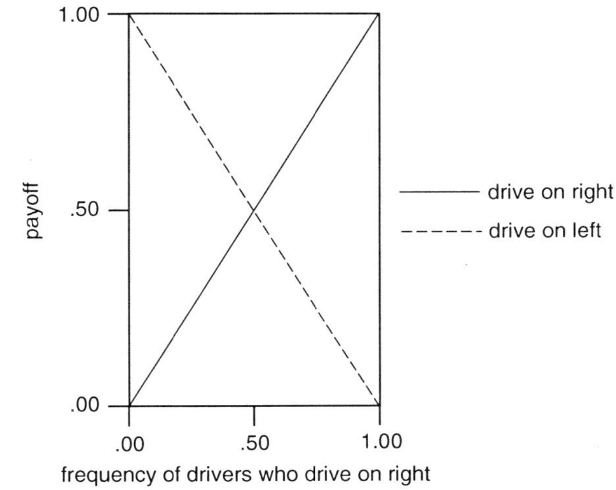

Fig. 10.1 Payoff structure for the driving game. Payoff to individuals adopting either strategy (drive left, drive right) is simply the frequency of others who do the same thing

advantageous. As Trivers (1971) showed, if individuals act as reciprocal altruists, meaning they provide the benefit initially but continue providing it only if partner reciprocates in kind, evolution will stabilize (make an adaptive peak of) "tit-for-tat" reciprocity when the long-term net benefit of reciprocating is greater than the one-time benefit of cheating (failing to reciprocate), i.e., $t(b - c) > b$, where b and c are defined as above, and t is the expected number of interactions. Thus, cheaters receive the one-time benefit b, and reciprocators the lesser net benefit of $b - c$, each time they play a full round; if they play enough rounds (i.e., if t is large enough), it will pay to reciprocate (assuming, of course, $b > c$). In the repeated game, tit-for-tat stabilizes cooperation by a form of punishment—the withholding of future benefits that cheaters would have otherwise enjoyed. This does not mean that reciprocal altruism is likely in the real world, however, merely that it is possible, i.e., it is a stable equilibrium.

An often missed point is that while tit-for-tat reciprocity is a stable equilibrium, it is not the only one on this evolutionary landscape. As it turns out, in the reciprocity game (and for that matter, in any repeated game in which individuals strongly influence each other's payoffs) any pattern of behavior can be stabilized by mutual self-interest (Boyd and Richerson 2005:135). Every schoolchild is familiar with the ultimatum: "You play by my rules or I'm taking my ball and going home." In such cases, mutual self-interest can produce beneficial cooperation, but it need not. As every schoolchild knows, wanting to play means often having to play by silly or even unfair rules. More generally, in what is known as the "folk theorem" (because many game theorists had proved it without publishing it, none of them wanting to take credit for what everyone already knew) (Fudenberg and Maskin 1986), if pairwise interactions are repeated enough times, tit-for-tat type punishment can permanently maintain *any* kind behavior, cooperative or not. If enough of one's potential playmates are unwilling to let you play unless you play by their rules, the only sensible thing to do is play by those rules. Once they become common enough, any set of rules, say X, will be impossible to dislodge, because no one will have anything to do with any individual that does other than X. As Boyd and Richerson (1992) put it, punishment permits the evolution of cooperation—and anything else! This has two important implications.

First, once a convention of social interaction becomes common, it will be in everyone's interest to advertise the norm, so all will know how to behave. This will save well-meaning but uninformed individuals the cost of being punished and potential punishers the cost of punishing. When a stranger enters a foreign village to ask for information, for example, it will pay to know how villagers expect to be addressed and the kind of questions that might offend them; at the very least, it will pay to know they are playing by different rules. Since virtually every kind of social interaction (e.g., greeting strangers, treatment of in-laws, treatment of the dead, etc.) has a very large number of stable solutions (i.e., equilibria or local adaptive peaks), separate groups will frequently settle on very different conventions. As McElreath argues (2003) (see also Boyd and Richerson 1987), when such groups live close enough that they regularly encounter each other, evolution will favor the development of marker traits that signal the local convention. It will also favor forms of social transmission in which both the marker and various behaviors integral to the coordination are transmitted as package (e.g., rule of driving on the right is bundled with cars with steering wheels on the left, etc.) (e.g., Bettinger and Eerkens 1999). In short, evolution will favor the development of clearly marked ethnic units, as Mason (this volume), Prentiss (this volume), and Chatters (this volume) argue. Put more plainly, ethnic groups are not figments of eighteenth-century preconception. Human ultra-sociality virtually guarantees them. On this view cultures can be thought of as pseudo-species (Bettinger 1994:53–54), roughly analogous to biological species, which implies their evolution and competition between them requires a macroevolutionary perspective on a par with that needed to understand species selection in biology.

The second implication is as important as the first. As just noted, the idea that social interactions define adaptive landscapes with multiple stable equilibria does not imply that all equilibria are of equal fitness; their payoffs are likely to be quite different. Because that is so, the force of natural selection is going to act more on the differences between groups than on the differences between individuals within groups, where coordination and punishment suppress individual variation. This, in turn, will favor groups with superior conventions, and especially conventions that cause individuals to engage in costly behaviors that generate group-level benefits. This is illustrated in Fig. 10.2 where individuals can either be selfish or cooperate, and payoffs are a function of coordination *and* a group-level benefit generated by cooperators.

As in Fig. 10.1, the coordination contribution to each payoff is just the strategy frequency. When everyone is selfish (frequency selfish = 1, frequency cooperators = 0), selfish individuals receive 1 and cooperators receive 0; and when everyone cooperates (frequency cooperators = 1, frequency selfish = 0), cooperators receive

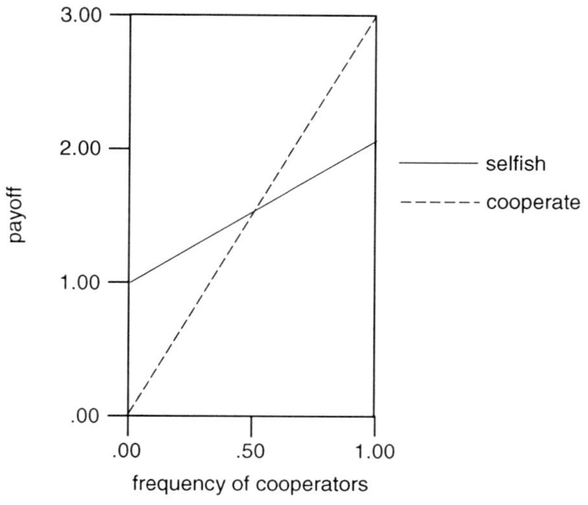

Fig. 10.2 Payoff structure for individuals in a simple cooperation game. Payoff for individuals following either strategy (selfish, cooperate) is the frequency of others doing the same thing *plus* one share of the pool of public good to which only cooperators contribute (2 units apiece)

1 and selfish individuals 0. In Fig. 10.2, however, all payoffs are augmented from a common pool to which only cooperators contribute, each individual cooperator adding 2 units to this group benefit pool. The group benefit is small when there are few cooperators, because it is split between all group members, most of whom are selfish, and increases directly in proportion the frequency of cooperators to a maximum of 2 (i.e., when everyone cooperates, each contributing 2 units to the common pool). As shown in Fig. 10.2, the group-level benefit causes both selfish and cooperator payoffs to increase with the frequency of cooperators, but the cooperator payoff rises more rapidly owing to coordination; when everyone cooperates the selfish payoff is entirely this group-level benefit. As in the driving game, the payoff structure will make either behavior (selfish, cooperate) stable once it becomes common enough (>0.50), but evolution will strongly favor groups that cooperate owing to higher payoffs (3 vs. 1). Note that while the group advantage depicted here is due to altruistic cooperation, the group selection effect modeled by McElreath et al. (2003) merely requires that competing forms of social coordination have different payoffs. For example, if existing technology or human physiology made "all drive right" a slightly better solution than "all drive left" (90% of humans *are* right-handed), selection would favor groups that settled on the "all drive right" solution over groups that settled on "all drive left."

Punishment deserves special attention in social interactions such as these because while the group effect does not require it, altruistic cooperation obviously confers major advantages to groups where it manages to evolve, and theory suggests that no kind cooperation is likely to be stable in the absence of punishment (Henrich and Boyd 2001:194). This prediction is strongly supported by empirical work indicating that punishment, including third-party punishment (where I punish you if you cheat someone else) is near-universal, its intensity varying directly with observed altruistic cooperation (Henrich et al. 2006; see also Ostrom 1990). It follows that much of cultural evolution is likely to center on innovations in punishment, e.g., making it cheaper. There is a good deal of room for innovation on this count, because punishment is costly to the punisher and is therefore a form of cooperation, which as just noted cannot evolve without punishment. Thus an evolutionary Catch 22. To be stable, cooperation requires people who punish non-cooperators. However, because punishing is costly to the punisher, people will not punish non-cooperators unless there are people who will punish them if they do not. And people will not punish non-punishers unless there are people who will punish them if they do not - and so on. Thus, natural selection alone will never favor the development of punishment because at the margin non-punishers always do better (Henrich and Boyd 2001), neither will payoff-biased social transmission (copy the successful) for the same reason (higher payoffs to non-punishers). However, if cooperation is common to begin with, even weak conformist transmission (copy the majority) can overcome the cost of punishment. That both conformist and payoff-biased transmission are involved makes sense because people weigh new behaviors using a variety of cues: how sensible it seems (direct bias), how many people are doing it (conformist transmission), the prestige of the people doing it (indirect bias), etc. In the case of punishment, when most individuals want to cooperate, failure to cooperate, and thus the need for punishment, will be mainly the result of rare mistakes: individuals who want, but inadvertently fail, to cooperate. Similarly, failure to punish these rare defectors will also be the result of rare mistakes: individuals who intend, but inadvertently fail, to punish rare non-cooperators. . .and so on. As a result, the need for, and cost of, punishment will decrease as one proceeds up the punishment chain. Under reasonable conditions, even small amounts of conformist social transmission will eventually override the cost of punishment and stabilize it somewhere up this chain, and this will stabilize punishment at lower levels. For example, if being punished is costly, once everyone punishes individuals who do not punish non-cooperators, everyone will punish non-cooperators because it will be too costly to do otherwise. And if everyone punishes non-cooperators, then everyone will cooperate, again because it will be too costly to do otherwise. Thus, once conformist transmission stabilizes punishment at any higher level, it will stabilize punishment at successively lower levels and eventually cooperation itself (Henrich and Boyd 2001). All that is needed is something that makes cooperation common to begin with—a mixture of punishment and conformist transmission will do the rest.

Simulation suggests that some stable solutions will consist of a mixture of individuals, a majority that cooperate under threat of punishment (reluctant cooperators) and a minority that cooperate and punish (cooperator-punishers). Boyd and Richerson (1992) jokingly call this the "big man" equilibrium—outwardly similar to social organizations observed in New Guinea—and speculate about chiefdom-like arrangements where there are reluctant cooperators and individuals who punish non-cooperators but do not cooperate themselves. In these cases, being punished is usually costly so that it pays to cooperate, but under quite plausible conditions punishment can stabilize cooperation *even when the cost of being punished is less than the cost of cooperating*, i.e., non-cooperation pays more. Groups that somehow manage to find these superior solutions will enjoy selective advantages over groups where punishment (and thus the cost of punishing) is more costly and eventually replace them. Many of Rosenberg's speculations about the origin of social stratification and complexity, and particularly his linking of it to peaceful, within-group conflict resolution, are consistent with these models. Conflicts between individuals are often resolved with threat of punishment (i.e., settlement is a form of cooperation), and Rosenberg's peaceful resolution scenario implies that the cost of punishing has been minimized so that disputants settle even when the cost of settling is greater than the cost of being punished in the event they do not.

History Matters

Unfortunately, while evolutionary theory can tell us much about the relative worth of different cultural configurations (whether they are stable, whether they can resist invasion by other configurations, etc.), it provides few reliable clues about shifts from one adaptive peak to another. Again, this is because under normal circumstances, "You can't get there from here." For example, where selfish behavior is common, coordination and related forces will prevent group beneficial behavior from evolving unless some chance event intervenes to make it common instead. Any number of historical circumstances might do this, abrupt climate change, for example (Price et al. 1993). Suppose that selection has favored the transmission of two behaviors, X and Y, as a package so that when X changes, Y changes with it, and conversely, as is common in cultural transmission. Further, suppose that climate strongly affects the values of X but not the values of Y, which show two contrasting optima, one that is group beneficial, the other not (the selfish default state), giving an adaptive landscape with two adaptive peaks. Normally, climate change acting on X that pulls Y off its selfish peak as a consequence of the XY correlation will be countered by natural selection acting on Y, pulling both of them back, again as a consequence of the XY correlation. However, if the XY correlation is strong enough and climate change is severe enough, X can drag Y from its selfish to its group-beneficial peak. The more fitness declines as Y passes between peaks, the more likely Y will linger at the first until a severe climate change acting on X drags it rapidly and abruptly to the other (Price et al. 1993). Here the peak shift from selfish to group beneficial behavior is a historical accident triggered by climate change. Y does not respond to climate—it responds only to changes in X due to the XY correlation, which has nothing to do with climate.

The evolutionary archaeologists, then, are certainly right; evolutionary change that entails peak shifting is going to have this serendipitous, historical character. The most revolutionary change in Great Basin plant procurement, for example, seems likely the result of an entirely unrelated change in hunting technology—the introduction of the bow (Bettinger and Eerkens 1999). The much less accurate atlatl had formerly required widespread sharing in relatively large social units. This reduced subsistence risk but made freeloading common enough to discourage individual investment in costly activities, including intensive plant procurement and storage. When bow technology made smaller, more closely related social units possible, the freeloading problem was eliminated, and plant procurement rapidly intensified. On this view, there is nothing about plant procurement *per se* that can explain plant intensification and storage—its explanation lies elsewhere, in social relationships that affect, but are not determined by, plant procurement or the obvious advantages of intensive plant use and storage in the Great Basin environment.

Several contributors to this volume grapple with this difficult issue of peak shifting. Rosenberg argues at length about the emergence of social stratification and its likely connection with intra-group conflict. Spencer wonders what circumstances might have motivated Zapotec leaders to centralize settlement while diversifying control at upper levels of management—the innovation that resulted in the first Mesoamerican state. Archaeologists accustomed to simple cost-benefit explanations (e.g., prey choice, inclusive fitness), and particularly processualists who favor broad stroke forces (e.g., population growth and environment), are bound to find such accounts unwieldy precisely because they are historical and therefore do not necessarily generalize from one case to another. Yet I disagree with the contention that it is pointless to seek and test various hypotheses about these historical events, however difficult that might be. Spencer's speculations about Zapotec settlement change and Rosenberg's about social conflict and stratification are testable—at least they have eminently testable implications.

Finally, there are issues of degree. *Serendipitous* does not mean *random*. A great many scientific discoveries are serendipitous—experimenters finding things they did not expect to find—but the context of such discoveries is hardly random. Trial and error human learning works much the same way. A flint knapper who, while in the process of refining a traditional point shape to improve his hunting success, stumbles upon a fundamentally new and better point shape, is not acting randomly in the usual sense. Merely on that count, human intention cannot be ignored as a macroevolutionary force (Rosenberg 1990).

Macroevolution and Cultural Transmission

Multi-peaked adaptive landscapes, social coordination, and group selection cannot be all there is to cultural macroevolution. They are probably enough to explain some basics, for example, why the worth of specific behaviors will often be missed if not viewed in relation to the coordinated patterns they serve, why much human behavior cannot be explained with reference to individual self-interest, and why some culture change is abrupt and progressive rather than random and gradual. In the last analysis, however, group selection and adaptive landscapes fail to account for what is arguably the dominant feature of macroevolutionary culture change: the quickening pace with which ever more complex and successful forms of social organization have appeared since the advent of the Holocene. Selection acting on groups is not fast enough to explain this.

In the most closely modeled case of cultural group selection (Soltis et al. 1995), a stable, group-beneficial arrangement arises due to historical circumstance and is maintained by a combination of payoffs, social coordination and conformist transmission, as outlined above. It then replaces less group-beneficial arrangements at a rate determined by its relative fitness advantage. Cultural group selection is faster than genetic group selection, because only groups go extinct (fission); their members join other groups, adopting their conventions. Even so, ethnographic data from New Guinea suggest that among simple horticulturalists it will take something like 1000 years for the group-beneficial arrangement to replace lesser arrangements completely (Soltis et al. 1995) marshal. Some cultural change is that slow, of course, but a good deal of it is much faster, as in the rapid appearance of secondary states on the heels of the initial development of Spencer's Zapotecan state. Neither does simple cultural group selection account for the quickening pace of cultural evolution—the evident tendency for more complex cultural systems to develop relatively quickly following the appearance of key behaviors in simpler systems. That complex systems more likely developed by recombining behaviors already present in simpler systems rather than simply from scratch seems clear (e.g., Zeder, this volume) but does not explain why these component behaviors seem to recombine in less time than they took to develop in the first place. For group beneficial arrangements to spread rapidly and rapidly recombine into more complex social systems requires the kind of social transmission modeled by Boyd and Richerson (2002; Henrich 2004).

Boyd and Richerson assume a region with different groups whose local composition determines the payoffs of selfish and group beneficial behaviors (see Fig. 10.2). Both traits are stable when common. That is, coordination and related forces guarantee that when the frequency of cooperation is below some critical value, being selfish pays better than being cooperative, and that above that being cooperative pays better than being selfish. By definition, however, payoffs are higher in groups with more cooperators than groups with fewer cooperators. The model assumes that cultural transmission is payoff-biased; individuals differentially imitate other individuals with higher payoffs. Individuals mostly encounter and imitate members of their own group, but have some chance of meeting, and thus of imitating, members of other groups.

In the default condition, selfish behavior is common in all groups, so payoffs are uniformly low. Strangers are sometimes imitated, but since mean payoffs are the everywhere the same, these chance events roughly balance out, so neighboring groups imitate each other at about the same rate. However, if something pushes the frequency of cooperation above its critical value in just one group, the frequency-dependent payoff structure will reward individuals who cooperate in that group and make cooperation locally stable. Once that happens, the mean payoffs of individuals in this one cooperator group will be higher than payoffs of individuals in the other groups, which do not cooperate. Payoff-biased transmission will then increase the frequency with which members of the cooperator group are imitated when encountered as strangers by members of neighboring non-cooperator groups, where payoffs are lower. The force of this differential imitation can then increase the group beneficial behavior above its tipping point and stabilize it in the nearest neighboring group of non-cooperators and eventually in the region as a whole. Since the trajectory here is driven by social transmission rather than selection, this can happen very rapidly, approximately as fast as an individually beneficial trait can spread within a local group. This kind of rapid cultural transmission seems necessary to account for the spread of state-level organization in Mesoamerica (Spencer), of complex adaptive strategies (or more awkwardly, "resource management strategies," which amount to the same thing) in the Northwest (Prentiss), social stratification (Rosenberg), and perhaps the establishment of Arctic *Baupläne* (Mason). An extension of this process leads to the prediction that existing simpler arrangements will recombine into new more complex ones faster than they evolved in the first place.

Suppose there are two traits representing cooperation in different areas (e.g., subsistence and warfare), and individuals acquire these traits separately by the kind of payoff-biased imitation just described. As before, each cooperative trait separately produces higher payoffs than its selfish alternative in groups where its frequency is above a given tipping point and lower payoffs when its frequency is below that. It follows that when both cooperative traits are common, individuals with both traits will receive the highest payoff; conversely, they will be doubly penalized when selfish behavior is common, making it unlikely that chance events will make both cooperation in subsistence and cooperation in warfare common simultaneously in any single group. It is much more likely that cooperation in subsistence and cooperation in warfare will develop and then spread independently by the simple process of differential imitation just described. It is the spreading of these two fronts that creates conditions promoting rapid recombination. Where they meet, payoff-biased imitation will generate a boundary group in which cooperation in subsistence and cooperation in warfare are both relatively common. In the simplest case, half the population (0.50) will cooperate in subsistence but not warfare, the other half (0.50) will cooperate in warfare but not in subsistence. What happens now depends on the tipping point, i.e., the frequency at which cooperation is common enough to be favored. If the tipping point is less than or equal to one-half ($\leq$0.50), individuals that manage to acquire both cooperative traits will receive higher payoffs than individuals with just one or neither; thus, two- trait cooperation will become stable in this boundary group and subsequently spread to other groups by payoff-biased imitation. If the tipping point is greater than one-half (>0.50), individuals with both cooperative traits will suffer only a slight disadvantage relative to individuals with other combinations because each cooperative trait is relatively common (0.50). Under these circumstances only a small chance fluctuation in trait frequency will be enough make the two-trait cooperation payoff superior. Indeed, the force needed to make both group beneficial traits common enough to recombine in a boundary group will be less than the force that was needed to push the frequency of either trait past its tipping point initially, i.e., make it common enough to be viable in the first place. This is simply because both cooperative traits are more common in the boundary group (0.50), than either was initially (0.00). Thus, on average it will take less time for group beneficial traits to recombine than to evolve. To the extent that cultural evolution is driven by group beneficial social arrangements (and coordinated social arrangements with similar properties), more complex organizations will evolve by recombination with increasing ease. The valleys between adjacent peaks will become shallower, facilitating leaps from less to more optimal adaptive peaks. As more groups cooperate, group size and population density will grow, increasing the likelihood that cultural transmission will be able to maintain complex cultural arrangements once they evolve (Henrich 2004).

The more complex *Baupläne* that are the subject of so many contributions to this volume must have developed by this recombining of existing arrangements. Niche construction (Kuijt and Prentiss) and landscape learning (Rockman) have this quality. The kind of niche shattering discussed by Kuijt and Prentiss might well increase chances for recombination through constant recycling; niches and cultural landscapes are built up by recombination, shattered by externalities including climate deterioration, and recombined in new and different ways when climate ameliorates. A process like recombination would seem to explain why Childe's Neolithic revolution took so long if we mark it as starting with the development of its earliest individual components, each its own developmental history, as described by Zeder. Recall here that traits are more likely to recombine than they are to develop initially and that the development of individual traits proceeds independently: Traits that eventually recombine may well develop at different times for different reasons, some much earlier than others. Perhaps individual Neolithic components (e.g., ceramics) produced viable payoffs only when coupled with certain beneficial social arrangements, the Neolithic whole being built up from these. Alternatively, components may have persisted simply by virtue of chance bundling with beneficial social arrangements. Behaviors that seem to arise from whole cloth without these developmental histories, the Natufian sedentism discussed by Kuijt and Prentiss, for example, may be the consequences of recombination—beneficial behavioral innovations made possible or necessary as a consequence of a novel juxtaposition of preexisting elements.

Implications

I intended to conclude with a powerful argument for cultural macroevolution but elected not to. By now most readers have pretty much decided where they stand. If you have come along for the ride this far, you are probably already convinced that cultural macroevolution is worth thinking about. If you are not, I am unlikely to make much of a dent in the short space remaining; the best I can do is thank you for coming and let you off here. For the rest of us, the problem is charting an agenda for future macroevolutionary archaeology.

The success of different strains of microevolutionary archaeology is instructive. These programs matured by first linking evolutionary theory to just a few models with testable implications and then accumulating whole batteries of case studies to show exactly how these models could be applied to the archaeological record—the exemplars of Kuhn (1970) and Giere (1988). Evolutionary ecologists operationalized fitness via optimal foraging theory (Winterhalder 1981), especially the diet breadth model and developed a variety of ways to detect its archaeological signatures (Grayson 1991; Grayson and Delpeche 1998). Evolutionary archaeologists did the same, first via the style-function dichotomy (Dunnell 1978, 1980; Neiman 1995) and more recently via cladistics and cultural phylogeny (Darwent and O'Brien 2006; Lipo et al. 2006). A workable agenda for macroevolutionary archaeology will have the same components—theory, models, and case studies. The contributions to this volume add to all three of these areas, even if as a whole they understandably fail to display the conceptual coherence that signifies scientific maturity. For that to happen, macroevolutionary archaeologists will have to settle on a few "pet" models that articulate key evolutionary processes, are easy to understand, and have archaeologically testable implications. Sewall Wright's (1932) landscape model ought to be one of these. It is widely accepted by evolutionary theorists and the contributions to this volume by Rosenberg, Prentiss and Lenert, Chatters, Rockman, Prentiss, and most notably Spencer, have shown its potential for application to the archaeological record. I suspect that most, if not all, of the remaining contributions would have benefited from the interpretive perspective of Wright's model. The key test implication of the landscape model is "jerky" evolution, the stasis-change-stasis sequence of Kuijt and Prentiss: Change will not proceed smoothly and may entail wholesale cultural reorganization.

The landscape model is only part of the puzzle. Because cultural macroevolution acts at levels above the individual, some of its models will have to grapple with social processes and cultural transmission in particular. The models I have outlined above are obvious candidates. The good thing about the culture transmission models is that their statistical signatures should be detectable in the archaeological record. McElreath's (2003) (see also Boyd and Richerson 1987) coordination model, for example, implies that correlation between individual traits and trait variables should vary directly with social contact, being highest at the points of contact or boundaries between adjacent social units; this is ripe for testing with a variety of archaeological media, ceramics for example (actually much of the New Archaeology work on ceramics is premised on much the same idea (Deetz 1965; Whallon 1968). McElreath's model, of course, deals with traits that are governed by culturally biased transmission, a proposition that is also testable. Henrich (2001) has shown that culturally biased transmission produces distinctly s-shaped (logistical) historical trajectories. This is mainly because the force of culturally biased transmission hinges on variation in the population. New behaviors spread slowly at first (because there are few people with the behavior to copy = low variance), then faster (because there lots of people who have the behavior and lots of people who do not = high variance), and finally slower at the end (because there are relatively few people who do not have the behavior = low variance). In short, the behavior will grow at an increasing, then decreasing rate, the switch (inflection point) occurring when the frequency of the new behavior is exactly 0.50—when half the population has the trait and half do not (Fig. 10.3 right). This is in contrast to behaviors that increase as the result of individual learning, which display r-shaped trajectories that are fastest at the start and monotonically decrease with time (Fig. 10.3 left), the so-called "steep

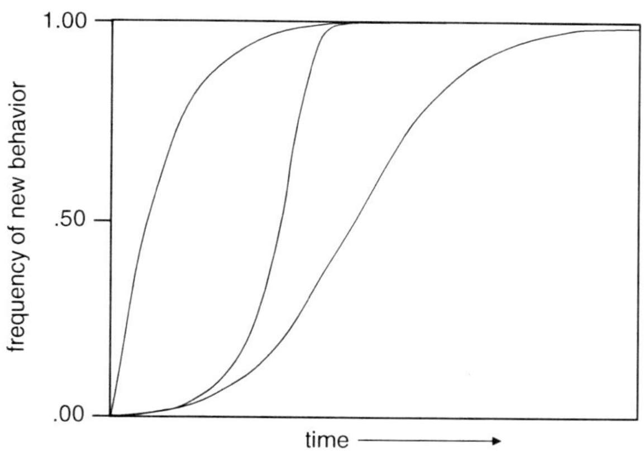

Fig. 10.3 Trajectory of behavioral change under different processes. Individual learning produces the r-shaped trajectory on the *left*. Payoff-biased social transmission with constant payoffs produces the symmetrical s-shaped trajectory on the *right*. Where payoffs change as the result of coordination and group beneficial effects, payoff-biased social transmission produces the asymmetrical s-shaped curve between the two

learning curve," sometimes misread as implying that learning a new task is initially slow when in fact it is fast. If payoffs are constant, traits spread by payoff-biased social transmission will display symmetrical s-trajectories—the shape of the curve to the left of the inflection point will mirror the shape of the curve to its right. However, if payoffs change as a consequence of coordination (McElreath et al. 2003) or group beneficial effects (Boyd and Richerson), payoff-biased social transmission will display distorted, nonsymmetrical s-trajectories because as the spreading behavior becomes more common within each group, the local payoff structure shifts to favor it, ever-hastening its adoption. This will increase the frequency at which its rate of increase shifts from positive to negative (i.e., the inflection point), compressing the s-trajectory into one that is more j-shaped (Fig. 10.3 center). That conformist (copy the majority) transmission also produces long left-tailed j-trajectories (Henrich 2001) adds a degree of complication but in the final analysis amounts to the same thing because conformist transmission has essentially the same implications for group selection as coordination, both implying that the differences between groups will be greater than differences within them. In any event, the point remains that cultural transmission (or dual inheritance) theory provides the macroevolutionary archaeologist an array of simple models that are eminently testable with the kinds of data that archaeologists routinely acquire, and recent advancements in precision radiocarbon dating (e.g., Waters and Stafford 2007) make it possible to discriminate between alternative hypotheses about the processes driving evolutionary change on the basis of their temporal trajectories.

In addition, somewhere down the line macroevolutionary archaeology will need to settle on some analytical conventions that streamline the casting and testing of hypotheses. No research program can advance without polite agreement about appropriate shortcuts that reduce the chaos of the empirical world to a manageable caricature that captures processes of interest; at minimum, there has to be agreement about what should be measured and how it should be measured. Optimal foraging theorists, for example, have done this in choosing to measure subsistence benefit in calories (kcals) when they know full well that subsistence is not merely a matter of calories, that things like vitamins and trace minerals also matter. Since evolutionary analysis almost always comes down to change, for starters it seems essential to adopt a standard measure of change in dimensionless units, allowing for comparison across widely varying phenomena. My colleague Pete Richerson has always argued that archaeologists would benefit from measuring change in Darwins, d, the unit proposed by Haldane (1949), $d = (\ln X_2 - \ln X_1)/y$, where X_1 is the initial value of the trait being measured, X_2 is its final value, and y is time in millions of years. Darwins can be measured in any unit, units canceling out during calculation. One Darwin is equivalent to an e-fold (approximately 2.718) change in 1 million years. While obviously designed to measure change over very long periods, the Darwin is perfectly applicable to change over much shorter ones.

Lastly there is that issue of parsimony. Evolutionary theory allows a place for both individual and group selection, but individual selection entails fewer assumptions. Doesn't it follow that the more parsimonious individual selection hypothesis ought to be tested and rejected before entertaining the less parsimonious group selection hypothesis? An interesting question, but since most of the time the two can be tested simultaneously, the issue really amounts to this: When individual and group selection hypotheses provide equally compelling accounts of a given cultural behavior, shouldn't the more parsimonious individual selection hypothesis be accepted? The answer naturally lies in what one takes to be the prior probabilities of individual and group selection, i.e., the likelihood that one is more common, therefore more likely beforehand. This issue slices neatly the difference between microevolutionary and macroevolutionary archaeologists. If one sees human evolution as an undifferentiated segment of the broader picture of biological evolution, where group/species selection is comparatively rare, then individual selection will always get the nod. If, on the other hand, one sees human culture as a revolutionary development on a par with the rise of multicellularity, one that made humans an essentially new grade of organism (Maynard Smith and Szathmary 1995), genetically predisposed to cooperate (Richerson et al. 2003:263–264), then group selection and the macroevolutionary processes it feeds should perhaps enjoy the benefit of the doubt. The time has long passed when group selection explanations could be dismissed out of hand—there is too much empirical evidence (think Stockholm syndrome), supporting theory, and models. The task for macroevolutionary archaeology is to craft these pieces into a workable program of research that tackles the big problems that have interested anthropologists from the beginning.

References

Bettinger, R. L. (1978). Alternative adaptive strategies in the prehistoric Great Basin. *Journal Anthropological Research* 34(1), 27–46.

Bettinger, R. L. (1980). Explanatory-predictive models of hunter-gatherer behavior. *Advances in Archaeological Theory and Method* 3 (pp. 189–255). New York: Academic Press.

Bettinger, R. L. (1991). *Hunter-gatherers: archaeological and evolutionary theory*. New York: Plenum Press.

Bettinger, R. L. (1994). How, when, and why Numic spread. In Madsen, D. B. and Rhode, D. R. (eds.), *Across the West: Human Population Movement and the Expansion of the Numa* (pp. 44–55). Salt Lake City: University of Utah Press.

Bettinger, R. L. (1999). What happened in the Medithermal? In Beck, C. (ed.), *Models for the Millennium: Great Basin Anthropology Today* (pp. 62–74). Salt Lake City: University of Utah Press.

Bettinger, R. L. and Eerkens, J. (1999). Point typologies, cultural transmission, and the spread of bow and arrow technology in the prehistoric Great Basin. *American Antiquity* 64(2), 231–242.

Bettinger, R. L., Winterhalder, B. and McElreath, R. (2006). A simple model of technological intensification. *Journal of Archaeological Science* 33(4), 538–545.

Boyd, R. and Richerson, P. J. (1985). *Culture and the Evolutionary Process*. Chicago: University of Chicago Press.

Boyd, R. and Richerson, P. J. (1987). The evolution of ethnic markers. *Cultural Anthropology* 2, 65–79.

Boyd, R. and Richerson, P. J. (1992). Punishment allows the evolution of cooperation (or anything else) in sizable groups. *Ethology and Sociobiology* 13, 171–195.

Boyd, R. and Richerson, P. J. (2002). Group beneficial norms can spread rapidly in a structured population. *Journal of Theoretical Biology* 215, 287–296.

Boyd, R. and Richerson, P. J. (2005). *The Origin and Evolution of Cultures*. Evolution and Cognition. Oxford: Oxford University Press.

Clark, D. L. (1968). *Analytical Archaeology*. London: Methuen.

Coyne, J. A., Barton, N. H. and Turelli, M. (1997). Perspective: a critique of Sewall Wright's shifting balance theory of evolution. *Evolution* 51(3), 643–671.
Coyne, J. A., Barton, N. H. and Turelli, M. (2000). Is Wright's shifting balance process important in evolution? *Evolution* 54(1), 306–317.
Darwent, J. and O'Brien, M. J. (2006). Using cladistics to construct lineages of projectile points from northeastern Missouri. In Lipo, C. P., O'Brien, M. J., Collard, M. and Shannon, S. J. (eds.), *Mapping Our Ancestors: Phylogenetic Approaches in Anthropology and Prehistory* (pp. 185–208). New York: Aldine.
Darwin, C. (1859). *On the Origin of Species by Means of Natural Selection, or the Preservation of Favoured Races in the Struggle for Life* (1st ed). London: J. Murray.
Deetz, J. (1965). *The Dynamics of Stylistic Change in Arikara Ceramics*. Illinois Studies in Anthropology 4. Urbana: University of Illinois Press
Dunnell, R. (1978). Style and function: a fundamental dichotomy. *American Antiquity* 43, 192–202.
Dunnell, R. (1980). Evolutionary theory and archaeology. *Advances in Archaeological Method and Theory* 3, 35–99. New York: Academic Press.
Eldredge, N. (1989). *Macroevolutionary Dynamics: Species, Niches, and Adaptive Peaks*. New York: McGraw-Hill.
Fudenberg, D. and Maskin, E. (1986). The Folk Theorem in repeated games with discounting or with incomplete information. *Econometrica* 54, 533–556.
Ghiselin, M. T. (1995). Perspective: Darwin, progress, and economic principles. *Evolution* 49(6), 1029–1037.
Giere, R. N. (1988). *Explaining Science: A Cognitive Approach*. Chicago: University of Chicago.
Gingerich, P. D. (1983). Rates of evolution: effects of time and temporal scaling. *Science* 222(4620), 159–161.
Goodnight, C. J. and Wade, M. J. (2000). The ongoing synthesis: a reply to Coyne, Barton, and Turelli. *Evolution* 54(1), 317–324.
Grayson, D. K. (1991). Alpine faunas from the White Mountains, California: adaptive change in the late prehistoric Great Basin? *Journal of Archaeological Science* 18, 483–506.
Grayson, D. K. and Delpeche, F. (1998). Changing diet breadth in the early Upper Paleolithic of southwestern France. *Journal of Archaeological Science* 25, 1119–1129.
Haldane, J. B. S. (1949). Suggestions as to quantitative measurement of rates of evolution. *Evolution* 3, 51–56.
Henrich, J. (2001). Cultural transmission and the diffusion of innovations: adoption dynamics indicate that biased cultural transmission is the predominate force in behavioral change. *American Anthropologist* 103(4), 992–1013.
Henrich, J. (2004). Demography and cultural evolution: how adaptive cultural processes can produce maladaptive losses – the Tasmanian case. *American Antiquity* 69(2), 197–214.
Henrich, J. and Boyd, R. (2001). Why people punish defectors: weak conformist transmission can stabilize costly enforcement of norms in cooperative dilemmas. *Journal of Theoretical Biology* 208, 79–89.
Henrich, J., McElreath, R., Barr, A., Ensminger, J., Barrett, C., Bolyanatz, A., Cardenas, J. C., Gurvey, M., Gwako, E., Henrich, N., Lesorogol, C., Marlowe, F., Tracer, D. and Ziker, J. (2006). Costly punishment across human societies. *Science* 312(23), 1767–1770.
Joshi, A. (1999). The shifting balance theory of evolution. *Resonance* 4 December, 66–75.
Kennett, D. J. and Winterhalder, B., Eds. (2006). *Behavioral Ecology and the Transition to Agriculture*. Berkeley: University of California Press.
Kuhn, T. (1970). *Structure of Scientific Revolutions*. 2nd ed. Chicago: University of Chicago Press.
Leonard, R. D. and Jones, G. T. (1987). Elements of an inclusive evolutionary model for archaeology. *Journal of Anthropological Archaeology* 6, 199–219.
Lipo, C. P., O'Brien, M. J., Collard, M. and Shennan, S. J., (eds.) (2006). *Mapping our ancestors: phylogenetic approaches in anthropology and prehistory*. New York: Aldine.
Maynard Smith, J. and Szathmary, E. (1995). *The Major Transitions in Revolution*. Oxford: W.H. Freeman/Spectrum.

McElreath, R., Boyd, R. and Richerson, P. (2003). Shared norms and the evolution of ethnic markers. *Current Anthropology* 44(1), 122–129.

Neiman, F. D. (1995). Stylistic variation in evolutionary perspective: inferences from decorative diversity and interassemblage distance in Illinois Woodland ceramic assemblages. *American Antiquity* 60, 7–36.

Ostrom, E. (1990). *Governing the Commons: The Evolution of Institutions for Collective Action*. Cambridge: Cambridge University Press.

Price, T., Turelli, M. and Slatkin, M. (1993). Peak shifts produced by correlated response to selection. *Evolution* 47(1), 280–290.

Richards, R. J. (2005). Darwin and Progress. *New York Review of Books* 52 (20).

Richerson, P. J., Bettinger, R. L. and Boyd, R. (2005). Evolution on a restless planet: were environmental variability and environmental change major drivers in human evolution. In Wuketits, F. M. and Ayala F. J. (eds.), *Handbook of Evolution,* Vol. 2: Evolution of Living Systems (Including Hominids) (pp. 223–242). Weinheim: Wiley-VCH Verlag GmbH & Co KGaA.

Richerson, P. J. and Boyd, R. (2005). *Not by Genes Alone: How Culture Transformed Human Evolution*. Chicago: University of Chicago Press.

Richerson, P. J., Boyd, R. and Henrich, J. (2003). The cultural evolution of human cooperation. In Hammerstein, P. (ed.), *The Genetic and Cultural Evolution of Cooperation* (pp. 357–388). Cambridge, MA: MIT Press.

Rosenberg, M. (1990). The mother of invention: evolutionary theory, territoriality, and the origins of agriculture. *American Anthropologist* 92(2), 399–415.

Simpson, G. G. (1944). *Tempo and Mode in Evolution*. New York: Columbia University Press.

Soltis, J., Boyd, R. and Richerson, P. J. (1995). Can group-functional behaviors evolve by cultural group selection? *Current Anthropology* 36(3), 473–494.

Trivers, R. (1971). The evolution of reciprocal altruism. *Quarterly Review of Biology* 46, 35–57.

Waters, M. R. and Stafford, T. W. Jr. (2007). Redefining the age of Clovis: implications for the peopling of the Americas. *Science* 315(5815),1122–1126.

Whallon, R. J. (1968). Investigations of late prehistoric social organization in New York State. In Binford, S. R. and Binford, L. R. (eds.), *New Perspectives in Archaeology* (pp. 223–244). Chicago: Aldine.

Winterhalder, B. (1981). Optimal foraging strategies and hunter-gatherer research in anthropology. In Winterhalder, B. and Smith, E. A. (eds.), *Hunter-Gatherer Foraging Strategies: Ethnographic and Archaeological Analyses* (pp. 13–35). Chicago: University of Chicago Press.

Wright, S. (1931). Evolution in Mendelian populations. *Genetics* 16, 97–159.

Wright, S. (1932). The roles of mutation, inbreeding, crossbreeding and selection in evolution. *Proceedings of the VI International Congress of Genetrics* 1, 356–366.

Chapter 11
Material Cultural Macroevolution

Niles Eldredge

Introduction

Ever since at least the days of Herbert Spencer, notions of "cultural evolution" have dotted the intellectual landscape. And if cultural evolution has not dominated cultural anthropology and archaeology to the degree to which it may be said to have dominated biology (and even there the impact and relevance of evolutionary ideas has been more checkered than is usually conceded), nonetheless the legitimacy and importance of notions of evolution in human cultural history have by and large seemed obvious to most observers (save perhaps British structural anthropologists).

Much of the work in cultural evolution in recent decades has tended to mirror—and often downright mimic—the approaches of hard-core neo-Darwinists in biological evolutionary theory. Boyd and Richerson's (1985) influential volume, for example, comes to mind in this regard—as of course does Dawkins' (1976) notion of "memes"—his parallel construct to his perhaps more elemental notion of the "selfish gene." A strong, reductive selectionism—a sort of "one size fits all"—is the prevailing core concept in such endeavors. Nor is the paradigm mere analogy—as it is not uncommon to find notions of "fitness" incorporated into discussions of cultural evolution, even to the point of allegations that cultural evolutionary change enhances the "fitness" of the developers/exponents of cultural novelty. Build a stronger bow, and make more babies! This extreme line of thinking puts cultural evolution in the invidious position of being a form of biological evolution after all. Talk about reductionism!

This present volume, along with another recent effort by archeologists (Lipo, et al., 2006), by and large takes the ontologically far more compelling and realistic position that cultural systems—be they material cultural or social systems—are more complicated than a single "population," where selection winnows each generation's standing crop of variation. Selection works in such domains—but the larger-scaled systems and patterns that archeologists (like their biological

N. Eldredge (✉)
The American Museum of Natural History, NY, USA
e-mail: epunkeek@amnh.org

A.M. Prentiss et al. (eds.), *Macroevolution in Human Prehistory*,
DOI 10.1007/978-1-4419-0682-3_12, © Springer Science+Business Media, LLC 2009

counterparts—paleontologists) typically confront present degrees of complexity that prevent (or at least *should* prevent) the straightforward application of simple, linear selection models. Simple selection models will not work for entire species divided into semiautonomous demes (the picture of species structure hinted at in Darwin and developed explicitly by Sewall Wright). I doubt they work any better for complex social systems and will demonstrate below that they certainly are not appropriate for complex material cultural systems, such as those typical of manufactured goods in industrialized nations[1].

It is not ipso facto wrong to seek parallels between biological and material cultural evolution in the attempt to clarify the underlying ontological structure and causalities within each system. Reciprocal illumination can work—especially if the underlying intent is not to derive—or simply import—a theory of evolutionary process for one system directly from the other. Before such comparisons can be made, however, it is helpful to start with a definition of "evolution" that is both suitable and appropriate to both such systems.

What is Evolution?

"Change through time" vaguely fits the bill for a general definition of evolution—though of course stability of systems (in some contexts known as "stasis") is not directly embraced. In general this definitional phrase is too weak to be of much service.

I prefer "the long-term fate of transmissible information" as a definition of evolution in both biological and material cultural systems. Stronger yet would be the more restrictive "long-term fate of transmissible information in an economic context." The objection to adding economics to the definition is simply that both genetic and material cultural information can be added or lost or modified strictly through the processes of transmission themselves—as when sexual selection modifies reproductive features of organisms for reasons having nothing to do with organismic relative economic success; or perhaps when information about making stone tools is modified as a simple by-product of errors in transmission of the original information (as in the parlor game when a message is invariably changed when serially whispered into enough ears).

Another objection to adding "economics" to the definition is that it is possible simply to chart the actual historical fate of information in both systems: This is what "phylogenetics" is all about in biology—with some schools of thought notorious for their aversion to considerations of underlying causal theory. A cladogram will suffice—yet cladograms are undeniably approximations of actual evolutionary history. Such is also the case for descriptions of the histories of material cultural systems: spear point design—or the design history of cornets—can be described in and of itself, with no further analysis of the causal factors underlying that history.

On the other hand, most evolutionary biologists seek to achieve an enhanced understanding of underlying evolutionary process based upon—and transcending—their phylogenetic diagrams. And such considerations almost automatically necessitate a consideration of the economic ("ecologic") context in which organisms lead their lives. Likewise, I really cannot think of material culture manufacture without considering the economic context of the objects being manufactured.

At the very least, the vast majority of long-term stasis and change in genetic information pertains to the economic side of organism's lives: the conformation of the soma, the adaptations for the basic physiological functions like respiration, digestion, etc. Darwin's classic formulation of natural selection boiled down to the effects that relative economic success (among conspecific organisms in a population) have on the relative reproductive success of those same organisms—given the competition for resources that limit the ultimate size of populations. And, of course, most things made by humans serve some sort of purpose, playing more-or-less obvious roles in the economics of daily life.

Returning to "fitness" for a moment, we can ask if there is an exact parallel between relative fitness—the statistical correlation between reproductive and economic success in biological populations[2]—in the biological realm, with a similar phenomenon in material cultural evolution. I have already commented that equating stasis and change in material cultural systems with patterns of relative reproductive success of humans (and even groups of humans) involved with designing, manufacturing, and using such products is tenuous at best.

The obvious difficulty of linking human reproductive success with the relative "success" of items of material culture in the marketplace highlights what at first glance at least is the beauty of Dawkins' concept of "memes." As with the parallel (and antecedent, really) notion of the "selfish gene," what's really going on in both biological and cultural systems is that it is the unit of information—the gene, the meme—whose "reproductive success" is really at stake. Were this really the case, we would have no difficulty in agreeing with Dawkins and those who see memes as the underlying basis for understanding cultural evolution.

But is it rational to see evolution in either system as a matter of bits of information competing with one another to be relatively better represented in the next succeeding generation? Despite Dawkins' (1982) own distinction between the information in a system, on the one hand and the "vehicles" that carry the information on the other, to follow Dawkins, one has to suppose that it is the bits of information themselves that are really the active competitors for enhanced representation as time goes by.

Such teleology fails because it is impossible to imagine how a piece of information—be it ensconced in a genome, a patent drawing, or in an actual spear point accompanied with an oral tradition specifying its manufacture—could itself be in competition with alternative forms of the "same" information. Squirrels may compete for acorns, and people may compete for recognition as the greatest cornet player on earth—but the information underlying squirrel behavior and anatomy is only a passive player in the game. And the information underlying various different cornet models—which may well play a role in the contest between their human players to see who is indeed the best—is likewise confined to a passive response to various forms of competition. Cornet models grow and diminish in popularity through time, but it is conceptually vacuous to describe such waxings and wanings in the borrowed jargon of genetics.

Thus there is no proper analogue of biological concepts of "fitness" in the material cultural realm—both because of the exiguous linkage between manufactured objects and human reproductive success, *and* because reducing natural selection to the rubric of the selfish gene, and producing a parallel—the "selfish meme"—into considerations of cultural evolution, forces information to play active competitive roles instead of the passive, driven-by-the-fates roles it actually plays in both systems. Think of it this way: Information doesn't give a damn whether it exists at all, let alone whether it will still be around in greater or diminished frequencies as time goes by. As I have found in my own database on the history of the cornet, the reasons why cornet designs (whether in their entirety, or in piecemeal, "meme" fashion) become more popular sometimes, or die out in others, are sometimes a reflection of relative efficiency ("selection" to be sure—but without the baggage of "fitness"; sometimes it *is* a matter of building a better mousetrap); but it just as often seems to be a matter of esthetics of a fickle buying public—who may be following the voguish dictates of professional players arbitrarily choosing among a spectrum of functionally equally good designs. I have found that some makers made better instruments than others—even if the designs were indistinguishably the same—and even when the makers of the better instruments themselves switched designs, or added alternative designs to their catalogues. A surprisingly high percentage of details of cornet design history cannot be attributed to actual improvements in the "playability" of the instruments.

Information in Biological and Material Cultural Systems

It has long been appreciated that the modes of transmission of genetic and human cultural information are very different. Genes are evanescent corporeal bodies; the information encoded on them is variably transcribed or copied before the actual gene dies. There are no such analogues in the cultural realm—"memes" being the information associated with, or underlying, a specifiable piece of cultural behavior—or, more concretely, an attribute of a physical artifact. Thus the "bore size" of a cornet (i.e. diameter of the air passage measured at the valves) might indeed usefully be construed as a "meme." But bore size is an abstract category of a cornet's property—and the bore size of any given cornet is merely an expression of the underlying information, whether from a shop drawing, or a physical mold or mandrel. Memes are the information lying in the "memotype"—and knowledge of them comes strictly from the phenotype. Genes, however, are themselves physical entities and can be studied as such in their own right.

And it is of course widely appreciated that, with the exception of bacterial systems and (some) plant taxa, genetic transmission is entirely *vertical*—i.e., from parent(s) to offspring. Not so with the transmission of human cultural information—neatly exemplified by the simple act of reading this sentence. Most considerations of the differences in how information is spread in biological (genetic) systems and human cultural systems emphasize the potential for far greater speed in the latter—on occasion engendering further thoughts of potentially faster rates of evolution.[3]

But there is another direct consequence of the disparities in mode of information transmission: For the evolutionary histories of complex material cultural systems can be notoriously difficult to analyze. Using a portion of my cornet data, Tëmkin and Eldredge (2007) showed that the memetic information—the 17 variables used to describe cornet "morphology" in my data base—is nearly entirely "unlinked" in cornet history, meaning that makers were free to mix and match parts. Openly stealing from another's designs, virtually all permutations and combinations were possible—including "retrofitting" older designs with new ideas. Bottom line: Information can easily spread back across separate branches (lineages) after they are already established as phylogenetically distinct. That is the equivalent, e.g., of innovations in dogs (Canidae) being able to spread back over to cats (Felidae).

That fact alone makes the straightforward application to cultural systems of the sort of phylogenetic reconstruction algorithms that are routinely applied to biological systems virtually impossible: The results are almost invariably gibberish. Tempering the raw computer output with what can be deduced about cornet history on other criteria (historical sequence of innovations in cornet morphology, patents, advertising, etc.) yields the sort of results in Fig. 11.1.

One mitigating fact is that in relatively simpler systems—such as Tëmkin's data on the Baltic psaltery (including some archaeological specimens!) or for that matter relatively simple lithic industries confined to single area cultural traditions—fare better when it comes to their analysis through biological phylogenetic techniques (Fig. 11.2). Such histories are perforce more linear—with innovations added seriatim, relatively few "branch points"—and thus a minimum of exchange of information across the branches of the tree.

As if this were not enough to establish the inherently greater complexity of the evolutionary histories of material cultural systems vs. biological evolutionary systems, consider the "Hannah Principle" (Eldredge 2006; Tëmkin and Eldredge 2007). The cornerstone assumption of phylogenetic analysis in biological systems is that evolution proceeds in greatest measure by the successive transformation of homologous traits[4]—meaning the "same" feature undergoes different forms of transformation in collateral lineages, and the sequence of such transformations can be resurrected through phylogenetic analysis. Of course, in deep phylogenetic history, entirely novel features appear (though even these ultimately have a progenitor). Thus a straightforward goal of phylogenetic analysis is to distinguish homologous similarities (synapomorphies) from the false signals of nonhomologous adventitious similarity (e.g., through convergent evolution; in general, "homoplasy").

And so we can ask: Is all change in the evolutionary histories of material cultural systems a matter of the transformation of preexisting states of homologous features? For clearly some of it must be: When the bell of cornets came to be lengthened (arguably in response to the market inroads of the rival trumpet—beginning in the early decades of the twentieth century), it was a progressive modification of a preexisting structure.

But consider the replacement of the original cornet piston valve by what has remained the "modern" valve. The original Stölzel valve, in which the airway ran

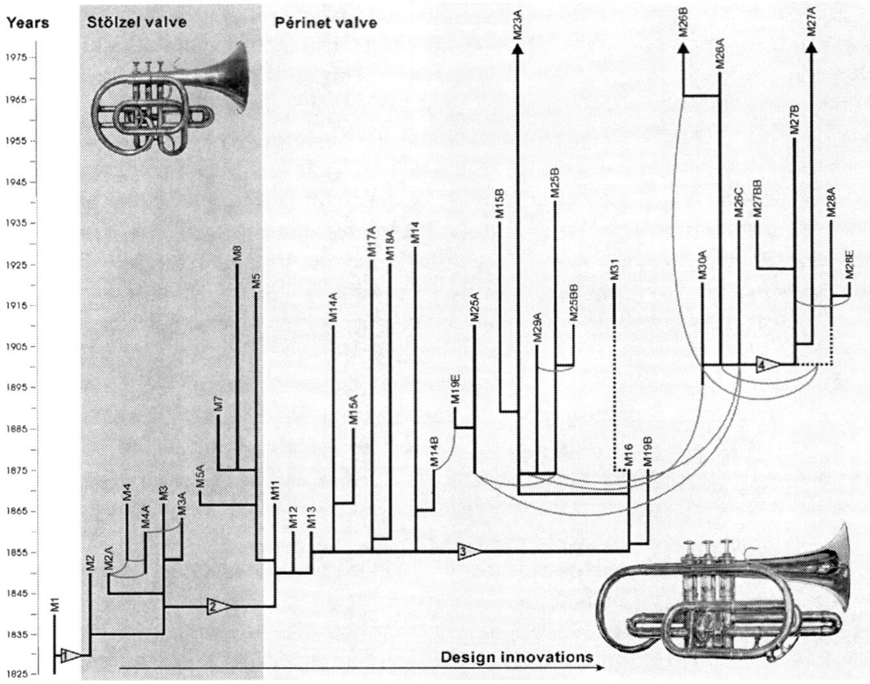

Fig. 11.1 Evolution of cornets. The *vertical bars* correspond to periods of manufacture of particular models (M). Shaded *left* and *right* areas correspond to instruments equipped with Stölzel and Périnet valves, respectively. *Curved lines* represent reticulations. *Triangles* mark several key innovations in cornet design: (1) valve number, (2) shifting of the second valve slide and valve alignment, (3) changing of bell exit position and bell placement, and (4) alteration of bell shape ("trumpetization"). From Tëmkin and Eldredge (2007), Fig. 11.2; for further information on analytic procedures underlying the generation of this diagram, see the caption for Tëmkin and Eldredge (2007), Fig. 11.2. Downloadable data files are available via the online version of this paper

vertically along the inside of the valve, was "challenged" and eventually replaced by the newer Périnet valve—which transmits the airway only across the valve. The first piston valves (ca. 1825) were all Stölzel valves; the oldest known Périnet-valved cornet was built by Adolphe Sax in 1843. It took until at least World War I for the Stölzel valve entirely to disappear—though it was largely replaced by the Périnet valve by the mid-1850s in all but the cheapest of instruments. The "transition" in the 1840s and 1850s even included some instruments with a mixture of both types of valves.

At issue here is the simple question: Did the Périnet valve "evolve" in some meaningful sense from the earlier Stölzel valve? Answer: No—not in the sense that the Périnet valve could be construed as a direct derivative, a descendant of the Stölzel valve. Rather, clearly it is an alternative design, meant to remove the sharp angles of the direction of the airway, as claimed in Périnet's patent of 1838.

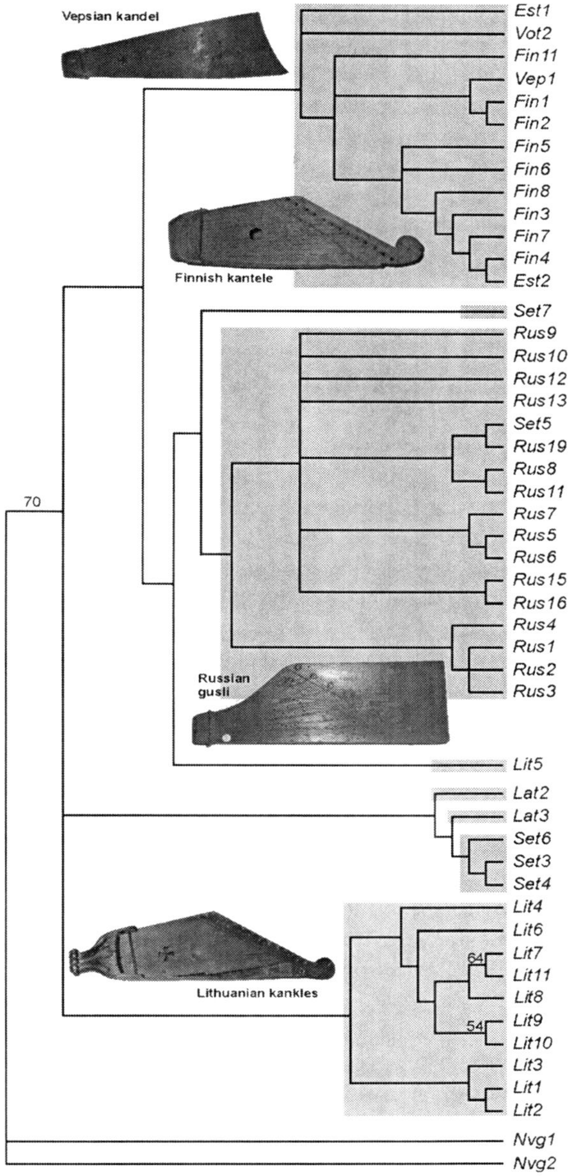

Fig. 11.2 Phylogeny of the Baltic psaltery. Shaded branches of the cladogram represent linguistic groups. Bootstrap values shown above branches (2000 replicates). Baltic group: *Lat*, Latvian; *Lit*, Lithuanian; Ural-Altaic group: *Est*; Estonian; *Fin*, Finnish (including Karelian); *Set*, Setu; *Vep*, Vepsian; *Vot*, Votic; Slavic group: *Rus*, Russian. For further details on analytic procedures, see Tëmkin and Eldredge (2007), Fig. 11.1 (caption)—from which this figure is derived

Homologous, perhaps, as tubular piston valves—but the newer valve was Périnet's way of thinking around what he saw as deficiencies in the structure—hence function—of the older Stölzel valve.

This is the "Hannah Principle" (Eldredge 2006; see Tëmkin and Eldredge 2007): In the "evolution" of material culture, new ideas, conceived as better ways to build an artifact (or at the very least to get around patent restrictions). Such considerations broach the topics of "directed variation" and indeed a sort of instantaneous selection. What is relevant here is that changes in material cultural systems that do *not* reflect simple modification of preexisting structures[5] further complexify the problem of historical reconstruction—in this case especially through the routine application of the kinds of phylogenetic algorithms developed and used for biological systems. Once again, we see how evolution in material cultural systems is inherently more complex than it is in biological systems.

Hierarchies and Patterns

Phylogenetic reconstruction perforce focuses solely on the information-based system variably conserved and modified in the passage of time in both biological and material cultural systems. Folding in the economic context in which patterns of stasis and change occur allows analysis of the dynamics of such pattern generation through time—the heart and soul of true evolutionary theory.

As other authors (e.g., Rosenberg especially) have discussed in this volume, the hierarchical structure of social systems is manifest—as it is in biological systems, and, for that matter, material cultural systems. Analyzing the hierarchical structure of biological systems (e.g., Eldredge and Salthe 1984; see Eldredge 2008 for a concise review) has proved fruitful in developing theories of the biological evolutionary process on a variety of spatiotemporal scales—from "micro-" to "macro-evolution."

The basis of the dual biological hierarchy scheme followed here rests on the ontological supposition that organisms engage in two—and only two—classes of processes: matter/energy transfer processes ("economics")—that allow differentiation, growth, and maintenance of the soma (aka "staying alive"), on the one hand, and reproduction on the other. Both sets of processes generate hierarchical systems of which organisms are a part: Organisms are parts of local economic systems ("avatars"), which form parts of local ecosystems. Local ecosystems are connected geographically with adjacent ecosystems, across the boundaries of which matter and energy commonly flow. This is the "ecological hierarchy." (Fig. 11.3).

Reproduction sets up a parallel hierarchy: Sexually-reproducing organisms are parts of demes, which themselves are parts of species. Species also "reproduce"—in the sense that speciation occurs, creating lineages of ancestral and descendant species that are themselves hierarchically arrayed as subsets within larger sets: the variously ranked taxa of the Linnaean hierarchy. This is the "evolutionary" hierarchy (Fig. 11.4).

Note how Darwin's original formulation of natural selection virtually falls out of the mere act of setting the two hierarchies side by side. Organisms are of course the only entities in both systems: Natural selection is the statistical effect that relative economic success has on reproductive success among conspecific organisms (members of the same avatar/deme)—given the presence of heritable variation.

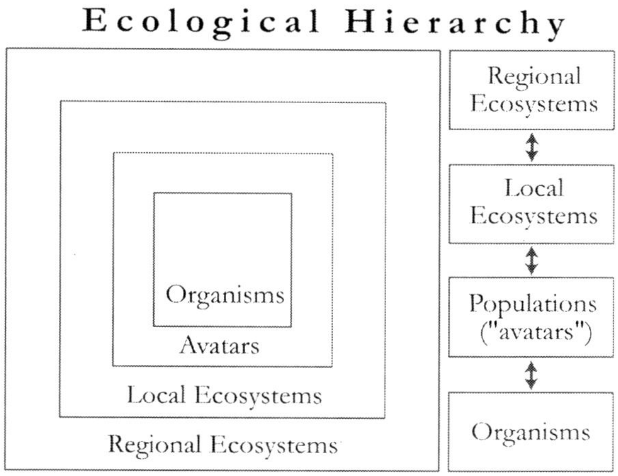

Fig. 11.3 The ecological hierarchy

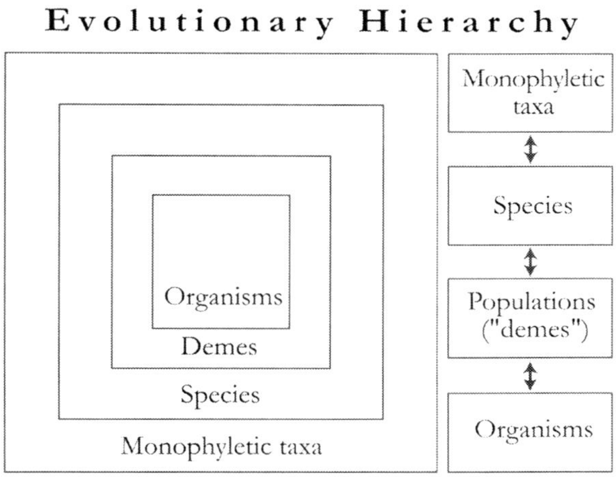

Fig. 11.4 The evolutionary hierarchy

The "sloshing bucket" (Eldredge 2003a, 2008) notion of (macro-) evolution—i.e., at spatiotemporal scales larger/higher than within-population microevolution through natural selection and drift—is derived simply by mapping classes of evolutionary events onto the dual-hierarchy conceptual structure. Lower levels of ecological disturbance characteristically trigger recovery through ecological succession on the scale of months/years/decades. Recruitment comes from adjacent demes—and little or no discernible morphological evolutionary change accrues.

On the upper end of the spectrum, the five or six global mass extinctions that have occurred in the past .5 billion years are sufficiently devastating—accounting for the loss of so many species—that entire clades disappear. Recovery entails considerable evolution, as surviving clades often radiate (usually after a lag of several million years). The most famous example is the Paleocene and Eocene multiphasic radiation of mammals after the demise of terrestrial dinosaurs in the end-Cretaceous mass extinction event. Mammals had been present since the Triassic—evolving at roughly the same time as had the terrestrial dinosaurs (and other collateral reptilian kin clades). Yet, ecologically speaking, at least, mammals did not diversify to any major degree until after the dinosaurs had become extinct.

Thus radiations, involving extensive amounts of evolution occurring in bursts, are as a rule typically "decoupled" from the actual origin of major clades—marked as they often are by major adaptive "innovations." Three middle ear bones, the presence of hair, lactation, and (later, but still in a pre-Tertiary world) the acquisition of placentation—mammalian hallmarks all—did not themselves trigger the bursts of evolution that produced most of the extant orders of mammals.

The spatiotemporally intermediate level of the "sloshing bucket" is in many ways the most interesting: When climate change or other environmental factors provoke extinction events of sufficient severity, covering sufficiently large spatial areas that the geographic ranges of entire species are affected, species begin to go extinct. Relatively rapidly occurring waves of extinction are common in the history of life, always followed by waves of speciation—mini-versions of the more notorious and easily appreciated mass extinction events. These are the "turnovers" of Vrba (1985). There is growing consensus among paleontologists that most speciation events are concentrated in pulses following regional extinction events (e.g. see van Dam 2006, on rodent evolution in Spain over the past 24.5 million years). And, if it is indeed the case that stasis characterizes most of the histories of most species in the history of life (Eldredge and Gould 1972), the implication is that most morphological evolutionary change occurs in conjunction with speciation events—a conclusion now gaining a measure of support from recent results in molecular biology (see Pagel, et al., 2006).

Most speciation events associated with turnovers produce only small-scale, relatively minor morphological change; rarely does the truly new appear in turnover-related speciation events. Yet, if the macroevolutionary larger scaled radiations following mass extinctions themselves usually fail to produce major-large-scale adaptive innovations that are justly considered to constitute the other half of what is usually meant by the phrase "macro-evolution"—it must be the case that many such "major" morphological changes in fact do occur in speciation events in the course of relatively modest "turnover pulse" events. Here, perhaps, is where relatively minor changes in the genetic regulatory apparatus, mediated by selection, produce the sorts of changes that, in hindsight at least, are recognized as the major large-scale changes of macroevolution.

Hierarchies in Material Cultural Systems

What would a hierarchy scheme of material cultural systems look like? Asking myself that question specifically as applied to cornets, I derived the hierarchical scheme of Fig. 11.5. At a glance, and once again, it appears that material cultural

Material Culture Evolution in a Hierarchical/Systems Context

Pure Economics	Maker: Economics+Information		Pure Information
Global Market	Universe of Makers		Universe of Artifacts
Regional Market	Regional Makers		$Subset_1$ of Universe
Local Market	Focal Maker		$Subset_2$ of Universe
	Eco	*Info*	Focal Set of Artifacts
	manufctre←------→design		
	[materials	[plans	
	payroll/ovrhd	prototype	
	markting]	othr mkrs models	
		patents]	
Selection II	**Selection I**		**Stasis & Change**
Larger-scale Eco forces	**Intentional/directed variation**		

Fig. 11.5 Hierarchical structure of complex material cultural systems in an evolutionary context. For discussion, see text

systems are at least an order of magnitude more complex than the much simpler dual hierarchy scheme that describes biological systems. Nor does the scheme of Fig. 11.5 display all the true complexities of the material cultural realm.

Like the biological systems, there are "pure" economic and informational hierarchies in material cultural systems: The regionally structured "marketplace" encapsulates the economic side—related to, but different from, the complexities of the world of music—where the instruments are of course put to use. On the other hand, like Plato's chair, there is the "pure information"—the hierarchically structured world of artifacts, with their underlying information. Unsurprisingly, given the discussion above, these hierarchies are by no means as neat and unambiguous as are the hierarchies of relationships among sexually reproducing organisms. Cornets, for example, are members of the lip-blown "family" of musical instruments—often referred to as "brasswinds." But unambiguous classifications of musical instruments are notoriously difficult to produce: Is a 4.5' long brasswind equipped with a cornet-style mouthpiece, but with a slide instead of valves, to be called a "slide cornet" or a "soprano trombone?" Both terms have been used—depending to some extent on the use to which the instrument is put. It is a trombone in a trombone quartet; it was a slide cornet when Louis Armstrong used one occasionally with his jazz group in the 1920s.

Thus the information and economic hierarchies of biological systems do seem to have direct analogues in material cultural systems—though they are seemingly more complex in the latter than in the former. What is truly new about material cultural systems (vis à vis biological systems) is the dual hierarchy scheme lodged *between* the information and the marketplace: the manufacturers—themselves typically arrayed in local and regional hierarchies.

Minimally, of course, a manufacturing system requires a single maker—unlikely in the industrialized world of manufactured goods, but less unlikely in smaller, nonindustrialized social contexts—such as hunter-gatherer bands. No real hierarchy there, if only one person is involved; but there is still an interplay between information/ideation and the manufacture of a given object. Along with other lines of evidence, the work sites of stone tool traditions are identified as such primarily from the discarded shards produced in manufactory, as well as unused raw materials (which may have been rejected as not optimal—and, of course, as tools that did not turn out well enough to be used). Even if only one maker is involved, there is an interplay between the idea (i.e. the underlying information of the desired product and the steps known to be necessary in its successful production) and the actual production itself—which includes obtaining the raw materials and the actual labor involved.

In postindustrial manufacturing systems, many makers, employing several to hundreds of variously skilled people, are found in local and regional settings. My piston-valved cornet database lists some 125 makers (between the years 1825-Present), primarily located in France, Belgium, England, and the United States (northern and eastern European cornets have historically typically been fitted with rotary valves). Surviving instruments, advertising, patents, and written records help pinpoint the history of design innovation and the spread of information within and among manufacturers throughout cornet history (a far better record than anything I have seen in paleontology; fossils do not come with serial numbers!).

I have utilized 17 variables to describe cornet morphology; these variables are sufficient to describe all but a fraction of known cornet variation throughout the 180+ years of its "evolution." As already noted, these 17 variables are almost completely "unlinked": the "bell" (flaring portion emitting the sound) can be on either side of the valve cluster for example—and the configuration of the valve tubing can be fitted to any variant version of valves. As shown especially clearly in Fig. 11.1, the earlier Stölzel-valved cornets (often called "cornopeans") were retrofitted with later innovations associated initially with later Périnet-valved cornets—including placing the bell to the left side of the valve cluster and moving the tubing from the second valve to the right side of the instrument; both conditions have long been invariant throughout later cornet history.

It is possible to see innovations arising in one atelier and being applied to other models under production in the same shop. The oldest known cornet with the modern configuration of the bell on the left-hand side of the valves was made by the Parisian maker Antoine Courtois in 1855. Called the "nouveau modèle," shortly thereafter the same maker moved the bell to the left side on several different models—including one with a combination of Stölzel and Périnet valves, and on another two models that differed from the original configuration of the 1855 instrument. These latter two models rapidly became stabilized as the twin iconic models for the remainder of the nineteenth century. All four were called the "new model"—signifying the rapid spread of the idea within the shop itself.

The reason for this spread was the commercial success of these French-made cornets in London. England had used mostly German-designed (and/or made) Stölzel-valved cornets through the 1840s, but a switch to French-made instruments was already underway, imported by the Distin family, John Pask and perhaps others (As a youth, William Thomson, the future Lord Kelvin, who annoyed Darwin by insisting on a relatively young age of the earth, played a John Pask French-made Stölzel-valved cornet, possibly made by Antoine Courtois). Courtois' "New Model" cornets (so stamped in English on London-sold instruments) were an almost instant hit in London. Played and endorsed by the virtuoso Hermann Koenig, star soloist (arguably history's first instrumentalist "rock star") in Jullien's orchestra that mostly played in London, by the late 1850s cornets with the new "English bell" had all but replaced the older, traditional, retrospectively dubbed "modèle français"—"French model"—with its bell on the right side of the valve cluster.

Other makers began to catch on but not until nearly an entire decade had passed. Henry Distin had begun making his own instruments, and Courtois' great Parisian rival Gustave Besson had opened a factory in 1858 in London: Both were making English bell instruments by the mid-late 1860s for sale to the English market. But this "new model" was much slower to become popular in France and other countries in continental Europe, not fully supplanting the modèle français until World War I.

An important feature of the hierarchy of makers that stands between, and connects, the hierarchies of "pure" information and economics ("marketplace") is that it, *in itself*, is actually a dual structure of information and economics (Fig. 11.5). Part of an operation is concerned with innovation and production of existing models using plans (e.g. shop drawings and specifications, prototypes, patents, other makers' models, and the experience of the artisans); the other part is concerned with procuring materials, equipping the factory, paying the workers, and advertising and marketing of the product in general. These two facets of course interact—as when decisions are made (based at least in part on sales results) on which models to produce, including whether or not to produce a relatively greater or narrower variety, and of differing quality and "price points."

And, of course, information is transferred among makers—locally and regionally—driven in large measure by relative success of models in the marketplace. I have already mentioned the origin and spread of the English bell model of cornets—from regional hit in London to today's worldwide design standard. Looking for all the world like the spread of an evolutionary novelty across a Sewall Wright-style landscape (e.g. Spencer, this volume), the theft of ideas in material culture (between makers, as well as between models in a single maker's atelier) is an underlying dynamic of cornet evolution with ramifications that transcend the mere muddying of phylogenetic diagrams. Though there was a measurable lag in the spread of English bell designs, the mere fact that the information could be readily co-opted (and there were no known patents protecting the design in any country) accounts for a very different evolutionary history of the cornet than would have occurred had that design been restricted (like genetically based innovations) to a single small French company.

Figure 11.5 points to two distinct forms of "selection" in material cultural evolution. One is inherent in the very act of design novelty: A designer can conceive of a novelty and instantly reject it as impractical for a variety of reasons—without bothering to build a prototype, let alone testing it in the marketplace. The history of cornet design as seen in patents is very different from the actual, realized history—as makers routinely patented ideas that never got beyond the conceptual stage, or prototype stage, all the way to the marketplace. There is an element of selection to the very act of design creation—which otherwise is a directed form of variation, an ad hoc solution to a perceived problem, perhaps, that, if it seems plausible, might be further put to the test of prototype construction, then to limited production and marketing, etc. A form of "selection" enters in each step of the way (all subsumed under "Selection I" in Fig. 11.5). There would seem to be no analogue of "fitness" here—the "selection" merely being the evaluation of the designer(s).

The marketplace presents another, very different form of selection, where a form of "fitness" measured in sales, numbers of instruments—or even numbers of different makers producing the same model—might constitute a plausible analogue of the forms of fitness encountered in population genetics. Certainly continued existence—and expansion—of makers heavily depends upon their success in the marketplace. This—plus the unpredictable fluctuations of regional and global economies—are subsumed under "Selection II" in Fig. 11.5. But a *caveat* is necessary here: "fitness" in this sense has to do with the survival of business entities and not, in any formal, explicit way, with the relative reproductive success of owners, managers, and workers in those businesses.

Omitted from that diagram is another form of selection—as when makers decide to borrow ideas from one another. Makers compete with one another, but they also cooperate (some makers buy parts from others; price rigging is not unknown, etc.). What gets made, and what never makes it to the marketplace, is in part an inter-maker form of selection as well. When it comes to humanly made artifacts, selection is a multifaceted "process" that far transcends in its totality of complexity even the multiplicity of generally accepted forms of "selection" in biological evolutionary theory.

Material Cultural Evolution and the "Sloshing Bucket"

Is there an analogue to the "sloshing bucket," where we can match classes of events in cornet design history to various spatiotemporal levels of this hierarchical structure of material culture? And, if so, are major evolutionary events initiated more by external perturbations in the economic arena, or are they a reflection of innovation and selection or a combination of factors from both the informational and economic sides of the hierarchical ledger?

Elsewhere I have documented the major features of cornet "evolution"—including examples of a number of evolutionary patterns, including gradual linear change, stasis, turnovers (of varying scales) etc. (Eldredge 1997, 2000, 2002). Here I will just briefly mention five major episodes in cornet history and see how they connect with the hierarchical scheme outlined above:

- Valved Cornet Origins—ca. 1825. Brass wind instruments lacking valves or other means of changing the length of the tubing (e.g. slides, keys) can only play the overtone series: Limited by the principles of acoustical physics, it is impossible to play a chromatic scale except in the very highest reaches of the instrument attainable only by virtuosi. Made possible by advances in metallurgy, various devices (such as valves) that minimized leakage of air allowed inventors to try out a multiplicity of mechanisms that could change the length of the instrument—thus allowing full chromaticism (especially when a third valve was added a few years after the earliest two-valved cornets were marketed). This was an achievement driven by the needs of Western music that was enabled by general scientific and engineering advances. It was an instant success—though the valveless bugle maintains its role as a military signaling device.
- Replacement of the Stölzel valve (invented ca. 1825) by the Périnet valve (invented ca. 1838). The alleged superiority (in production of sound qualities) of the Périnet valve over the Stölzel valve was not immediately apparent to musicians. Manufacturers continued to produce Stölzel-valved instruments alongside Périnet-valved instruments (as well as a "hybrid" form with a central Périnet valve flanked by two Stölzel valves) well into the 1850s. The "hybrid" form allowed the classic deep-bodied form of the older Stölzel-valve cornopeans to be retained. Tests with restored instruments reveal no discernible superiority in sound qualities or playing characteristics. The best explanation of the eventual replacement of Stölzel-valved instruments by Périnet instruments is that makers evidently found them to be a sturdier form of construction and less prone to wear and tear (Robb Stewart, pers. comm.). Moreover, there has been no experimental or experiential confirmation of the near-constantly made claims of rival manufacturers that they had devised a type of Périnet valve superior to all others in acoustical properties. The public, choosing from the variety of styles, including valve types, seems to have made its choice on reputation of maker and overall style (i.e., appearance) of an instrument.
- Such may be the explanation for the instant success of the "English bell" in England: the novel look of the instrument. However, with the bell on the left side of the valves, there is a more secure purchase as the player grips the valves with the left hand.
- 1900–1905: Major turnover in cornet design. The iconic Victorian double waterkey designs of Courtois (copied by many other makers) almost completely disappeared, virtually overnight—replaced by a slight modification of a more conservative design dating from the late 1860s/early 1870s. Nothing about the replacement models were alleged to be superior to the Victorian models they so abruptly replaced in the public's esteem. Nor were there global depressions, changes in music or anything else of a large-scale nature that could be reasonably supposed to have caused this turnover, except one thing—the change in the calendar from 1899 to 1900. Out with the old—Victorian style—in with the new! Advertising of all manner of manufactured goods reflected the need for novelty for the new century. In the United States (unsurprisingly), Conn, York, Holton, White (King), Buescher, and other makers outdid themselves producing many

newfangled, short-lived models—most pronounced to be superior in their playing aspects (these included still more experimental valve designs). Planned obsolescence was part of this, as many models (especially by Conn) were continuously "improved" with each passing year. What won out—the dominant cornet model still in use today—was however a simple style invented by Besson in the late 1860s—with some later, minor modifications.

- Near extinction of the cornet—1920s. Cornets, especially in the United States, had become gradually longer, more trumpet-like, throughout the first two decades of the twentieth century. Sometime in the early to mid-1920s, the modern form of the Bb trumpet—invented by cornet makers (probably the aforementioned Parisian maker Besson) in the late 1870s/1880s through an experimental trial-and-error phase—suddenly became popular not only in jazz but in commercial and symphony orchestras and town bands. Bb trumpets had never been sold by Sears, Roebuck until the early 1920s. Why the chief symbol of this switch—Louis Armstrong—himself dropped the cornet for the trumpet in the 1920s is not really known: Some authors feel that the exigencies of recorded music might have had something to do with it. Trumpets are brighter instruments, capable of producing louder, more piercing sounds than the traditional more-horn-like, mellower cornets—with their deeper mouthpieces. But I agree with Lewis (1991) who has pointed to the "trumpetization" of cornets for several decades prior to the switch—a reflection (he plausibly argues) of the preference for lighter, brighter upper register sounds from the brass section that can be traced back at least as far as the bands and music of the American Civil War.

Thus the major features of cornet evolutionary history seem to overwhelmingly reflect marketplace fickleness—an unanticipated (i.e., by the makers) rejection of what had been "the usual" in favor of something else already available out there. The exceptions are the relatively few true "improvements" in design history: addition of valves for chromaticism; less dramatically (and consequently more slowly) adoption of the "English bell." But the near-extinction of Victorian double-waterkey cornets at the turn of the past century; and the switch to Bb trumpets that nearly drove cornets completely extinct (especially when trumpets are considered a derivative clade off cornets in the first place) reveal a pattern of external perturbation not driven by functional considerations of musical worth, or indeed by changes in musical style, so much as by shifting moods in the marketplace. The end-century switch-over seems to have reflected nothing more (nor less) than a desire to have something more modern—even though what was chosen was basically a tinkered version of an alternative model that had already existed for three decades. And the sudden adoption of the Bb trumpet was a culmination of a trend in that direction—the "trumpetization of the cornet"—finally just going the final step, and adopting something that had been available, but little used, for half a century. If it reflected musical taste, it was the taste of the musicians and the public they played for and not of the composers of any new form of music that called for brighter sounds over the mellower cornet sounds of old.

Conclusions

Like biological macroevolution, then, material cultural evolution seems seldom driven by "key innovations"—the better mousetraps that the consumer world is famously supposed to be beating a path to. It seems more driven by stylish fads—which sometimes reflect a true design change (as when clamshell flip phones surprised industry giant Nokia, losing them a major market share). Yet even here, nobody thought the clamshell phones worked better *as phones* than the older, more traditionally telephone-like cell phone models: That, after all, depends on the electronic guts of the phone—and on the carrier which it connects to. But the flip phone is cooler.

As long as a tool works, it stays in the ambient culture. True innovations that actually improve tools are very rare—some of the best cornets I have ever played were made in the 1850s. If the look, the feel—the *gestalt*, the *jizz*—of a tool can be changed without compromising its function, that will happen as unanticipated rapid turnovers.

And, of course, it is at the very beginning of the existence of classes of material cultural objects that the interplay between the creative mind of the true innovator and the successful adoption and spread of the innovation is most clearly seen. It is not the better mousetrap, but the initial fact of a mousetrap. And even the invention of the truly new—like computers or even desktop computers—may take a while to catch on.

These isolated events taken from cornet history are merely suggestive. Do they constitute actual *patterns*—classes of events that permeate the history—not just of cornets, but of other musical instruments, and for that matter, all the evolution of *all* material culture? For focusing on the evolutionary history of one well-delineated system over its entire history, informative as it may be, would be like deriving an entire theory from a single example. The "sloshing bucket" approach to macroevolutionary history could not be derived in full from the history of the Class Trilobita: Patterns of disturbance-induced extinction and subsequent evolutionary diversification within surviving lineages need to be timeless (e.g. seen throughout at least the Phanerozoic history of multicellular life) and affecting at least potentially all forms of life—not just a single clade.

Even granted the complex hierarchical system underlying the manufacture and sale of cornets to be potentially general for industrially manufactured goods, the critical question is: Were there similar events affecting other material cultural products at the same time and place? My claim, for example, that the turn of the century in itself led to the shift away from the ornate Victorian models to more conservative, yet more streamlined, models predicts that cornets are but one example of a general *fin-de-siècle*/brave-new-world mentality in the marketing world/cultural life of Western Europeans and Americans generally. There literally have to be more examples.

And what about the near extinction of cornets as the Western world (save the British Brass bands) nearly universally dropped the cornet in favor of the Bb trumpet (which itself hardly underwent any further "evolution"—it was ready to go right off the design shelf). This sounds like it could be a unique event, but what else happened in the Western cultural world of musical instruments in the 1920s? Vague allusions to unspecified acoustical demands of the recording (and radio) industries, unsatisfying as they are, would imply that perhaps other classes of instruments were similarly affected. There was the famous "saxophone craze" in the 1920s: There are as many old saxophones in attics in the United States as there are violins—and cornets. Maybe more. And the possibility arises that this sudden mania for saxophones (which had been around—virtually unchanged, but little used—since Adolphe Sax invented them in the 1840s) and the sudden rise of the Bb trumpet had the same underlying cause: if not the demands of the electronic era, then perhaps the marketing genius of Carl Greenleaf, who took over the floundering Conn corporation in 1916 and, faced with the radio-induced demise of the local town bands (there had been over 80,000 such bands in the late nineteenth century according to Hazen and Hazen, 1987), invented the "school band movement" that saved his company—and gave a shot in the arm to the entire industry. Perhaps the saxophone craze and sudden demise of the cornet in favor of the Bb trumpet represent successful outcomes of a deliberate marketing scheme.

Evolutionary patterns—like stasis, gradual change, and turnovers at various spatiotemporal scales—do seem to be common to both biological and material cultural evolutionary realms (see also Prentiss this volume; Kuijt and Prentiss this volume). The greater complexity of material cultural systems derives in large measure from the greater frequency of horizontal information transfer in material cultural systems, and from the violation of the principle of homology engendered by the "directed variation" of ad hoc solutions dreamt up by a fertile creative designing mind—the "Hannah Principle."

And both sorts of systems are manifestly hierarchically arranged—but the presence of those creative minds establishes a dual information/economic hierarchy in between the "pure" hierarchies of information and economics in human material cultural systems. The "sloshing bucket" notion of biological evolution—which relates disturbances and extinctions at varying spatiotemporal scales to magnitude of evolutionary events—may have its analogues in the evolution of material cultural systems.

The "sloshing bucket" further suggests that evolutionary rebounds after particularly large-scale extinction events are not necessarily where the more profound adaptive innovations in biological evolutionary history necessarily arise—further implying that those novelties in themselves seldom act immediately as "key innovations" triggering prodigious amounts of rapid diversification. The same appears to be true in material cultural systems (Chatters this volume; Zeder this volume): Most innovation in cornet design history come early in the histories of the ateliers where they first arose. If "English bell" cornets were an instant hit in England (though not in France where they arose in Courtois' shop in the mid-1850s), the success of the Besson *desideratum* (forerunner to the modern cornet, invented in the 1860s but not market-dominant until the twentieth century) and the piston-valved Bb trumpet (derived from the cornet in the 1870s but not market-dominant until the 1920s) are the more typical cases in point.

But all this remains to be seen as material cultural systems over the last 2.5 million years are looked at with these possibilities in mind.

Notes

1. In this paper, I shall be drawing on my database on the history of piston-valved cornets—lip-blown brasswind instruments manufactured and sold between 1825 and the Present, predominantly in France, England and (somewhat later) the United States. The database consists of 17 variable vectors that describe cornet morphology and comprises nearly 200 makers and over 120 distinct "models" (i.e. unique vectorial combinations).
2. not excepting humans—though culture has radically altered the straightforward usual correlation between economic and reproductive success, with exceptions that invariably delight sociobiology's proponents, robber barons, and the just plain wealthy typically have fewer children than do people in lower socioeconomic classes—a point well developed, of course, by Karl Marx.
3. Though long ago George Simpson denied that differences in generation time neatly transcribes into differences in the rate of evolution, when he maintained that elephants, with their long generation times, appear to have evolved more morphological disparity than have the far shorter-generation mice—over the same lengths of geological time.
4. Molecular biology has recently—and very interestingly—muddied the waters on the very concept of homology. It turns out the same genetic basis for the vertebrate eye—the *Hox* gene—is also what is responsible for the development of arthropod eyes. Traditionally, vertebrate and arthropod eyes have been considered excellent examples of analogous (nonhomologous) structures—what with their manifestly separate phylogenetic development and utterly different conformations. Yet now we hear they are, at some even deeper genetic level, homologous! Fun!
5. Mme. Besson, highly successful heir of her late husband Gustave's musical instrument-making atelier, patented yet another form of the Périnet valve in Paris in 1874. The patent text is fascinating—as Mme. Besson gives a (largely accurate) history of the Périnet valve from its inception through her present patent design (Eldredge 2003b). Tellingly, though she traces the development (I would say "evolution") of the valve through four antecedent stages, she pointedly does not refer to the Stölzel valve—clearly to her an utterly different valve form altogether.

References

Boyd, R. and P.J. Richerson (1985). *Culture and the Evolutionary Process*. Chicago: University of Chicago Press.

Dawkins, R. (1976). *The Selfish Gene*. New York and Oxford: Oxford University Press.

Dawkins, R. (1982). *The Extended Phenotype.* San Francisco: H. Freeman and Co.
Eldredge, N. (1997). Evolution in the marketplace. *Structural Change and Economic Dynamics* 8, 385–398.
Eldredge, N. (2000). Biological and material cultural evolution: Are there any true parallels? *Perspectives in Ethology* 13, 113–153.
Eldredge, N. (2002). An overview of piston-valved cornet history. *Historic Brass Society Journal* 14, 337–390.
Eldredge, N. (2003a). The Sloshing Bucket: How the physical realm controls evolution. In J. Crutchfield and P. Schuster (eds.), *Evolutionary Dynamics. Exploring the Interplay of Selection, Accident, Neutrality, and Function. SFI Studies in the Sciences of Complexity Series* (pp. 3–32). New York: Oxford University Press.
Eldredge, N. (2003b). Mme. F. Besson and the early history of the Périnet valve. *The Galpin Society Journal*, LVI, 147–151.
Eldredge, N. (2006). Foreword. In C. Lipo, M.J. O'Brien, M. Collard and S.J. Shennan (eds.), *Mapping Our Ancestors. Phylogenetic Approaches in Anthropology and Prehistory* (pp. xiii–xvi). New Brunswick: Transaction Publishers.
Eldredge, N. (2008). Hierarchies and the sloshing bucket: Toward the unification of evolutionary biology. *Evolution: Education and Outreach,* 1:10–15.
Eldredge, N. and S.J. Gould (1972). Punctuated equilibria: An alternative to phyletic gradualism. In T.J.M. Schopf (ed.), *Models in Paleobiology* (pp. 82–115). San Francisco: Freeman, Cooper and Co.
Eldredge, N. and S. N. Salthe (1984). Hierarchy and evolution. *Oxford Reviews in Evolutionary Biology* 1, 182–206.
Hazen, M. H. and R. M. Hazen (1987). *The Music Men. An Illustrated History of Brass Bands in America, 1800–1920.* Washington D.C.: Smithsonian Institution Press.
Lewis, H.M. (1991). How the cornet became a trumpet—the instruments and music of a transitional period in American music: 1880–1925. *ITG Journal* 16,17–26.
Lipo, C., M.J., O'Brien, M. Collard and S.J. Shennan (eds.) (2006). *Mapping Our Ancestors. Phylogenetic Approaches in Anthropology and Prehistory.* New Brunswick: Transaction Publishers.
Pagel, M., C. Venditti and A. Meade (2006). Large punctuational contribution of speciation to evolutionary divergence at the molecular level. *Science* 314, 119–121.
Tëmkin, I. and N. Eldredge (2007). Phylogenetics and material cultural evolution. *Current Anthropology* 48, 146–153.
van Dam, J.A. (2006). Long-period astronomical forcing of mammal turnover. *Nature* 443, 687–691.
Vrba, E.S. (1985). Environment and evolution: Alternative causes of the temporal distribution of evolutionary events. *South African Journal of Science* 81, 229–236.

Index

A.M. Prentiss et al. (eds.), *Macroevolution in Human Prehistory*,
DOI 10.1007/978-1-4419-0682-3, © Springer Science+Business Media, LLC 2009

Breinigsville, PA USA
06 October 2009
225324BV00004B/21/P
9 781441 906816